CONFIGURATIONAL MECHANICS

PROCEEDINGS OF THE CONFIGURATIONAL MECHANICS SYMPOSIUM HELD WITHIN THE 5TH EUROMECH SOLID MECHANICS CONFERENCE, 17–22 AUGUST 2003, THESSALONIKI, GREECE

Configurational Mechanics

Edited by

Vassilios K. Kalpakides
University of Ioannina, Greece

Gérard A. Maugin
Université Pierre et Marie Curie, France

A.A. BALKEMA PUBLISHERS LEIDEN / LONDON / NEW YORK / PHILADELPHIA / SINGAPORE

Published by: A.A. Balkema Publishers, a member of Taylor & Francis Group plc
www.balkema.nl and www.tandf.co.uk

ISBN 90 5809 667 X

Printed in the Netherlands

Configurational Mechanics, Kalpakides & Maugin (eds)
© 2004, Taylor & Francis Group, London, ISBN 90 5809 667 X

Table of Contents

Configurational Mechanics, Kalpakides & Maugin (eds)
© 2004, Taylor & Francis Group, London, ISBN 90 5809 667 X

Foreword

This volume is the outcome of a symposium on *configurational mechanics*, which was held during the 5th Euromech Solid Mechanics Conference (Thessaloniki, August 17–22, 2003). This was one event among many others that took place in 2003 stressing the increasing interest by the mechanics community in *configurational mechanics.*

A proceedings volume can never reproduce the vividness and the discussions atmosphere of a live meeting. Nevertheless, the reader's thirst for sensations can be compensated by the authors' efforts to expose their ideas in long and explanatory written texts. We hope that this will render the volume attractive not only to the specialists of the particular topic but also to those approaching from different perspectives this interdisciplinary field.

We take this opportunity to thank all the authors for their written contributions and for the efforts they produced in conforming to the spirit of the volume as well as following the complicated technical instructions of the Publisher. Also, we acknowledge the financial support of the University of Ioannina without which the present volume would not be finalized.

Last but not least, we thank the Ph.D. students Konstantinos Balassas and Eleni Agiasofitou for the generous technical aid they have provided during the editing of the volume.

December 15, 2003.
Vassilios K. Kalpakides, Gérard A. Maugin

Configurational Mechanics, Kalpakides & Maugin (eds)
© 2004 Taylor & Francis Group, London, ISBN 90 5809 667 X

A critical assessment of the theory and applications of material-configurational forces

G.A. Maugin
Laboratoire de Modélisation en Mécanique, Université Pierre et Marie Curie, Paris, France

ABSTRACT: Although it now finds applications in typical engineering settings such as the application of finite-element and finite-volume computational techniques, the theory of configurational – or material – forces smells good of its mathematical-physical origin. It is in the light of this evidence that we briefly examine some of the most recent fruitful developments of this theory. The newly entertained relationship with computational techniques based on mathematical formulations akin to general conservation laws (weak form such as the principle of virtual power, classical volume balance laws) is not so surprising. This contribution revisits some recent progress such as the generation of material "driving" forces in the spirit of the principle of virual power, and the allied application to numerical schemes such as those developed for studying the nonrectilinear progress of cracks (finite elements) and the progress of phase transition fronts (finite volumes).

1 INTRODUCTION

Both the deep physical meaning and the practical usefulness of the critical expressions (e.g., of thermodynamical driving forces) obtained within the present framework stem from the intimate relationship built from the start between these expressions and a field-theoretical invariance or lack of invariance. After all, symmetries and their eventual breaking are the physically most profound and intellectually puzzling tenets in modern physics. That engineering applications fit into this general picture is a comforting view of the mutual enrichment of pure science and modern engineering.

The field of continuum physics and mechanics called the theory of *configurational forces* or *material forces* can be viewed in two ways. First there are those, like the author, who claim that the theory of material forces is intimately related to a basic invariance of mathematical physics, the homogeneity or lack of homogeneity of the material (Maugin, 1993, 1995). This is not *per se* a new law of physics, but only the expression of the projection, onto the material manifold M^3, of a wellknown equation of balance of physics, that of linear momentum. All applications, known so far, can be related either to the local fully material expression of this equation when continuous fields are involved, or to the associated jump relation when discontinuous fields are involved across a singular surface, or else to an associated singular global integral when some field singularity is involved (Maugin, 1993; Kienzler and Herrmann, 2000). This unity and its broad range of applications contribute to both the practical strength and the irresistible beauty of this interpretation. Some strong minds, however, want to wrongly attach to this a mystic (Mueller, I., 1999) which they attribute to others while these "others" are fully pragmatic. The equations in question are just *canonical* projections onto the material manifold of wellknown equations. It happens that such projections facilitate the appearance and view of some effects that were invisible on the classical writing of the equation. A material force in that sense is less mystical that the force of gravity – see the first chapter of my book (Maugin, 1993). We can say that this viewpoint is certainly the one accepted by many prominent authors in the field (Rogula, 1977; Kienzler and Herrmann, 2000; Maugin and Trimarco, 2001; Steinmann, 2002, etc).

Another viewpoint is advocated by Gurtin (1999) and some of his co-workers, e.g., Podio-Guidugli (2001). It consists in viewing the balance of canonical or material momentum as a basic principle of physics, independently of the principle of balance of linear momentum, in a sense a new physical law! These authors are particularly interested in the treatment of singular surfaces. We do not agree with this viewpoint. In truth, there is no need to introduce a new law when everything is still explainable by the known ones. Also, like any modern physicist we believe that every conservation law is associated with an invariance principle. In the present context, the classical balance of linear momentum (with components in *physical* space) is related to the invariance under space translations in that space (invariance under *spatial* translations) in the absence of body forces (e.g., gravity which acts at the current physical point or *placement* x). This is called *homogeneity of space* (Landau and Lifshitz, 1986). Now, material space, i.e., matter *per se*, has no reason to be homogeneous. On the contrary, many materials are inhomogeneous in a smooth or more irregular or singular way. It is the canonical *material* balance of linear momentum which is generated by this invariance or lack of invariance (invariance under translations on the material manifold!) An ambiguity arises here because the actual placement x of a deformable material is in physical space while the usual space parametrization is the material point $\mathbf{X}$. The two are related by a time-parametrized mapping, the "motion" of the material. That is why one equation can be deduced from the other for *pure* mechanics and sufficiently smooth fields. This is not the case when the material body possesses other properties (cf. Maugin, 1995) and other degrees of freedom (cf. Maugin, 1997, 1998; Fomethe and Maugin, 1996) which are also parametrized by the material point and time. In that case the balance of canonical momentum captures **all** degrees of freedom, just in the same way as does the balance of energy. Accordingly, the balance of canonical momentum does not simply relate to a single degree of freedom but to them *all simultaneously*, a fact lightheartedly discarded by those not aware of the clear distinction between fields and parameters in a field theory. It is a pity that some respectable authors in mechanics and thermodynamics ignore this elementary fact.

Now what happens in a rigid body or in small deformations (when usually one does not introduce material coordinates)? The material in a rigid body is still subjected to the balance of canonical momentum since its properties (e.g., electromagnetic or heat-conduction ones) may still vary from one material point to another. In the deformable case, motion must be treated just like another field, and *no* more despite the fact that classical motion has the same geometrical and physical dimension as the spatial parametrization (in 3d). In small strains, one must be especially careful to distinguish between the placement and the coordinate frame as otherwise even the best authors may make errors of interpretation (see, Peierls, 1979).

In the case of a deformable solid exhibiting a singular surface such as a phase-transition front or a classical shock wave, it is clear that the material is *not* the same (or is not in the same phase or state) on both sides of the discontinuity surface which, therefore, breaks the translational symmetry on the material manifold. Accordingly, this dictates that the (co-vectorial) equation governing the change of material (phase) at the interface is none other than the jump relation associated with the balance of canonical momentum. It is such a method of typically mathematical-physics gesture than we used in all cases to construct with success the relevant equation and not an equation issued from some new physical law as Gurtin would like us to believe. Yes, we know that Eshelby stress is the enthalpy stress tensor, a fact known since Bowen's (1967) theory of mixtures and Grinfeld's (1991). But what counts above all is the wide landscape covered by the general concepts and more inclusive notions such as the generation of the Eshelby stress tensor by means of local structural rearrangements (Epstein and Maugin, 1990, 1997). In order to substantiate our view we complement previously published works with a brief presentation of a series of new developments that easily conform the wellfoundedness of our approach.

2 NOTION OF LOCAL STRUCTURAL REARRANGEMENT AND APPLICATIONS

We first recall a decisive step taken by Epstein and Maugin (1990) who established that in a continuously *inhomogeneous* elastic body the bulk *equilibrium* equation projected onto M^3 could be written as

$$div_R \mathbf{b} = \mathbf{b}:\Gamma \tag{1}$$

in the absence of given body force. Here the divergence operator is the referential one, **b** is the quasi-static material Eshelby stress and Γ is the material connection based on the uniformity map **K** so that, according to these authors

$$\mathbf{b} = -\frac{\partial \overline{W}}{\partial \mathbf{K}}\mathbf{K}^T \quad , \quad \Gamma = \mathbf{K}^{-1}.(\nabla_R \mathbf{K})^T \tag{2}$$

or, more classically,

$$\mathbf{b} = W\mathbf{1}_R - \mathbf{T}.\mathbf{F} \quad , \quad W = W(\mathbf{F};\mathbf{X}) = \overline{W}(\mathbf{F};\mathbf{K}(\mathbf{X})) \quad , \quad \mathbf{T} = \frac{\partial W}{\partial \mathbf{F}}, \tag{3}$$

where W is the elastic energy per unit referential volume, **F** is the deformation gradient, and **T** is the first Piola-Kirchhoff stress which satisfies the equilibrium equation $div_R\, \mathbf{T} = \mathbf{0}$ at all regular material points inside the body. Here $\mathbf{K}^{-1}$ maps the reference configuration K_R at material point **X** to a locally defined prototypical configuration called the *reference crystal.* Whenever the skewsymmetric part of the connection Γ is interpreted as the dislocation density tensor of a *continuously dislocated elastic body* in the manner of E.Kroener and others, then eqn. (1) is the equation that relates the notion of dislocation density and the true material mechanics.

Among the most recent advances in the material mechanics of materials, we should emphasize the applications of the concepts of uniformity map or local structural rearrangement **K** – see eqn. (4) below – to various complex behaviors of matter. It was first realized by Epstein and the author (1995) – see also Maugin (1994) – that the Eshelby stress tensor could be viewed as the (bulk) driving force behind finite-strain plasticity. This was facilitated by the identification of the second part of the material tensor **b**, as the Mandel stress $\mathbf{M} = \mathbf{T} \cdot \mathbf{F}$ of finite-strain anelasticity, perhaps to be considered in the so-called intermediate configuration. In this case the above introduced **K** is none other than the inverse of the plastic deformation component $\mathbf{F}_p$ of the celebrated multiplicative decomposition $\mathbf{F} = \mathbf{F}_e \cdot \mathbf{F}_p$. Indeed,

$$\mathbf{K}^{-1} \equiv \mathbf{F}_p \quad , \quad \mathbf{FK} = \mathbf{F}.\mathbf{F}_p^{-1} \equiv \mathbf{F}_e. \tag{4}$$

The deformation $\mathbf{F}_p$ defines locally an *elastically released configuration* at material point **X**. This is also called a (local) *intermediate configuration* K_i (cf. Mandel, 1971; Sidoroff, 1976; Teodosiu and Sidoroff, 1978). Use can be made of formula (4) to find out the thermodynamic force associated with the anelastic internal variable of state $\alpha \equiv \mathbf{F}_p$ through the geometrical description. Introducing the second Piola-Kirchhoff stress **S** and the "Mandel" stress **M** by

$$\mathbf{S} = \mathbf{T}.\mathbf{F}^{-T} \quad , \quad \mathbf{M} = \mathbf{T}.\mathbf{F} = \mathbf{S}.\mathbf{F}^T.\mathbf{F} = \mathbf{S}.\mathbf{C}, \tag{5}$$

we obtain that

$$\mathbf{b} = W\mathbf{1}_R - \mathbf{M} \quad or \quad \mathbf{M} = W\mathbf{1}_R - \mathbf{b}. \tag{6}$$

As a consequence, the thermodynamic force associated with α is given by (cf. Maugin, 2003)

$$A = (\mathbf{M} - W\mathbf{1}_R).\mathbf{F}_p^{-T}. \tag{7}$$

Simultaneously, the *intrinsic dissipation* Φ^{intr} takes on the following form:

$$\Phi^{intr} = A\dot{\alpha} = tr\left[(\mathbf{M} - W\mathbf{1}_R).\mathbf{F}_p^{-T}.\dot{\mathbf{F}}_p^T\right], \tag{8}$$

or

$$\Phi^{intr} = tr\left[\left(\mathbf{M} - W\mathbf{1}_R\right).\left(\mathbf{L}_R^p\right)^T\right], \tag{9}$$

where

$$\mathbf{L}_R^p = \dot{\mathbf{F}}_p.\mathbf{F}_p^{-1} \tag{10}$$

is the "plastic" finite-strain rate in the reference configuration K_R. If the plastic deformation is assumed to be incompressible, then $tr\ \mathbf{L}_R^p = 0$, and (9) reduces formally to

$$\Phi^{intr} = tr\left[\mathbf{M}.\left(\mathbf{L}_R^p\right)^T\right]. \tag{11}$$

In plain words this means that the *Mandel stress is the driving force behind plasticity*. The relation (11) can be as well expressed with geometrical objects pushed forward to the intermediate configuration K_i, a more usual formulation. Obviously, the Mandel stress is only one part of the Eshelby stress. The first relationship between finite-strain plasticity and the notion of Eshelby stress in an intermediate configuration was established by Maugin (1994). More on Eshelby stress and finite-strain elastoplasticity in Tigoiu and Maugin (2000). No wonder then that the Eshelby stress **b** may also be used as the *resolved shear stress* activating dislocation systems (Kh. Chau Le, 1999).

Since volumetric *material growth* is very much akin to plasticity save, obviously, for the bulk behavior – i.e., it is a special kind of local structural rearrangement in which more matter of the same kind is squished in at each material point – it was natural to develop a "material" theory of material growth on the same bases (Epstein and Maugin, 2000) emphasizing the role of the Eshelby stress in such processes. In other words, the process of volumetric growth can be seen as an evolution of material-point neighborhoods in a fixed reference configuration, an evolution that finds its expression in *temporal changes of density* and concomitant distorsions of material-point neighborhoods. While preserving the uniform material properties, the growth process will in general cause the developments of *inhomogeneities*, responsible for *residual stresses* in the body. Then mass density, even in the reference configuration, can evolve in time. Accordingly, discarding diffusive effects, the global mass balance equation reads locally on the material manifold

$$\left.\frac{\partial \rho_0}{\partial t}\right|_{\mathbf{X}} = \Pi . \tag{12}$$

It was shown (Epstein and Maugin, 2000) that, for this theory, the first Piola-Kirchhoff stress **T**, and the Eshelby stress tensor **b** are left formally unchanged while a *material growth force* $\mathbf{f}^{gr}$: $= \Pi\ \mathbf{V}$ appears in such a way that the balance equation of canonical momentum – essentially eqn. (1) – then reads

$$\left.\frac{\partial P}{\partial t}\right|_{\mathbf{X}} - div_R\mathbf{b} = \mathbf{f}^{gr} \tag{13}$$

where P is the so-called material momentum, **b** is the dynamical generalization of the Eshelby stress, and **V** is the material velocity defined by taking the partial time derivative, at fixed actual placement **x**, of the inverse motion mapping $\mathbf{X} = \chi^{-1}(\mathbf{x}, t)$. The Clausius-Duhem inequality is also left formally unchanged compared to the case of thermoelasticity, i.e.,with usual symbols it reads

$$-\left(\dot{W} + S\dot{\theta}\right) + tr\left(\mathbf{T}\left(\nabla_R\mathbf{v}\right)^T\right) - \theta^{-1}\mathbf{Q}.\nabla_R\theta \geq 0. \tag{14}$$

Now we introduce the inhomogeneity mapping **K** called *transplant* in this growth theory with physiological applications. Let J_k = det **K**, and ρ_c the truly invariant mass density in the crystal reference so that

$$\rho_0 = J_K^{-1} \rho_c \,. \tag{15}$$

By time differentiation this shows that *growth* (increase in density) corresponds to $\partial(det\,\mathbf{K})/\partial t < 0$ and *resorption* (decrease in density) corresponds to the opposite condition. With a simplified functional dependence of W taken as $W = \overline{W}(\mathbf{F}, \theta, \mathbf{X}, t)$ we can write $W = J_K^{-1} W_c(\mathbf{F} \cdot \mathbf{K}(\mathbf{X}, t), \theta)$. We also introduce the following two transplant velocity gradients:

$$\mathbf{L_K} = \dot{\mathbf{K}}.\mathbf{K}^{-1}, \; \overline{\mathbf{L}}_\mathbf{K} = \mathbf{K}^{-1}.\dot{\mathbf{K}} = -\dot{\mathbf{K}}^{-1}.\mathbf{K} \,. \tag{16}$$

On account of (15) we have

$$\left.\frac{\partial \rho_0}{\partial t}\right|_\mathbf{X} = -\rho_0 tr \mathbf{L_K} = \Pi \,. \tag{17}$$

This shows that there is no need of an evolution equation for the volumetric source, once we know the evolution of **K**. This must be ruled by the second law of thermodynamics. As a matter of fact, **T** and S receiving their classical form as state laws, the reduced dissipation reads as follows:

$$\Phi = tr(\mathbf{M}.\mathbf{L_K}) - \theta^{-1}\mathbf{Q}.\nabla_R \theta \geq 0 ,$$

where **M** is the above introduced Mandel stress. The system of constitutive equations must now be closed by a heat conduction law (e.g., Fourier's) and the evolution equation for the transplant **K**. The latter can be envisaged as the following general relationship between various arguments:

$$\varphi(\mathbf{K}, \dot{\mathbf{K}}, \mathbf{M} \; or \; \mathbf{b}, \mathbf{F}, \dot{\mathbf{F}}) = 0 , \tag{18}$$

reflecting thus some kind of *nonlinear viscoelasticity*, because of the obvious presence of one or several time scales. Using invariance arguments it can be shown (Epstein and Maugin, 2000), that this can be reduced to the following relationship:

$$\varphi(\overline{\mathbf{L}}_\mathbf{K}, \mathbf{b}_c, \mathbf{C}_e, \dot{\mathbf{C}}_e) = 0 , \tag{19}$$

where $\mathbf{C}_e = \mathbf{F}_e^T \cdot \mathbf{F}_e$ and $\dot{\mathbf{C}}_e$ come from $\mathbf{F}_e = \mathbf{F} \cdot \mathbf{K}$ – the elastic deformation gradient, $\overline{\mathbf{L}}_\mathbf{K}$ comes from an invariant combination of **K** and $\dot{\mathbf{K}}$, and $\mathbf{b}_c$ is given by $\mathbf{b}_c = J_K \mathbf{K}^{-1} \cdot \mathbf{b} \cdot \mathbf{K}$.

As a simple example of isotropic behavior for the symmetric part $\overline{\mathbf{L}}_S$ of $\overline{\mathbf{L}}_\mathbf{K}$ we have the evolution equation

$$\overline{\mathbf{L}}_S = \varphi_0 \mathbf{1} + \varphi_1 \mathbf{b}_c + \varphi_2 \mathbf{b}_c^2 , \tag{20}$$

where the coefficients are functions of the fundamental invariants of $\mathbf{b}_c$. Such a law, akin to *non-linear viscoelasticity*, is exploited in computations of growth by Imatani and Maugin (2003) who also account for *adaptation* via the evolution of a possible anisotropy. This receives applications in the biomechanics of soft tissues.

Finally, with a view to studying the biomechanics of objects like remodelled bones it was necessary to envisage a theory of binary mixtures including both a solid with pores and a fluid saturating the latter. Since the material view is essentially one where the most relevant kinematics is that of the inverse motion instead of the traditional "direct" motion $\mathbf{x} = \chi(\mathbf{X},t)$ – the question naturally arises of the choice of which material manifold should the motions of several co-existing media be pulled back to; the answer is simple and logical, both solid and fluid deformations are pulled back to the *solid* material manifold, i.e., on the skeleton in terms of porous-media mechanics. This is exemplified by the work of Quiligotti *et al* (2003).

3 BODY MATERIAL FORCES CAN BE COMPUTED AND ARE USEFUL!

It is usually, and correctly, remarked that at all regular material points, the balance of pseudo-momentum is "equivalent" to the balance of physical momentum. The reason for this is that at all such points, the former can be deduced from the latter by some algebraic and analytic operations allowed by the smoothness of the fields. That is, at such points the balance of pseudo-momentum is an *identity.* However, it is not exactly true that both equations of momentum are *operationally* equivalent, being then only two different "projections" on different manifolds, of the same equation, because they *cannot* play the same role in problem solving. This is due to the fact that, whether we like it or not, applied forces – which are real physical forces in the Newtonian-Euler-Cauchy sense – are prescribed in physical space (a world to which they belong by definition) so that they intervene in problem solving in the equations (field equations and boundary conditions) expressed in that space. Therefore, initial-value boundary-value problems are, of necessity, solved, analytically or numerically, in physical space. The question then arises of the usefulness of the balance of pseudo-momentum at regular material points, its role in capturing singularities being clear at singular points. This usefulness resides in the redundancy between the two (essentially vectorial) equations as it can be exploited thus. While the balance of physical momentum – and other equations governing other fields than deformation – are used in the direct problem solving, the balance of pseudo-momentum (or its "equilibrium" form where this applies) is to be exploited either in the form of a *criterion* to be satisfied by the already obtained point-wise field solution (e.g., in fracture, phase-transition, anelasticity, etc), or as a way to checking globally the accuracy and quality of the obtained solution. This is very similar to the exploitation of the energy balance in the numerics of hyperbolic systems where the accuracy of the adopted scheme is measured by a global energy criterion (i.e., does the scheme conserve energy?). Herein below we focus attention on such an application of the balance of pseudo-momentum in various numerical techniques, noting in passing that the notion of momentum is vectorial, and therefore "directional", and thus richer than that of energy (a scalar) which addresses only magnitude. Three numerical techniques are briefly examined in the light of recent progresses. Before, however, a simple reminder of basic equations is useful.

In the absence of material inhomogeneities and at all regular material point, eqn. (1) reduces to

$$div_R \mathbf{b} = \mathbf{0}, \tag{21}$$

a pure identity deduced from the classical equilibrium equation (with components in physical space in the present configuration)

$$div_R \mathbf{T} = \mathbf{0}. \tag{22}$$

Equation (21) is trivially deduced from (22) by right-composition of (5) with **F** and manipulations on account of the elastic energy dependency on **F**, and only on **F**. A boundary-value problem of nonlinear elasticity is solved on the basis of (22) accompanied by appropriate boundary conditions at the boundary of the material volume. Once eqn. (22) is solved numerically by, e.g., a finite-element scheme, eqn. (21) should be checked. In most cases this is not exactly checked and, as rewritten in a continuous framework, one shall obtain an eqn. (21) with a source term, i.e.,

$$div_R \mathbf{b} = \mathbf{f}^{comp}, \tag{23}$$

where $\mathbf{f}^{comp}$ is a spurious "material force" induced by an incorrect meshing to the problem (this quantity in fact is computed at the nodes of the FEM scheme). It provides a *directional* (since it is a co-vector on the material manifold M^3) *indicator*, in direction and intensity, of how the FEM nodes should be moved to obtain a better approximation. The latter should be such as to minimize the potential energy. This was first noticed by Braun (1997) and these ideas pursued and put to practice by Maugin (2000a), Mueller *et al* (2001), and Steinmann (2002). Moreover, this practice having shown how easy it is to evaluate these material forces, the computation of such quantities as the *succion force* (Mueller and Maugin, 2002) or *driving force* on the tip of a crack in general physical circumstances has become a rather easy matter. That is, the co-vectorial driving force (not a simple scalar such as the *J-integral*) can be evaluated numerically on account of the numerical solution of a complete problem, not only of homogeneous elasticity, but also in the presence of thermal, anelastic and smooth or abrupt inhomogeneity effects. The quasi-static non-collinear progress of a crack being attracted by an inhomogenity using a simple kinetic law of progress has thus been spectacularly illustrated by Mueller and Maugin (2002). Steinmann and co-workers (2003) have shown the practicality of such computations in more general thermal-anelastic cases. All these computations are based on a volume integration of the balance of material momentum (generalization of eqn. (1)) and the energy equation, the first of which reading then (cf. Maugin, 1997)

$$div_R \mathbf{b} = -\left(\mathbf{f}^{inh} + \mathbf{f}^{th} + \mathbf{f}^{anel} + \mathbf{f}^{inert}\right), \quad \mathbf{f}^{inert} = -\frac{\partial P}{\partial t}, \tag{24}$$

where in the right-hand-side of $(24)_1$, using a revealing notation, we have contributions from true material inhomogeneities, and thermal, anelastic, and inertial effects, all in the form of "pseudo-inhomogeneity" material forces. All these may be computed at each node and at each time step, although not perhaps so simply. Note that the volume of integration for (24) evolves with time due to the irreversible progress of the crack in the material body and this is what brings in the picture, by duality, the notion of driving force acting on the tip of the crack.

Examples of applications in the finite-element (FE) method: In the case where eqn. (21) applies at all regular points, considering hyperlasticity of the neo-Hookean type, Mueller and Maugin (2002) have first considered the 2d problem of a block of rectangular material loaded by a displacement on the top side, assuming plane-strain state. The corresponding discrete material forces are readily computed (Figure 1b). Particularly large material forces appear at the boundary while no material forces should appear in the interior. The material forces here appear because the FE approximation is not smooth with respect to strains (stresses). Material forces on the boundary appear as reaction forces to the constraint imposed by the fact that the boundary is fixed in the reference configuration. Question: does there exist a "better" internal arrangement of the nodes for which the material forces vanish? This is a finite dimensional optimization problem which may

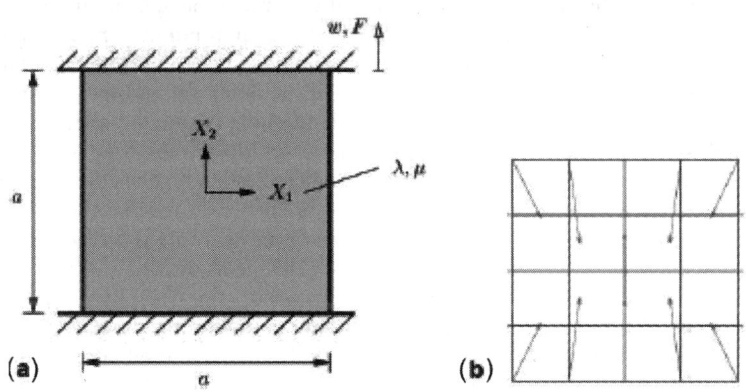

Figure 1. (a) Block loaded by displacement *w*; (b) network of material forces.

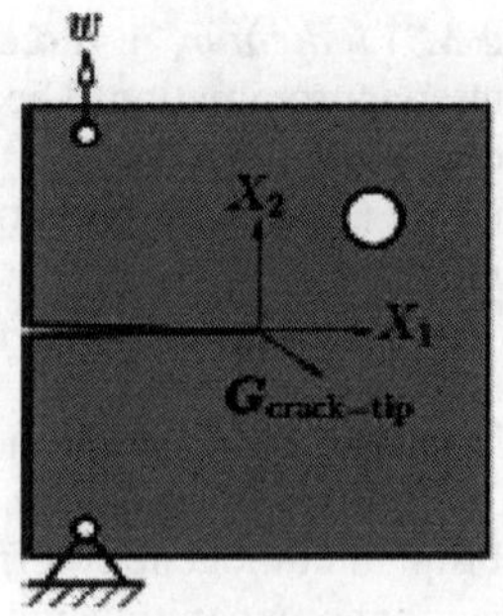

Figure 2. Specimen with a crack and a hole loaded by a dead displacement w.

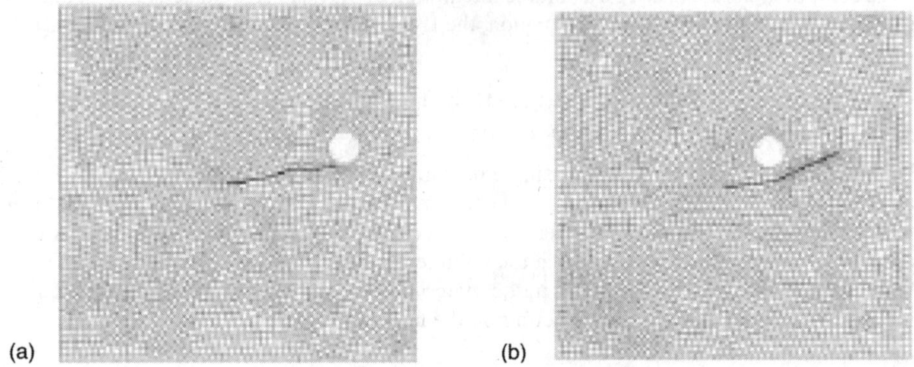

Figure 3. (a) Trajectory of crack tip: hole far enough from starting point; (b) Trajectory of crack tip: hole too close to starting point and the hole is missed.

not be convex. The material force $\mathbf{f}^{comp} = \mathbf{G}$ (an energy-release rate) at a nodal point $\mathbf{X}$ allows one to use a simple "evolution criterion" (however here there is no time scale) such that the new location of the interior node $\mathbf{X}$ will be given by the mapping

$$\mathbf{X} \to \mathbf{X} - c\mathbf{G}, \tag{25}$$

where c is an appropriate constant which is chosen small enough to achieve convergence and to avoid "unhealthy" mesh distorsion. This is an *updating technique* which parodies the progress criterion of structural defects in the "material" mechanics of materials. Indeed, another illuminating example consists in studying the possible progress of the crack tip in Figure 2 in a 2d simulation of a specimen loaded by a dead displacement w, a circular hole being here introduced in the specimen so as to study the interaction of the material force at the crack tip and the material forces caused by the hole (a material inhomogeneity by itself). If the hole is situated relatively far from the crack tip we may say in the archetry language that the crack tip has the time to adjust to this target because it is attracted to the vacuum of the hole. The crack then reaches the hole by gently changing its direction, but the simulation is stopped just before the crack reaches the hole, because in this situation the automatic mesh generation fails. If, however, the hole is already too close to the crack tip, although out of axis, then the crack tip does not change its direction sufficiently fast to reach the hole, and it misses it while the crack tip trajectory is still significantly perturbed by the hole (see Figures 3a and 3b). In this simple but impresssive simulation the crack tip propagation is governed by a simple (evolution but without time scale) rule such that (compare to eqn. (25))

$$\mathbf{X}^{new}_{crack-tip} \to \mathbf{X}^{old}_{crack-tip} - c\mathbf{G}_{crack-tip}, \tag{26}$$

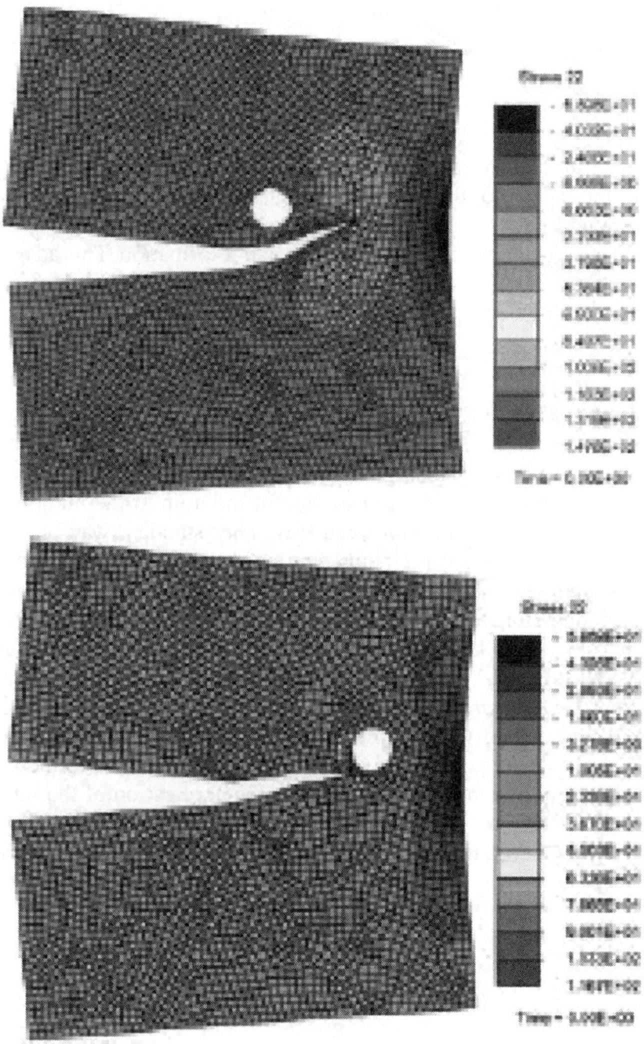

Figure 4. (a) The crack misses the hole (cf. Figure 3b). (b) The crack hits the hole (cf. Figure 3a).

where $\mathbf{G}_{crack\text{-}tip}$ is the material force (energy-release rate) at the crack tip and c is an approriately chosen constant so that the crack propagation has to be sufficiently small (incremental viewpoint in a rate-independent situation) and at the same time large enough to ensure a proper mesh generation for the new geometry (see also Figures 4a and 4b). This illustration having only pedagogical purpose no threshold value of crack initiation is here introduced although a value *à la* Griffith could be introduced (so-called energy criterion). Other examples can be found in Mueller and Maugin (2002). The question is naturally raised of the adaptivity of the FE method using the Eshelby-stress tensor notion. The magnitude of the discrete material forces which are computed in the post-processing of the solution of the material force balance – eqn. (23) written in a weak form – may be used as an indicator for mesh refinement (here a scalar indicator). The main concern is that of boundaries where, as noted before, important material forces are developed as reaction forces (see Mueller, Gross and Maugin, 2003). The said magnitude is used to check the discretization and assign new mesh sizes. If h^I is the smallest edge length of all elements adjacent to the node I, then in the adaptive scheme the new discretization is generated from a new distribution of mesh sizes.

This, introducing refinement factors r^I, is schematized by the new modal values of the mesh size given by the expression

$$h^I_{new} \leftarrow h^I_{old}\, r^I \quad , \tag{27}$$

where the r^I are chosen so as to make the magnitude of the discrete material forces, perhaps summed over the whole of the interior of the body (i.e., all internal nodal points) decrease. This is not ideal because only the magnitude of the material forces is taken into account while each of these, by its very nature of being a co-vector, also carries directional information. This adaptivity question is currently debated as shown by the discussions at the EUROMECH Colloquium held in Kaiserslautern (Steinmann and Maugin, 2003).

4 APPLICATION TO THE PROGRESS OF PHASE-TRANSITION FRONTS

To conclude our presentation of the large field of applications of materials mechanics and material forces, we mention the possibility to use this formulation in the numerics of the progress of phase-transition fronts, but in a **non-standard way**. The "standard way" consists in complementing the system of field equations valid in any continuous phase and the system of jump (or transition) relations at the phase front Σ – a special type of shock waves – by a *kinetic relation* relating the local normal propagation velocity of the wave front to the computed driving force at the same singular point and using a standard FEM scheme. This follows works by Truskinowsky, Abeyaratne, Knowles and others. But we have shown (Maugin, 1997) within the logics of viewing the presence of the front as a *breaking of translational symmetry* on the material manifold that the driving force $\mathbf{f}_\Sigma$ was generally a material co-vector computed from the jump of the quasi-static Eshelby stress across the front. Again, this in general is a *directional indicator* of the further progress of the front, which progress should not be in contradiction with the application of the second law of thermodynamics across the front. That is, the source of entropy, if any, at the wave front, should not be negative. We have shown that, in fact, for homothermal fronts, the local dissipation at the front is such that

$$\Phi_\Sigma = \mathbf{f}_\Sigma . \overline{\mathbf{V}} \geq 0 \ . \tag{28}$$

The usual *pure normal-growth* kinetic law then is a generally nonlinear relationship

$$V_N := \overline{\mathbf{V}}.\mathbf{N} = g\left(f_\Sigma = \mathbf{N}.\mathbf{f}_\Sigma\right), \tag{29}$$

where $\mathbf{N}$ is the unit normal to Σ. Such a relation must be provided in an up-scaling approach (the natural one for scale effects in homogenization problems) by a lower-level description essentially one where the front Σ has a non-zero thickness across which dissipative processes take place (compare Maugin, 2000b). Then the progress of the transition front could be treated in an incremental way, using a finite-element scheme in space.

The "non-standard way" devised by Maugin and Berezovski (2000) is based on the initial remark that the *finite-volume element method* (for short FVM) – favored in many problems of fluid dynamics – considers at once all equations of the continuum problem as balance laws over each cell-volume element. Furthermore, in its discretization it happens to become identical to the thermodynamics of so-called *Schottky* (discrete) *thermodynamical systems*, the exchanges with neighboring system being readily identified to the fluxes at the boundaries of the cells, which boundaries are simple flat surfaces. Accordingly, a problem of phase-transition front propagation is numerically treated in this way, the additional co-vectorial balance equation provided by the balance of canonical momentum (24) in which the associated flux is none other than the Eshelby stress, being logically incorporated in the scheme and the change of the components of the Eshelby stress at the phase change being duly accounted for. In this type of approach two important ingredients are the introduction of a generating function f_g at any point $\mathbf{X}$ in the numerical scheme by the definition (Berezovski and Maugin, 2003)

$$f_g = -\mathbf{N}.\mathbf{b}_S.\mathbf{N} + f_0 \ , \tag{30}$$

where $\mathbf{N}$ is the unit normal to any oriented smooth surface $\Sigma_\mathbf{X}$ passing through the point $\mathbf{X}$, $\mathbf{b}_s$ is the quasi-static part of the Eshelby stress tensor, and f_0 is a constant such that its jump is zero across a true phase-transition front $\Sigma_\mathbf{X} = \Sigma$. If $\mathbf{X}$ happens to be situated on the front across which there holds the relation

$$f_\Sigma = \mathbf{N}.\mathbf{f}_\Sigma = \left[f_g \right]_\Sigma \quad , \tag{31}$$

where square brackets denote the jump of their enclosure.

Also, in the typical spirit of a finite-volume (FV) scheme one can establish as a basic equation of such a scheme the time evolution of the cell average of the quasi-static Eshelby stress tensor, e.g., in small strains (Maugin, 2000)

$$\begin{aligned} \frac{\partial}{\partial t}\int_V \mathbf{b}_S dV = &-\int_{\partial V} \mathbf{N}.\Sigma.\mathbf{V}\, dA \\ &+ \int_V \left(tr(\sigma.\nabla\mathbf{v})\mathbf{1} - \frac{\partial \sigma}{\partial t}(\nabla\mathbf{u})^T + (\nabla\sigma)\mathbf{v} \right) dV \end{aligned} \tag{32}$$

where V is the elementary cell in the FV scheme, Σ and $\mathbf{V}$ are the trace of the Cauchy stress and velocity at the boundary of this element (so-called contact quantities in Schottky's (discrete) thermodynamics), and $\boldsymbol{\sigma}$, $\mathbf{u}$, and $\mathbf{v}$ are the standard Cauchy stress, mechanical displacement, and physical velocity, respectively.

Together with the definition of the threshold of stability between phases in the form of relationships between thermodynamic derivatives, the quantities defined in (30) and (32) and the standard equations of conservation and constitution written in the FV formalism, allow one to simulate the progress of phase-transition fronts in thermoelasticity, especially in such problems as the impact-induced displacement of a phase front (Berezovski, Engelbrecht and Maugin, 2003).

Just for the sake of completeness we mention the rôle played by the application of the global balance of material or canonical momentum in the finite-difference (FDM) schemes used in the simulation of nonlinear wave-propagation problems for strongly localized solutions (e.g., solitons, or quasi-solitonic solutions). It can obviously be used again as a check of the accuracy of the scheme: both energy and total canonical momentum must be conserved in the absence of material inhomogeneities and dissipative processes or forcing. In case the latter effects are present and clearly modify the dynamical solution that is no longer "inertial", then this equation can be exploited both analytically and numerically to treat the gross perturbation caused by these effects (alteration in the total canonical momentum with time due to the presence of a "driving force"). We refer the reader to older works for this aspect of FDM applications (Maugin, 1999; Maugin and Christov, 2001).

5 CONCLUSION

Referring to all above considered cases we must emphasize the clearly distinguishable roles played by the field equations, *per se* – e.g., eqn. (22) – and the canonical equation of momentum. The latter proves to be extremely useful, and in fact the only tool available, to devise *criteria of progress* on a phenomenological scale. This applies not only to physical theories such as those of elasto-plasticity, growth, and phase-transition that involve known driving forces in terms of the Eshelby stress or one of its associates, but also to all numerical methods examined, whether of the finite-element, finite-difference, or finite-volume type. These applications succeed in granting a clearly engineering-like flavor to concepts, those of material momentum and Eshelby stress, whose origin is to be found in abstract field theory and the geometry of the material manifold or material space.

ACKNOWLEDGEMENT

Some of the work presented was developed within the European Network TMR-98-0229. The author benefits form a *Max-Planck Award* for International Cooperation (2001–2005).

REFERENCES

Berezovski, A., Engelbrecht, J. & Maugin, G.A. 2003. Numerical Simulation of Moving Phase Boundaries in Solids. *Interfaces and Free Boundaries* (in the press).

Berezovski, A. & Maugin, G.A. 2003. On the thermodynamic conditions at moving phase-transition fronts in thermoelastic solids. *J. Non-Equilibr. Thermodynamics* (in the press).

Bowen, R. 1967. Toward a thermodynamics and mechanics of mixtures. *Arch. Rat. Mech. Anal.* 24: 370–403.

Braun, M. 1997. Configurational forces induced by finite-element discretization. *Proc. Est. Acad. Sci. Math. Phys.* 46: 4–31.

Cleja-Tigoiu, S. & Maugin, G.A. 2000. Eshelby's Stress Tensors in Finite Elastoplasticity. *Acta Mechanica* 139: 119–131.

Epstein, M. & Maugin, G.A. 1990. The energy-momentum tensor and material uniformity in finite elasticity. *Acta Mechanica* 83:127–133.

Epstein, M. & Maugin, G.A. 1995. On the geometrical material structure of anelasticity. *Acta Mechanica* 115: 335–348.

Epstein, M. & Maugin, G.A. 1997. Notions of material uniformity and homogeneity. In T. Tatsumi, E. Watanabe, and T. Kambe (eds) *Theoretical and Applied Mechanics* (ICIAM, Kyoto, 1996), pp. 201–215. Amsterdam: Elsevier.

Epstein, M. & Maugin, G.A. 2000. Thermomechanics of volumetric growth in uniform bodies. *Int. J. Plasticity* 16: 951–978.

Fomethe, A. & Maugin, G.A. 1996. Material forces in thermoelastic ferromagnets. *Continuum Mech. Thermodyn.* 8: 275–292.

Grinfeld, M. 1991. *Thermodynamic Methods in the Theory of Heterogeneous Systems.* Harlow, Essex: Longman.

Gurtin, M.E. 1999. *Configurational Forces as Basic Concepts of Continuum Physics*. New York: Springer-Verlag.

Imatani, S. & Maugin, G.A. 2003. A Constitutive modeling for growing materials and its applications to finite-element analysis. *ASME. Tran. J. Appl. Mech.* (in the press).

James, R.D. 2002. Configurational forces in magnetism with application to the dynamics of a small-scale ferromagnetic shape-memory cantilever. *Continuum Mech. Thermodyn.* 14: 55–86.

Kienzler, R. & Herrmann, G. 2000. *Mechanics in Material Space with Applications to Defect and Fracture Mechanics*. Berlin: Springer-Verlag.

Kienzler, R. & Maugin, G.A. (eds) 2001. *Configurational Mechanics of Materials* (CISM Courses and lectures No.427). Wien: Springer-Verlag.

Landau, L.D. & Lifshitz, E.M. 1986. *Theory of Fields*. Oxford: Pergamon.

Le, K.C. 1999. Thermodynamically based constitutive equations for single crystals. In G.A. Maugin (ed) *Geometry, Continua and Microstructure* (Proc.GCM1, Paris, 1997), pp.87–97. Paris: Hermann Editeurs.

Mandel, J. 1971. *Plasticité et viscoplasticité classique* (CISM Udine lectures). Wien: Springer-Verlag.

Maugin, G.A. 1993. *Material Inhomogeneities in Elasticity*. London: Chapman and Hall.

Maugin, G.A. 1994. Eshelby Stress in Plasticity and Fracture. *Int. J. Plasticity* 10: 393–408.

Maugin, G.A. 1995. Material forces: concepts and applications. *Appl. Mech. Rev.* 48: 213–245.

Maugin, G.A. 1997. Thermomechanics of inhomogeneous-heterogeneous systems: applications to the irreversible progress of two- and three-dimensional defects. *ARI* 50: 41–56.

Maugin, G.A. 1998. On the structure of the theory of polar elasticity. *Phil. Trans. Roy. Soc. Lond.* A356: 1367–1395.

Maugin, G.A. 1999. *Nonlinear Waves in Elastic Crystals*. Oxford: Oxford University Press.

Maugin, G.A. 2000a. Geometry of material space: its consequences in modern numerical means. *Technische Mechanik* (Proc. GCM3, Herrenalb, 1999) 20: 95–104.

Maugin, G.A. 2000b. Multiscale Approach to a Basic Problem of Materials Mechanics (Propagation of Phase Transition Fronts). In A. Sendig, W. Schielen et W.L. Wendland (eds), *Multifield Problems: State of the Art*, pp.11–22. Berlin: Springer-Verlag.

Maugin, G.A. 2003. Pseudo-plasticity and pseudo-inhomogeneity effects in materials mechanics. *J. Elasticity* (Truesdell Volume, in press).

Maugin, G.A. & Berezovski, A. 2000. Thermoelasticity of inhomogeneous solids and finite-volume dynamic computations. In B. Albers (ed) *Contributions to Continuum Theories* (K.Wilmanski's Festschrift), pp.166–173. Berlin: W.I.A.S.

Maugin, G.A. & Christov, C.I. 2001. Nonlinear waves and conservation laws: nonlinear duality between elastic waves and quasi-particles. In C.I. Christov and A. Guran (eds) *Selected Topics in Nonlinear Wave Mechanics*, pp.117–160. Boston: Birkhauser.

Maugin, G.A. & Trimarco, C. 1995. On material and physical forces in liquid crystals. *Int. J. Engng. Sci.* 33: 1663–1678.

Maugin, G.A & Trimarco, C. 2001. Elements of field theory in inhomogeneous and defective materials. In R. Kienzler & G.A. Maugin (eds). *Configurational Mechanics of Materials*, pp. 55–171. Wien: Springer Verlag.

Mueller, I. 1999. Eshelby tensor and phase equilibrium. *Theoretical and Applied Mechanics* (Belgrade) 25: 77–99.

Mueller, R., Gross, D. & Maugin, G.A. 2003. Use of the Eshelby-stress tensor in adaptive finite-element methods. *Computational Mechanics* (in press).

Mueller, R., Kolling, S. & Gross, D. 2002. On configurational forces in the context of the finite-element method. *Int. J. Num. Meth. Engng.* 53: 1557–1574.

Mueller, R. & Maugin, G.A. 2002. On material forces and finite-element discretizations. *Computational Mechanics* 29: 52–60.

Peierls, R. 1979. *Surprises in Physics*. Princeton, New Jersey: Princeton University Press.

Podio-Guidugli, P. 2001. Configurational balances via variational arguments. *Interfaces and Free Boundaries* 3: 223–232.

Quiligotti, S., Maugin, G.A. & Dell'Isola, F. 2003. An Eshelbian Approach to the Nonlinear Mechanics of Constrained Solid-Fluid Mixtures. *Acta Mechanica* 160: 45–60.

Rogula, D. 1977. Forces in material space. *Arch. Mech.* 29: 705–715.

Sidoroff, F. 1976. Variables internes en viscoé lasticité et plasticité. *State Doctoral Thesis in Mathematics*. Paris: Université Pierre et Marie Curie.

Steinmann, P. 2002. On spatial and material settings in hyperelastodynamics. *Acta Mechanica* 156: 193–218.

Steinmann, P. & Maugin, G.A. (eds) 2003. *Mechanics of Material Forces* (Proc. EUROMECH Kaiserslautern, 2003). In press.

Teodosiu, C. & Sidoroff, F. 1976. A theory of finite elastoplasticity in single crystals. *Int. J. Engng. Sci.* 14: 165–176.

Configurational Mechanics, Kalpakides & Maugin (eds)
© 2004, Taylor & Francis Group, London, ISBN 90 5809 667 X

On the impact of configurational mechanics on computational mechanics

Ellen Kuhl & Paul Steinmann
Chair of Applied Mechanics, University of Kaiserslautern, Kaiserslautern, Germany

ABSTRACT: This contribution is concerned with the impact of configurational mechanics on computational mechanics. To this end we derive a generic hyperelastic ALE formulation embedded in a variational framework. The governing equations follow straightforwardly from the Dirichlet principle for conservative mechanical systems. Thereby, the key idea is the reformulation of the total variation of the potential energy at fixed referential coordinates in terms of its variation at fixed material and at fixed spatial coordinates. The corresponding Euler–Lagrange equations define the spatial and the material motion version of the balance of linear momentum, i.e. the balance of spatial and material forces, in a consistent dual format. In the discretised setting, the governing equations are solved simultaneously rendering the spatial and the material configuration which minimise the overall potential energy of the system. The remeshing strategy of this FE formulation is thus no longer user-defined but objective in the sense of energy optimisation. As the governing equations are derived from a potential, they are inherently symmetric, both in the continuous case and in the discrete case. The characteristic features of the derived approach will be demonstrated by means of selected examples.

1 INTRODUCTION

The deformation of a body is essentially characterised through the balance of linear momentum. Typically, this balance of momentum is formulated in terms of the spatial deformation map mapping the material placements of particles in the material configuration to their spatial placements in the spatial configuration. Alternatively, we could formulate the balance of momentum in terms of the material motion map, see (Shield 1967, Maugin 1993, Maugin 1995, Gurtin 1995, Gurtin 2000, Kienzler and Herrmann 2000, Steinmann 2002a, Steinmann 2002b, Steinmann 2003, Kuhl and Steinmann 2003). Being the inverse of the spatial motion map, the material motion map characterises the mapping of particles in the spatial configuration to their material motion counterparts. While the former approach relates to the equilibration of the classical spatial forces in the sense of Newton, the latter lends itself to the equilibration of material forces in the sense of Eshelby. Thereby, spatial forces typically perform work over positional changes relative to their ambient space, whereas material forces perform work over positional changes relative to the ambient material. In the continuous case, the statement of spatial force equilibrium and material forces equilibrium are completely equivalent for homogeneous problems, i.e. they can be related to one another via a complete pull back onto the material manifold.

The finite element discretisation renders the discrete residual statement of the balance of momentum. In the classical sense, the solution of the finite element method corresponds to the vanishing residual of the spatial force balance. It is typically parameterised in terms of the discrete spatial motion map or rather in terms of the discrete spatial placements of the node point positions. Within the finite element context, this solution is related to the minimum of the potential energy. This minimum, however, is only a minimum with respect to fixed material placements.

The discrete solution of the spatial motion problem does not necessarily have to coincide with the discrete solution of the material motion problem, i.e. a vanishing residual of discrete spatial forces does not necessarily imply that the discrete material forces vanish equivalently, see (Braun 1997, Müller and Maugin 2002, Askes et al. 2003b, Kuhl et al. 2003, Askes et al. 2003a, Thoutireddy 2003). Non-vanishing discrete material forces indicate that the underlying discretisation is not yet optimal. An additional release of energy can take place when finite element nodes are moved in the direction opposite to the resulting discrete material node point forces.

In this respect, the proposed strategy can be interpreted as a particular version of an Arbitrary Lagrangian Eulerian formulation, see e.g. (Donéa et al. 1977, Belytschko and Kennedy 1978, Donéa 1980, Donéa 1983, Hughes et al. 1981, Belytschko et al. 2000, Donéa and Huerta 2003, Kuhl et al. 2003) for ALE formulations in general and (Yamada and Kikuchi 1993, Askes et al. 1999, Askes 2000, Askes and Sluys 2000, Askes and Rodriguez-Ferran 2001, Askes et al. 2003b, Rodriguez-Ferran et al. 2002, Armero and Love 2003) for their particular application to adaptive remeshing. In contrast to the existing ALE formulations, however, the selection of the mesh is no longer user-defined in our formulation. Rather, the optimal mesh follows straightforwardly from a variational principle. Unlike classical ALE formulations, the formulation derived herein is thus inherently symmetric.

In this contribution, we suggest to solve the spatial and the material motion problem simultaneously. We thus introduce both, the spatial and the material motion map as primary unknowns. The solution renders the spatial and the corresponding material configuration that are optimal in the sense of potential energy. In this sense, optimal meshes are defined through vanishing discrete material node point forces.

The present work is motivated by a one-dimensional model problem introduced in section 2. After reviewing the relevant kinematics in section 3, we then illustrate the variational framework of our formulation based on the Dirichlet principle in section 4. Section 5 finally treats its spatial discretisation parameterised in terms of the material and the spatial reference mapping. Some illustrative examples are given in section 6 before we close with a conclusion in section 7.

2 MOTIVATION

Let us motivate our work by considering a simple one-dimensional bar, clamped on both sides and loaded by a constant line load b_0 = const. The resulting displacement field can be determined analytically through the classical Dirichlet principle as the minimum of the potential energy $\mathcal{I}(\varphi)$. Hereby, the potential energy can be understood as the total potential energy density $U_\square$ integrated over the entire bar length as $\mathcal{I} = \int_{\mathcal{B}_\square} U_\square \, dV_\square$. The potential energy density $U_\square = W_\square + V_\square$ typically consists of an internal contribution $W_\square$, i.e. given as $W_\square = ¼\, E_\square\,[F^2 - 1 - 2\ln(F)]$ in the Neo-Hookean case, and an external contribution $V_\square$, given as $V_\square = -b_\square \varphi$ for the particular example considered herein. The potential energy is thus a function of the spatial placement φ and its gradient $F = \nabla_X \varphi$, of Young's modulus $E_\square$ and of the amount of loading $b_\square$. Its minimum is characterised through its vanishing variation with respect to the independent variables, in this case, the spatial placement φ.

$$\mathcal{I}(\varphi) = \int_{\mathcal{B}_\square} U_\square \, dV_\square \rightarrow \inf_{\varphi} \tag{1}$$

The above equation defines the spatial placement φ or rather the deformation u as the difference between material and spatial placement $u = X - \varphi$. The analytical solution is indicated through the grey curves in Figure 1. However, typically, the analytical solution is not easily available such that a numerical approximation has to be carried out instead. In the simplest case, we could discretise the clamped bar with two linear finite elements as illustrated in Figure 1, top. Since both

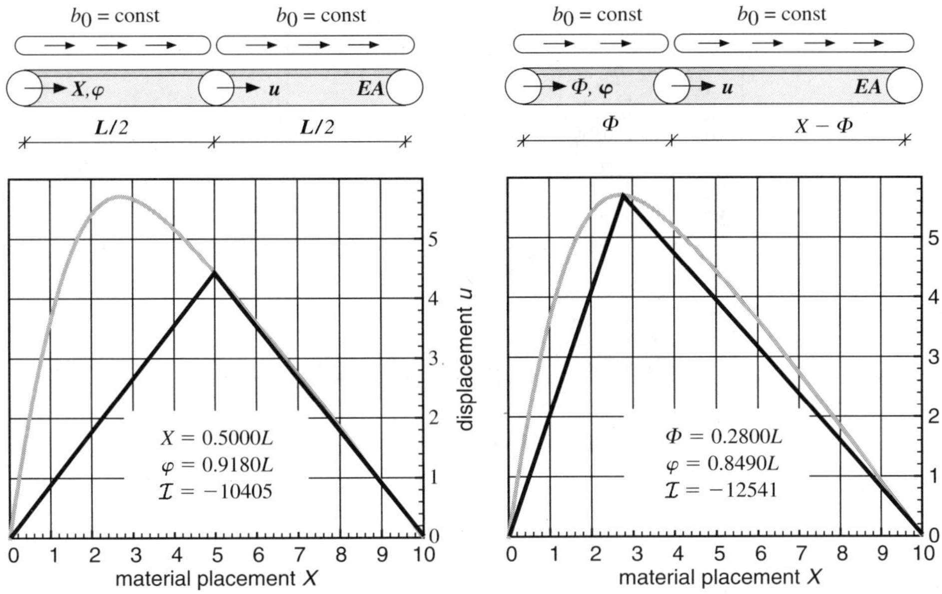

Figure 1. One-dimensional model problem – continuous vs. discrete lagrangian and discrete ale solution.

ends are clamped, we introduce only one degree of freedom for the midpoint node which is located right in the middle of the bar at $X = 0.5\ L$. The discrete potential energy $\mathcal{I}(\varphi^h)$ or rather its vanishing variation with respect to the unknown φ^h can thus be expressed as follows.

$$\mathcal{I}(\varphi^h) = X\,U_1^h + [L - X]\,U_2^h \to \inf_{\varphi^h} \tag{2}$$

The black lines in figure 1, top, illustrate the corresponding discrete solution for the displacement $u^h = X - \varphi^h$ of this classical lagrangian analysis. The area between the grey and the black curve represents a measure for the error of the discrete solution. A significant reduction of this area, and thus an improved approximation to the solution, can obviously be obtained by shifting the midnode to the left as illustrated in figure 1, bottom. Strictly speaking, we introduce the node point position or rather the material placement Φ^h as an additional degree of freedom next to the spatial placement φ^h. The corresponding discrete potential energy $\mathcal{I}(\varphi^h,\ \Phi^h)$ or rather its vanishing variation with respect to the unknowns φ^h and Φ^h can then be expressed in the following form.

$$\mathcal{I}(\Phi^h, \varphi^h) = \Phi^h U_1^h + [L - \Phi^h]U_2^h \to \inf_{\Phi^h, \varphi^h} \tag{3}$$

Figure 1 bottom, shows the solution for the material placement Φ^h and the displacement $u^h = \varphi^h - \Phi^h$ for this analysis, which can be interpreted as a particular form of an Arbitrary Lagrangian Eulerian analysis. The optimal node point position is found at $\Phi = 0.2800\,L$. The corresponding potential energy $\mathcal{I} = -12541$ indicates, that the potential energy $\mathcal{I} = -10405$ of the classical lagrangian analysis has been reduced by more than 20% through the relaxation of the node point position. Figure 2, which shows the potential energy $\mathcal{I}$ as a function of the material placements and the displacements supports these observations. For the classical lagrangian analysis, we look for the optimum of the potential energy at fixed material position. It is obvious, that the potential energy can be reduced considerably upon relaxation of this node point position. In what follows, we will present a general theoretical and numerical framework which allows to determine not only the spatial but also the material placements within a consistent variational framework.

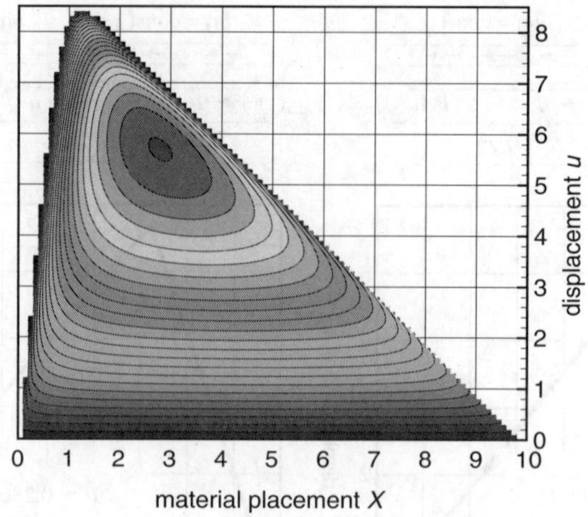

Figure 2. Potential energy $\mathcal{I}$ in terms of material placements X and displacements u.

3 KINEMATICS

Let us briefly summarise the fundamental kinematic relations of the present ALE formulation which is basically characterised through the introduction of an independent fixed reference domain $\mathcal{B}_\square$ next to the classical material and spatial domain $\mathcal{B}_0$ and $\mathcal{B}_t$, see e.g. (Donéa 1980, Donéa 1983, Hughes et al. 1981, Belytschko et al. 2000, Donéa and Huerta 2003, Askes et al. 2003b, Askes et al. 2003a, Kuhl et al. 2003). Thereby, any scalar- or tensor-valued quantity per unit volume in either $\mathcal{B}_\square$, $\mathcal{B}_0$ or $\mathcal{B}_t$ will be denoted as $\{\bullet\}_\square$, $\{\bullet\}_0$ or $\{\bullet\}_t$, respectively. The following considerations are essentially based on two independent mappings, i.e. the referential maps from the fixed referential configuration $\mathcal{B}_\square$ to the material configuration $\mathcal{B}_0$ and from the referential configuration $\mathcal{B}_\square$ to the spatial configuration $\mathcal{B}_t$ as illustrated in Figure 3.
Let $\tilde{\boldsymbol{\Phi}}$ with

$$\boldsymbol{X} = \tilde{\boldsymbol{\Phi}}(\boldsymbol{\xi}, t) : \qquad \mathcal{B}_\square \to \mathcal{B}_0 \tag{4}$$

denote the mapping of physical particles from a fixed position $\boldsymbol{\xi}$ in the referential domain $\mathcal{B}_\square$ to their material position $\boldsymbol{X}$ in the material domain $\mathcal{B}_0$. Its referential gradient $\tilde{\boldsymbol{f}}$

$$\tilde{\boldsymbol{f}} = \nabla_{\boldsymbol{\xi}} \tilde{\boldsymbol{\Phi}}(\boldsymbol{\xi}, t) \quad \text{with} \quad \tilde{j} = \det \tilde{\boldsymbol{f}} > 0 \tag{5}$$

defines the related tangent map from the referential tangent space $T\mathcal{B}_\square$ to the material tangent space $T\mathcal{B}_0$ with the related Jacobian $\tilde{j}$. With $\tilde{j} > 0$, we assure the existence of the inverse map $\tilde{\varphi}$ with $\boldsymbol{\xi} = \tilde{\varphi}(\boldsymbol{X}, t)$: $\mathcal{B}_0 \to \mathcal{B}_\square$ whereby $\tilde{\boldsymbol{F}} = \nabla_X \tilde{\varphi}(\boldsymbol{X}, t) = \tilde{\boldsymbol{f}}^{-1}$: $T\mathcal{B}_0 \to T\mathcal{B}_\square$ with $\tilde{J} = \det \tilde{\boldsymbol{F}} = 1/\tilde{j} > 0$. Moreover, we introduce the second independent mapping $\bar{\boldsymbol{\varphi}}$ from the referential domain $\mathcal{B}_\square$ to the spatial domain $\mathcal{B}_0$.

$$\boldsymbol{x} = \bar{\boldsymbol{\varphi}}(\boldsymbol{\xi}, t) : \qquad \mathcal{B}_\square \to \mathcal{B}_t \tag{6}$$

The related referential gradient $\bar{\boldsymbol{F}}$ and its Jacobian $\bar{J}$

$$\bar{\boldsymbol{F}} = \nabla_{\boldsymbol{\xi}} \bar{\boldsymbol{\varphi}}(\boldsymbol{\xi}, t) \quad \text{with} \quad \bar{J} = \det \bar{\boldsymbol{F}} > 0 \tag{7}$$

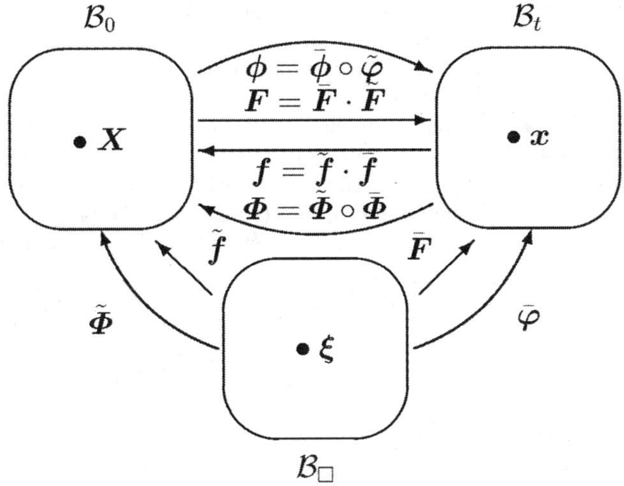

Figure 3. Reference domain $\mathcal{B}_\square$, material domain $\mathcal{B}_0$ and spatial domain $\mathcal{B}_t$.

characterise the corresponding tangent map from the referential tangent space $T\mathcal{B}_\square$ to the spatial tangent space $T\mathcal{B}_t$. The related inverse map $\bar{\boldsymbol{\Phi}}$ can be expressed as $\boldsymbol{\xi} = \bar{\boldsymbol{\Phi}}\,(\boldsymbol{x},\ t)$: $\mathcal{B}_t \to \mathcal{B}_\square$ with $\bar{\boldsymbol{f}} = \nabla_x \bar{\boldsymbol{\Phi}}(\boldsymbol{x}, t) = \bar{\boldsymbol{F}}^{-1}$: $T\mathcal{B}_t \to T\mathcal{B}_\square$ and $\bar{j} = \det \bar{\boldsymbol{f}} = 1/\bar{J} > 0$. The variation at fixed referential coordinate $\boldsymbol{\xi}$ denoted as $\delta\{\bullet\} = \delta\{\bullet\}|_{\xi\text{fixed}}$ will be referred to as total variation in the sequel. It can be expressed as the variation with respect to the spatial coordinates $\boldsymbol{x}$ at fixed material coordinates $\boldsymbol{X}$ denoted as $\delta_x\{\bullet\}$ plus a contribution with respect to the material coordinates $\boldsymbol{X}$ at fixed spatial coordinates $\boldsymbol{x}$ denoted as $\delta_X\{\bullet\}$.

$$\delta\{\bullet\} = \delta_{\boldsymbol{x}}\{\bullet\} + \delta_{\boldsymbol{X}}\{\bullet\} \tag{8}$$

The individual contributions $\delta_x\{\bullet\}$ and $\delta_X\{\bullet\}$ will be defined in the following subsections. With these definitions at hand, we can now formulate the kinematics of the spatial and the material motion problem. For the former we typically follow spatial placements $\boldsymbol{x}$ of physical particles through the ambient space at fixed material position $\boldsymbol{X}$ corresponding to a lagrangian viewpoint. For the latter, material placements $\boldsymbol{X}$ of physical particles are followed through the ambient material at fixed spatial position $\boldsymbol{x}$, the observer thus takes the Eulerian viewpoint. Recall, however, that neither the spatial nor the material deformation map represent an independent mapping in the present context since both can be reformulated in terms of the referential mappings defined above.

3.1 *Spatial motion problem*

In the spatial motion problem, the placement $\boldsymbol{x}$ of a physical particle in the spatial configuration $\mathcal{B}_t$ is described by the nonlinear spatial deformation map $\boldsymbol{\varphi}$

$$\boldsymbol{x} = \boldsymbol{\varphi}\,(\boldsymbol{X}, t) = \bar{\boldsymbol{\varphi}}\,(\boldsymbol{\xi}, t) \circ \tilde{\boldsymbol{\varphi}}\,(\boldsymbol{X}, t) \tag{9}$$

in terms of the placement $\boldsymbol{X}$ in the material configuration $\mathcal{B}_0$, see figure 3. It can thus be understood as a composition of the referential maps $\bar{\boldsymbol{\varphi}}$ and $\tilde{\boldsymbol{\varphi}}$. The related spatial motion deformation gradient $\boldsymbol{F}$, i.e. the linear tangent map associated with the spatial motion deformation map, is given as

$$\boldsymbol{F} = \nabla_{\boldsymbol{X}} \boldsymbol{\varphi}\,(\boldsymbol{X}, t) = \bar{\boldsymbol{F}} \cdot \tilde{\boldsymbol{F}} \quad \text{with} \quad J > 0 \tag{10}$$

with J denoting the related Jacobian. Note, that the spatial motion deformation gradient is essentially characterised through the multiplicative decomposition in terms of the referential gradients $\bar{\boldsymbol{F}}$ and $\tilde{\boldsymbol{F}}$. As typical strain measures of the spatial motion problem, we introduce the right and left spatial motion Cauchy-Green strain tensors $\boldsymbol{C} = \boldsymbol{F}^t \cdot \boldsymbol{F}$ and $\boldsymbol{b} = \boldsymbol{F} \cdot \boldsymbol{F}^t$, respectively. In what follows, the variation of a function $\{\bullet\}$ with respect to the spatial coordinates $\boldsymbol{x}$ at fixed material coordinates $\boldsymbol{X}$ will be denoted as $\delta x\{\bullet\} = \delta\{\bullet\}|_{X\text{fixed}}$. Consequently, the variation of the spatial motion deformation map and its gradient can be expressed as $\delta_x \boldsymbol{\varphi} = \delta\bar{\boldsymbol{\varphi}}$ and $\delta_x \boldsymbol{F} = \nabla_X \delta\bar{\boldsymbol{\varphi}}$. In the sequel, we will particularly make use of the expression $\delta_x\{\bullet\}_0\,(\boldsymbol{F}, \boldsymbol{\varphi}\,; \boldsymbol{X}) = \mathbf{D}_F\{\bullet\}_0 : \nabla_X\delta\bar{\boldsymbol{\varphi}} + \partial_x\{\bullet\}_0 \cdot \delta\bar{\boldsymbol{\varphi}}$ which relates the variation of a scalar-valued function $\{\bullet\}_0$ with a possible dependence on the deformation gradient $\boldsymbol{F}$, the deformation map $\boldsymbol{\varphi}$ and the material position $\boldsymbol{X}$ to its derivative with respect to the deformation gradient $\mathbf{D}_F\{\bullet\}_0$ and its explicit derivative with respect to the spatial coordinate $\partial_x\{\bullet\}_0$.

3.2 *Material motion problem*

In complete analogy to the spatial motion problem, we can introduce a material motion map $\boldsymbol{\Phi}$ defining the mapping of the placement $\boldsymbol{X}$ of physical particles in the material configuration $\mathcal{B}_0$

$$\boldsymbol{X} = \boldsymbol{\Phi}\,(\boldsymbol{x},t) = \tilde{\boldsymbol{\Phi}}\,(\boldsymbol{\xi},t) \circ \bar{\boldsymbol{\Phi}}\,(\boldsymbol{x},t) \tag{11}$$

in terms of the related placement $\boldsymbol{x}$ in the spatial configuration $\mathcal{B}_t$. As illustrated in figure 3, the material deformation map can thus be interpreted as a composition of the referential mappings $\tilde{\boldsymbol{\Phi}}$ and $\bar{\boldsymbol{\Phi}}$. The related linear tangent map from the spatial tangent space $T\mathcal{B}_t$ to the material tangent space $T\mathcal{B}_0$ is defined through the material motion deformation gradient $\boldsymbol{f}$ and its Jacobian j,

$$\boldsymbol{f} = \nabla_x \boldsymbol{\Phi}\,(\boldsymbol{x},t) = \tilde{\boldsymbol{f}} \cdot \bar{\boldsymbol{f}} \quad \text{with} \quad j > 0 \tag{12}$$

whereby $\boldsymbol{f}$ can be decomposed multiplicatively into the referential gradients $\tilde{\boldsymbol{f}}$ and $\bar{\boldsymbol{f}}$. The right and left material motion Cauchy-Green strain tensors $\boldsymbol{c} = \boldsymbol{f}^t \cdot \boldsymbol{f}$ and $\boldsymbol{B} = \boldsymbol{f} \cdot \boldsymbol{f}^t$ represent characteristic strain measures of the material motion problem. The variation of any function $\{\bullet\}$ with respect to the material coordinates $\boldsymbol{X}$ at fixed spatial coordinates $\boldsymbol{x}$ denoted as $\delta_X\{\bullet\} = \delta\{\bullet\}|_{x\text{fixed}}$ defines the variation of the material motion deformation map $\delta_X \boldsymbol{\Phi} = \delta\tilde{\boldsymbol{\Phi}}$ and its gradient as $\delta_X \boldsymbol{f} = \nabla_x \delta\tilde{\boldsymbol{\Phi}}$. Moreover, we obtain the following relation $\delta_X\{\bullet\}_t\;(\boldsymbol{f}, \boldsymbol{\Phi};\, \boldsymbol{x}) = \mathbf{d}_f\{\bullet\}_t \colon \nabla_x\delta\tilde{\boldsymbol{\Phi}} + \partial_X\{\bullet\}_t \cdot \delta\tilde{\boldsymbol{\Phi}}$ between the variation of a scalar-valued function $\{\bullet\}_t$ as a function of $\boldsymbol{f}$, $\boldsymbol{\Phi}$ and $\boldsymbol{x}$, the related derivative with respect to the material motion deformation gradient $\mathbf{d}_f\{\bullet\}_t$ and the explicit derivative with respect to the material coordinate $\partial_X\{\bullet\}t$.

Remark 1 (Spatial vs. material quantities) In the continuous case, the spatial and the material motion problem are related through the identity maps in $\mathcal{B}_0$ and $\mathcal{B}_t$ as $\boldsymbol{id}_{\mathcal{B}_0} = \boldsymbol{\Phi}\,(\boldsymbol{\varphi}\,(\boldsymbol{X}, t), t)$ and $\boldsymbol{id}_{\mathcal{B}_t} = \boldsymbol{\varphi}\,(\boldsymbol{\Phi}\,(\boldsymbol{x}, t), t)$ while the spatial and the material deformation gradient are simply related by their inverses. $\boldsymbol{f} = \boldsymbol{F}^{-1}$ and $\boldsymbol{F} = \boldsymbol{f}^{-1}$. Scalar-valued or tensorial quantities with material reference $\{\bullet\}_0$ and spatial reference $\{\bullet\}_t$ are related through the well-known transformation formulae $\{\bullet\}_t = j\{\bullet\}_0$ and $\{\bullet\}_0 = J\{\bullet\}_t$. Moreover, the equivalences of vector- or tensor-valued material surface fluxes $\{\Box\}$ through the boundary of the material domain $\partial\mathcal{B}_0$ and the corresponding spatial surface fluxes $\{\Diamond\}$ through the spatial boundary $\partial\mathcal{B}_t$ render the well-known Nanson's formulae $\{\Diamond\} = j\{\Box\} \cdot \boldsymbol{F}^t$ and $\{\Box\} = J\,\{\Diamond\} \cdot \boldsymbol{f}^t$. Later on, we will further invoke the following relations between the derivative of a scalar-valued function $\{\bullet\}_0$ or $\{\bullet\}_t$ with respect to the spatial and material motion deformation gradient

$$\begin{aligned}
\mathrm{D}_{\boldsymbol{F}}\{\bullet\}_0 &= \{\bullet\}_0 \boldsymbol{f}^t - J\boldsymbol{f}^t \cdot \mathrm{d}_{\boldsymbol{f}}\{\bullet\}_t \cdot \boldsymbol{f}^t \\
\mathrm{d}_{\boldsymbol{f}}\{\bullet\}_t &= \{\bullet\}_t \boldsymbol{F}^t - j\boldsymbol{F}^t \cdot \mathrm{D}_{\boldsymbol{F}}\{\bullet\}_0 \cdot \boldsymbol{F}^t
\end{aligned}$$

which has been given earlier in (Steinmann 2002a).

In what follows, we shall restrict ourselves to the hyperelastostatic case for conservative systems for which the Dirichlet principle defines the appropriate variational setting. In this context, we introduce the internal potential energy density $W_\square$ per unit volume in $\mathcal{B}_\square$, which is typically referred to as strain energy density. Conservative loading is characterised through the external potential energy density for conservative systems $V_\square$ per unit volume in $\mathcal{B}_\square$. The total potential energy density $U_\square$ per unit volume in $\mathcal{B}_\square$ can thus be expressed as the sum of the corresponding internal and external contribution $W_\square$ and $V_\square$.

$$U_\square = W_\square + V_\square \tag{13}$$

Then, the conservative mechanical system is essentially characterised through the total energy $\mathcal{I}$ defined through the integration of $U_\square$ over the reference domain $\mathcal{B}_\square$.

$$\mathcal{I}(\bar{\varphi}, \tilde{\boldsymbol{\Phi}}) = \int_{\mathcal{B}_\square} U_\square \, \mathrm{d}V_\square \to \inf \tag{14}$$

The infimum of the total potential energy $\mathcal{I}$ corresponds to a vanishing total variation, i.e. the variation at fixed referential coordinates $\boldsymbol{\xi}$.

$$\delta\mathcal{I}(\bar{\varphi}, \tilde{\boldsymbol{\Phi}}) = \int_{\mathcal{B}_\square} \delta U_\square \, \mathrm{d}V_\square = \delta_{\boldsymbol{x}}\mathcal{I} + \delta_{\boldsymbol{X}}\mathcal{I} \doteq 0 \tag{15}$$

According to equation (8), this total variation consists of a variation with respect to the spatial coordinates $\boldsymbol{x}$ at fixed material coordinates $\boldsymbol{X}$ plus variation with respect to the material coordinates $\boldsymbol{X}$ at fixed spatial coordinates $\boldsymbol{x}$ as illustrated in Figure 4. These variations will be elaborated in detail in the sequel. Thereby, we will particularly emphasise the striking duality of both formulations.

4.1 *Spatial motion problem*

The variation of the total energy $\mathcal{I}$ with respect to the spatial coordinates $\boldsymbol{x}$ at fixed material coordinates $\boldsymbol{X}$ can be expressed in the following form.

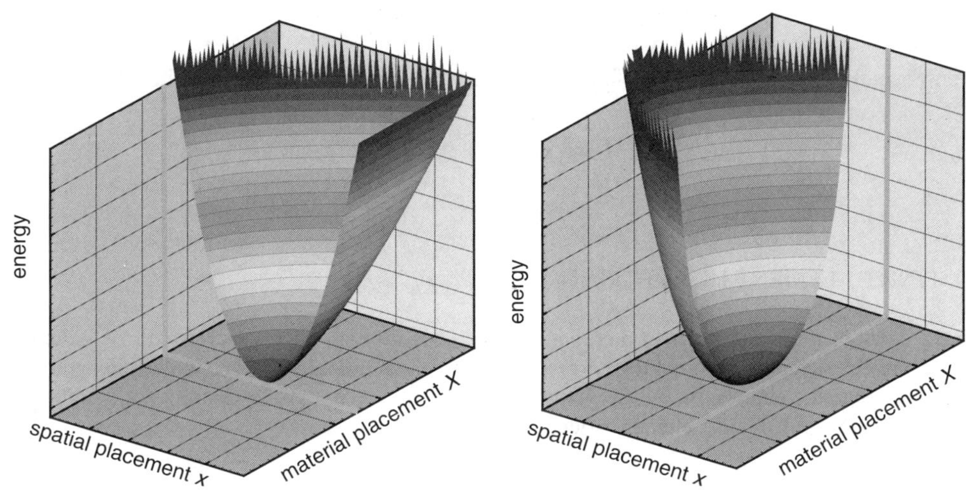

Figure 4. Variation of potential energy $\mathcal{I}$ with respect to spatial and material coordinates $\boldsymbol{x}$ and $\boldsymbol{X}$.

$$\delta_{\boldsymbol{x}}\mathcal{I} = \int_{\mathcal{B}_0} \nabla_{\boldsymbol{X}}\delta\boldsymbol{\varphi} : \mathrm{D}_{\boldsymbol{F}}U_0 + \delta\boldsymbol{\varphi}\cdot\partial_{\boldsymbol{x}}U_0\mathrm{d}V_0 \tag{16}$$

The above format motivates the introduction of the spatial motion momentum fluxes $\boldsymbol{\Pi}^t$ and $\boldsymbol{\sigma}^t = j\,\boldsymbol{\Pi}^t\cdot\boldsymbol{F}^t$, whereby the former denotes the first Piola-Kirchhoff stress tensor while the latter is typically referred to as Cauchy stress. Moreover, we introduce the corresponding momentum sources $\boldsymbol{b}_0$ and $\boldsymbol{b}_t$, i.e. the spatial volume force densities per unit volume in $\mathcal{B}_0$ and $\mathcal{B}_t$, respectively.

$$\begin{aligned} \boldsymbol{\Pi}^t &= \mathrm{D}_{\boldsymbol{F}}U_0 & \boldsymbol{b}_0 &= -\partial_{\boldsymbol{x}}U_0 \\ \boldsymbol{\sigma}^t &= U_t\boldsymbol{I} - \boldsymbol{f}^t\cdot\mathrm{d}_{\boldsymbol{f}}U_t & \boldsymbol{b}_t &= -\partial_{\boldsymbol{x}}U_t \end{aligned} \tag{17}$$

With the above definitions, the variation of the total energy (16) with respect to the spatial coordinates $\boldsymbol{x}$ takes the following more familiar format.

$$\delta_{\boldsymbol{x}}\mathcal{I} = \int_{\mathcal{B}_0} \nabla_{\boldsymbol{X}}\delta\boldsymbol{\varphi} : \boldsymbol{\Pi}^t\mathrm{d}V_0 - \int_{\mathcal{B}_0} \delta\boldsymbol{\varphi}\cdot\boldsymbol{b}_0\mathrm{d}V_0 \tag{18}$$

4.2 *Material motion problem*

The variation of total energy $\mathcal{I}$ with respect to the material coordinates $\boldsymbol{X}$ at fixed spatial coordinates $\boldsymbol{x}$ takes the following format.

$$\delta_{\boldsymbol{X}}\mathcal{I} = \int_{\mathcal{B}_t} \nabla_{\boldsymbol{x}}\delta\tilde{\boldsymbol{\Phi}} : \mathrm{d}_{\boldsymbol{f}}U_t + \delta\tilde{\boldsymbol{\Phi}}\cdot\partial_{\boldsymbol{X}}U_t\mathrm{d}V_t \tag{19}$$

In complete analogy to the spatial motion case, we can introduce the material motion momentum fluxes $\boldsymbol{\pi}^t$ and $\boldsymbol{\Sigma}^t$, a two-point stress tensor of Piola-Kirchhoff type and the classical Eshelby stress, respectively. Accordingly, the corresponding momentum sources, the material volume force densities per unit volume in $\mathcal{B}_t$ and $\mathcal{B}_0$, are introduced as $\boldsymbol{B}_t$ and $\boldsymbol{B}_0$.

$$\begin{aligned} \boldsymbol{\pi}^t &= \mathrm{d}_{\boldsymbol{f}}U_t & \boldsymbol{B}_t &= -\partial_{\boldsymbol{X}}U_t \\ \boldsymbol{\Sigma}^t &= U_0\boldsymbol{I} - \boldsymbol{F}^t\cdot\mathrm{D}_{\boldsymbol{F}}U_0 & \boldsymbol{B}_0 &= -\partial_{\boldsymbol{X}}U_0 \end{aligned} \tag{20}$$

Consequently, the spatial variation of the total energy (19) can be reformulated in the following form.

$$\delta_{\boldsymbol{X}}\mathcal{I} = \int_{\mathcal{B}_t} \nabla_{\boldsymbol{x}}\delta\tilde{\boldsymbol{\Phi}} : \boldsymbol{\pi}^t\mathrm{d}V_t - \int_{\mathcal{B}_t} \delta\tilde{\boldsymbol{\Phi}}\cdot\boldsymbol{B}_t\mathrm{d}V_t \tag{21}$$

5 FINITE ELEMENT DISCRETISATION

In the spirit of the finite element method, the reference domain $\mathcal{B}_\square$ is discretised in n_{el} elements which are characterised through the corresponding element domain $\mathcal{B}^e_\square$. The underlying geometry $\boldsymbol{\xi}$ is interpolated elementwise by the shape functions $N^i_{\boldsymbol{\xi}}$ in terms of the discrete node point positions $\boldsymbol{\xi}_i$ of the $i = 1, \ldots, n_{en}$ element nodes.

$$\mathcal{B}_\square = \bigcup_{e=1}^{n_{el}} \mathcal{B}^e_\square \qquad \boldsymbol{\xi}^h|_{\mathcal{B}^e_\square} = \sum_{i=1}^{n_{en}} N^i_{\boldsymbol{\xi}}\,\boldsymbol{\xi}_i \tag{22}$$

Following the isoparametric concept, the unknowns $\bar{\varphi}$ and $\tilde{\Phi}$ are interpolated on the element level with the same shape functions $N^i_{\bar{\varphi}}$ and $N^j_{\tilde{\Phi}}$ as the element geometry ξ. Similar shape functions are applied to interpolate the test functions $\delta\bar{\varphi}$ and $\delta\tilde{\Phi}$ according to the classical Bubnov–Galerkin technique.

$$
\begin{aligned}
\delta\bar{\varphi}^h|_{\mathcal{B}^e_\square} &= \sum_{i=1}^{n_{en}} N^i_{\bar{\varphi}}\, \delta\bar{\varphi}_i &&\in H^0_1(\mathcal{B}_\square)\\
\bar{\varphi}^h\;|_{\mathcal{B}^e_\square} &= \sum_{k=1}^{n_{en}} N^k_{\bar{\varphi}}\, \bar{\varphi}_k &&\in H_1(\mathcal{B}_\square)\\
\delta\tilde{\Phi}^h|_{\mathcal{B}^e_\square} &= \sum_{j=1}^{n_{en}} N^j_{\tilde{\Phi}}\, \delta\tilde{\Phi}_j &&\in H^0_1(\mathcal{B}_\square)\\
\tilde{\Phi}^h\;|_{\mathcal{B}^e_\square} &= \sum_{l=1}^{n_{en}} N^l_{\tilde{\Phi}}\, \tilde{\Phi}_l &&\in H_1(\mathcal{B}_\square)
\end{aligned}
\tag{23}
$$

Consequently, the related gradients of the test functions $\nabla_\xi\delta\bar{\varphi}^h$ and $\nabla_\xi\delta\tilde{\Phi}^h$ and the gradients of the primary unknowns $\nabla_\xi\bar{\varphi}^h$ and $\nabla_\xi\tilde{\Phi}^h$ take the following elementwise interpolation.

$$
\begin{aligned}
\nabla_\xi\, \delta\bar{\varphi}^h|_{\mathcal{B}^e_\square} &= \sum_{i=1}^{n_{en}} \delta\bar{\varphi}_i \otimes \nabla_\xi N^i_{\bar{\varphi}}\\
\nabla_\xi\, \bar{\varphi}^h\;|_{\mathcal{B}^e_\square} &= \sum_{k=1}^{n_{en}} \bar{\varphi}_k \otimes \nabla_\xi N^k_{\bar{\varphi}}\\
\nabla_\xi\, \delta\tilde{\Phi}^h|_{\mathcal{B}^e_\square} &= \sum_{j=1}^{n_{en}} \delta\tilde{\Phi}_j \otimes \nabla_\xi N^j_{\tilde{\Phi}}\\
\nabla_\xi\, \tilde{\Phi}^h\;|_{\mathcal{B}^e_\square} &= \sum_{l=1}^{n_{en}} \tilde{\Phi}_l \otimes \nabla_\xi N^l_{\tilde{\Phi}}
\end{aligned}
\tag{24}
$$

In particular, $\nabla_\xi\bar{\varphi}^h$ and $\nabla_\xi\delta\tilde{\Phi}^h$ denote the discrete deformation gradients $\bar{F}^h = \nabla_\xi\bar{\varphi}^h$ and $\tilde{f}^h = \nabla_\xi\tilde{\Phi}^h$, which define the discrete spatial and material deformation gradient as $F^h = \bar{F}^h \cdot \tilde{F}^h$ and $f^h = \tilde{f}^h \cdot \bar{f}^h$. With the above suggested discretisations at hand, the discrete residual of the balance of momentum of the spatial motion problem $\mathsf{R}^I_{\bar{\varphi}}$ and of the material motion problem $\mathsf{R}^j_{\tilde{\Phi}}$ can be expressed $\forall I, J = 1, \ldots, n_{np}$ as

$$
\begin{aligned}
\mathsf{R}^I_{\bar{\varphi}}\,(\bar{\varphi}^h, \tilde{\Phi}^h) &= \mathop{\mathsf{A}}_{e=1}^{n_{el}} \mathsf{r}^{ei}_{\bar{\varphi}} &= \mathbf{0}\\
\mathsf{R}^J_{\tilde{\Phi}}\,(\bar{\varphi}^h, \tilde{\Phi}^h) &= \mathop{\mathsf{A}}_{e=1}^{n_{el}} \mathsf{r}^{ej}_{\tilde{\Phi}} &= \mathbf{0}
\end{aligned}
\tag{25}
$$

with the element- and nodewise residuals

$$
\begin{aligned}
\mathsf{r}^{ei}_{\bar{\varphi}} &= \int_{\mathcal{B}^e_t} \nabla_{\boldsymbol{x}} N^i_{\bar{\varphi}} \cdot \boldsymbol{\sigma} - N^i_{\bar{\varphi}}\, \boldsymbol{b}_t \;\mathrm{d}V_t - \int_{\partial\mathcal{B}^{te}_t} N^i_{\bar{\varphi}}\, \boldsymbol{t}_t \;\mathrm{d}A_t\\
\mathsf{r}^{ej}_{\tilde{\Phi}} &= \int_{\mathcal{B}^e_0} \nabla_{\boldsymbol{X}} N^j_{\tilde{\Phi}} \cdot \boldsymbol{\Sigma} - N^j_{\tilde{\Phi}}\, \boldsymbol{B}_0 \,\mathrm{d}V_0 - \int_{\partial\mathcal{B}^{te}_0} N^j_{\tilde{\Phi}}\, \boldsymbol{T}_0 \,\mathrm{d}A_0
\end{aligned}
$$

and with the understanding that the operator $\mathbf{A}^{n_{el}}_{e=1}$ denotes the assembly over all $e = 1, \ldots, n_{el}$ element contributions at the $i, j = 1, \ldots, n_{en}$ element nodes to the global node point residuals at all I, $J = 1, \ldots, n_{np}$ global node points. The above derived discrete residual statements represent a highly nonlinear coupled system of equations which can be solved efficiently within the framework of a monolithic incremental iterative Newton-Raphson solution strategy. To this end, we perform a consistent linearisation of the governing equations

$$
\begin{aligned}
\mathsf{R}^{I\,k+1}_{\bar\varphi} &= \mathsf{R}^{I\,k}_{\bar\varphi} + \mathrm{d}\,\mathsf{R}^{I}_{\bar\varphi} \doteq \mathbf{0} \\
\mathsf{R}^{J\,k+1}_{\tilde\Phi} &= \mathsf{R}^{J\,k}_{\tilde\Phi} + \mathrm{d}\,\mathsf{R}^{J}_{\tilde\Phi} \doteq \mathbf{0}
\end{aligned} \tag{26}
$$

with $\mathrm{dR}^I_{\bar\varphi}$ and $\mathrm{dR}^J_{\tilde\Phi}$ denoting the iterative residua of the iteration $k+1$.

$$
\begin{aligned}
\mathrm{d}\mathsf{R}^{I}_{\bar\varphi} &= \sum_{K=1}^{n_{np}} \mathsf{K}^{IK}_{\bar\varphi\bar\varphi}\ \mathrm{d}\bar\varphi_K + \sum_{L=1}^{n_{np}} \mathsf{K}^{IL}_{\bar\varphi\tilde\Phi} \cdot \mathrm{d}\tilde{\boldsymbol{\Phi}}_L \\
\mathrm{d}\mathsf{R}^{J}_{\tilde\Phi} &= \sum_{K=1}^{n_{np}} \mathsf{K}^{JK}_{\tilde\Phi\bar\varphi}\ \mathrm{d}\bar\varphi_K + \sum_{L=1}^{n_{np}} \mathsf{K}^{JL}_{\tilde\Phi\tilde\Phi} \cdot \mathrm{d}\tilde{\boldsymbol{\Phi}}_L
\end{aligned} \tag{27}
$$

The iteration matrices $\mathrm{K}^{IK}_{\bar\varphi\bar\varphi}$, $\mathrm{K}^{IL}_{\bar\varphi\tilde\Phi}$, $\mathrm{K}^{JK}_{\tilde\Phi\bar\varphi}$ and $\mathrm{K}^{JL}_{\tilde\Phi\tilde\Phi}$ can finally be expressed as

$$
\begin{aligned}
\mathsf{K}^{IK}_{\bar\varphi\bar\varphi} &= \frac{\partial \mathsf{R}^I_{\bar\varphi}}{\partial \bar\varphi_K} = \mathop{\mathsf{A}}_{e=1}^{n_{el}} \int_{\mathcal{B}^e_0} \nabla_{\boldsymbol{x}} N^i_{\bar\varphi} \cdot \ \mathsf{c}\ \cdot \nabla_{\boldsymbol{x}} N^k_{\bar\varphi}\ \mathrm{d}V_0 \\
\mathsf{K}^{IL}_{\bar\varphi\tilde\Phi} &= \frac{\partial \mathsf{R}^I_{\bar\varphi}}{\partial \tilde{\boldsymbol{\Phi}}_L} = \mathop{\mathsf{A}}_{e=1}^{n_{el}} \int_{\mathcal{B}^e_t} \nabla_{\boldsymbol{x}} N^i_{\bar\varphi} \cdot \ \mathrm{d}_{\boldsymbol{f}}\boldsymbol{\sigma}^t \cdot \nabla_{\boldsymbol{x}} N^l_{\tilde\Phi}\ \mathrm{d}V_t \\
&\quad - \mathop{\mathsf{A}}_{e=1}^{n_{el}} \int_{\mathcal{B}^e_t} N^i_{\bar\varphi}\ \ \boldsymbol{b}_t \ \cdot \nabla_{\boldsymbol{X}} N^l_{\tilde\Phi}\ \mathrm{d}V_t \\
\mathsf{K}^{JK}_{\tilde\Phi\bar\varphi} &= \frac{\partial \mathsf{R}^J_{\tilde\Phi}}{\partial \bar\varphi_K} = \mathop{\mathsf{A}}_{e=1}^{n_{el}} \int_{\mathcal{B}^e_0} \nabla_{\boldsymbol{X}} N^j_{\tilde\Phi} \cdot \mathrm{D}_{\boldsymbol{F}}\boldsymbol{\Sigma}^t \cdot \nabla_{\boldsymbol{X}} N^k_{\bar\varphi}\ \mathrm{d}V_0 \\
&\quad - \mathop{\mathsf{A}}_{e=1}^{n_{el}} \int_{\mathcal{B}^e_0} \nabla_{\boldsymbol{X}} N^j_{\tilde\Phi} \cdot\ \ \boldsymbol{b}_0 \ \ N^k_{\bar\varphi}\ \mathrm{d}V_0 \\
\mathsf{K}^{JL}_{\tilde\Phi\tilde\Phi} &= \frac{\partial \mathsf{R}^J_{\tilde\Phi}}{\partial \tilde{\boldsymbol{\Phi}}_L} = \mathop{\mathsf{A}}_{e=1}^{n_{el}} \int_{\mathcal{B}^e_t} \nabla_{\boldsymbol{X}} N^j_{\tilde\Phi} \cdot \ \ \mathsf{C}\ \ \cdot \nabla_{\boldsymbol{X}} N^l_{\tilde\Phi}\ \mathrm{d}V_t
\end{aligned}
$$

in terms of the fourth order tensors c, $\mathrm{d} = \mathrm{d}_{\boldsymbol{f}}\boldsymbol{\sigma}^t$, $\mathrm{D} = \mathrm{D}_{\boldsymbol{F}}\boldsymbol{\Sigma}^t$ and C which take the following format.

$$
\begin{aligned}
\mathsf{c} &= [\,\boldsymbol{I}\ \overline{\otimes}\, \boldsymbol{F}\,] : \mathsf{A} : [\,\boldsymbol{I} \overline{\otimes}\, \boldsymbol{F}^t\,] \\
\mathsf{d} &= \boldsymbol{I} \otimes \boldsymbol{\pi}^t - \boldsymbol{I} \underline{\otimes} \boldsymbol{\pi} - [\,\boldsymbol{f}^t \,\overline{\otimes}\, \boldsymbol{I}\ ] : \mathsf{a} : [\,\boldsymbol{I} \overline{\otimes}\ \boldsymbol{I}\ ] \\
\mathsf{C} &= [\,\boldsymbol{I}\ \overline{\otimes}\, \boldsymbol{f}\ ] : \mathsf{a} : [\,\boldsymbol{I} \overline{\otimes}\, \boldsymbol{f}^t\,] \\
\mathsf{D} &= \boldsymbol{I} \otimes \boldsymbol{\Pi}^t - \boldsymbol{I} \underline{\otimes} \boldsymbol{\Pi} - [\,\boldsymbol{F}^t \,\overline{\otimes}\, \boldsymbol{I}\ ] : \mathsf{A} : [\,\boldsymbol{I} \overline{\otimes}\ \boldsymbol{I}\ ]
\end{aligned}
$$

The solution of the system of equations (26) finally renders the iterative update for the global unknowns $\bar{\boldsymbol{\varphi}}_I = \bar{\boldsymbol{\varphi}}_I + \mathrm{d}\bar{\boldsymbol{\varphi}}_I$ and $\tilde{\boldsymbol{\Phi}}_J = \tilde{\boldsymbol{\Phi}}_J + \mathrm{d}\tilde{\boldsymbol{\Phi}}_J$.

Remark 2 (Spatial vs. material quantities) It is important to recognise that the relations between spatial and material quantities derived in the previous sections are valid only for the continuous case. Upon discretisation, we introduce the spatial and the material motion problem as independent statements. In particular, a vanishing residual $\mathrm{R}^I_{\bar\varphi}$ of the spatial motion problem does not necessarily have to correspond to a vanishing residual $\mathrm{R}^J_{\tilde\Phi}$ of the material motion problem and vice versa, see (Askes et al. 2003a, Kuhl et al. 2003, Askes et al. 2003b).

Remark 3 (Spatial vs. material forces) The discrete residua $\mathrm{R}^{\mathrm{I}}_{\bar\varphi}$ and $\mathrm{R}^{\mathrm{J}}_{\tilde\Phi}$ introduced in equation (25) define the discrete spatial and material surface forces $\mathrm{f}^{Ih}_{\varphi sur}$ *and* $\mathrm{F}^{Jh}_{\Phi sur}$.

$$
\begin{aligned}
\mathsf{f}^{I\,h}_{\varphi sur} &= \mathop{\mathsf{A}}_{e=1}^{n_{el}} \int_{\mathcal{B}^e_t} \nabla_{\boldsymbol{x}}\ N^i_{\bar\varphi} \cdot \boldsymbol{\sigma} - N^i_{\bar\varphi}\ \boldsymbol{b}_t\ \mathrm{d}V_t \\
\mathsf{F}^{J\,h}_{\Phi sur} &= \mathop{\mathsf{A}}_{e=1}^{n_{el}} \int_{\mathcal{B}^e_0} \nabla_{\boldsymbol{X}}\ N^j_{\tilde\Phi} \cdot \boldsymbol{\Sigma} - N^j_{\tilde\Phi}\ \boldsymbol{B}_0\ \mathrm{d}V_0
\end{aligned} \tag{28}
$$

which are energetically conjugate to the spatial variations $\delta\bar{\boldsymbol{\varphi}}_{\mathbf{I}}$ and to the material variations $\delta\tilde{\boldsymbol{\Phi}}_{\boldsymbol{J}}$ of the node point positions. Recall, that material forces $\mathrm{F}^{Jh}_{\Phi sur}$ are typically associated with positional changes of defects, e.g. the motion of dislocations or the opening of a crack. In the context considered herein, the finite element discretisation can thus be interpreted as a source of inhomogeneity introducing discrete defects!

6 EXAMPLES

6.1 *Homogeneous block under tension*

The first example concerns a homogeneous block under tension. The geometry and loading conditions are adopted from (Müller and Maugin 2002). The elasticity constants are given as $\lambda_0 = 115385$ and $\mu_0 = 76923$. The finite element discretisation consists of 4×4 elements. For the spatial motion problem, the horizontal displacement of upper and lower edge, which are subjected to tension, are constrained. The specimen is loaded by displacement increments 0.0025 h, up to a final displacement of 0.5 h. Different boundary conditions have been taken for the material motion problem, in correspondence with the cases studied in (Müller and Maugin 2002). Firstly, keeping all nodes fixed a lagrangian description is obtained that serves to calculate material forces, as is depicted in Figure 5, bottom. The total energy for this configuration is given as $\mathcal{I}^{\text{lagrange}} = 11.6724$. Secondly, only the second and fourth midnode on the vertical axis of symmetry are allowed to move, and only in the vertical direction, which corresponds to 2 degrees of freedom in the material motion problem. The resulting material configuration $\mathcal{B}_0$ is shown in figure 5, middle. The material displacement $\tilde{\boldsymbol{\Phi}}$ of the considered second and fourth midnode is found as 0.0301 h, thus new vertical coordinates 0.2199 h and 0.7801 h are obtained, which agree perfectly with the results given in (Müller and Maugin 2002). For this set of boundary conditions, the total energy is $\mathcal{I}^{\text{ale02}} = 11.6717$, which is somewhat below the value obtained with the purely lagrangian analysis.

Finally, all internal nodes are allowed to move, we thus have 12 degrees of freedom in the material motion problem. The resulting mesh is shown in Figure 5, top. For this case, the total energy is $\mathcal{I}^{\text{ale12}} = 11.6688$, which is again lower than in the previous analysis. In short, $\mathcal{I}^{\text{ale12}} < \mathcal{I}^{\text{ale02}} < \mathcal{I}^{\text{lagrange}}$; that is, the more degrees of freedom are included in the material motion problem or the more nodes are allowed to move the more the total energy decreases. For a representative load step, the convergence of the Newton-Raphson iterations is illustrated in Table 1. As can be verified, a quadratic convergence towards an optimal mesh is obtained, which is an important improvement compared to the 500 iterations reported in (Müller and Maugin 2002) needed to obtain the optimal mesh.

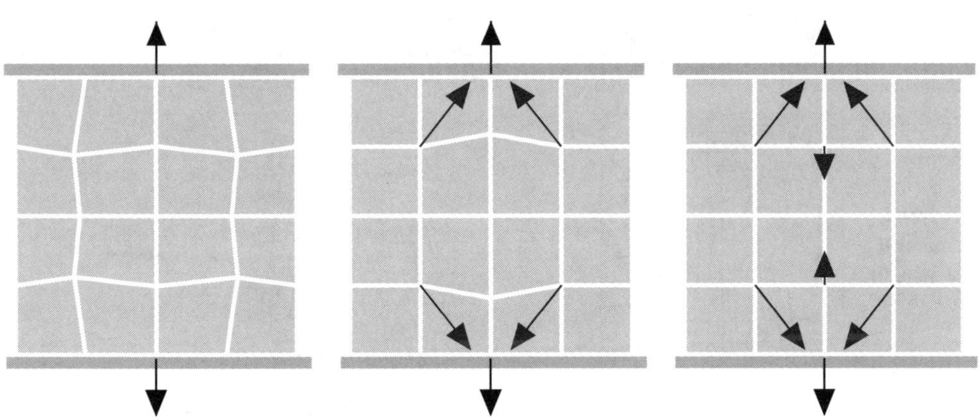

Figure 5. Homogeneous block under tension – zero, two and twelve material degrees of freedom.

Table 1. Homogeneous block under tension – typical quadratic convergence of Newton iteration.

Iteration	Norm of residual	Total energy $\mathcal{I}^{ale12}$
0	1.58114E−02	0.0000000000E+00
1	3.92275E+00	1.1672534150E+01
2	1.24279E+01	1.1669273828E+01
3	1.27318E+00	1.1668864896E+01
4	1.85232E−02	1.1668859084E+01
5	5.75864E−06	1.1668859082E+01
6	6.31087E−11	1.1668859082E+01

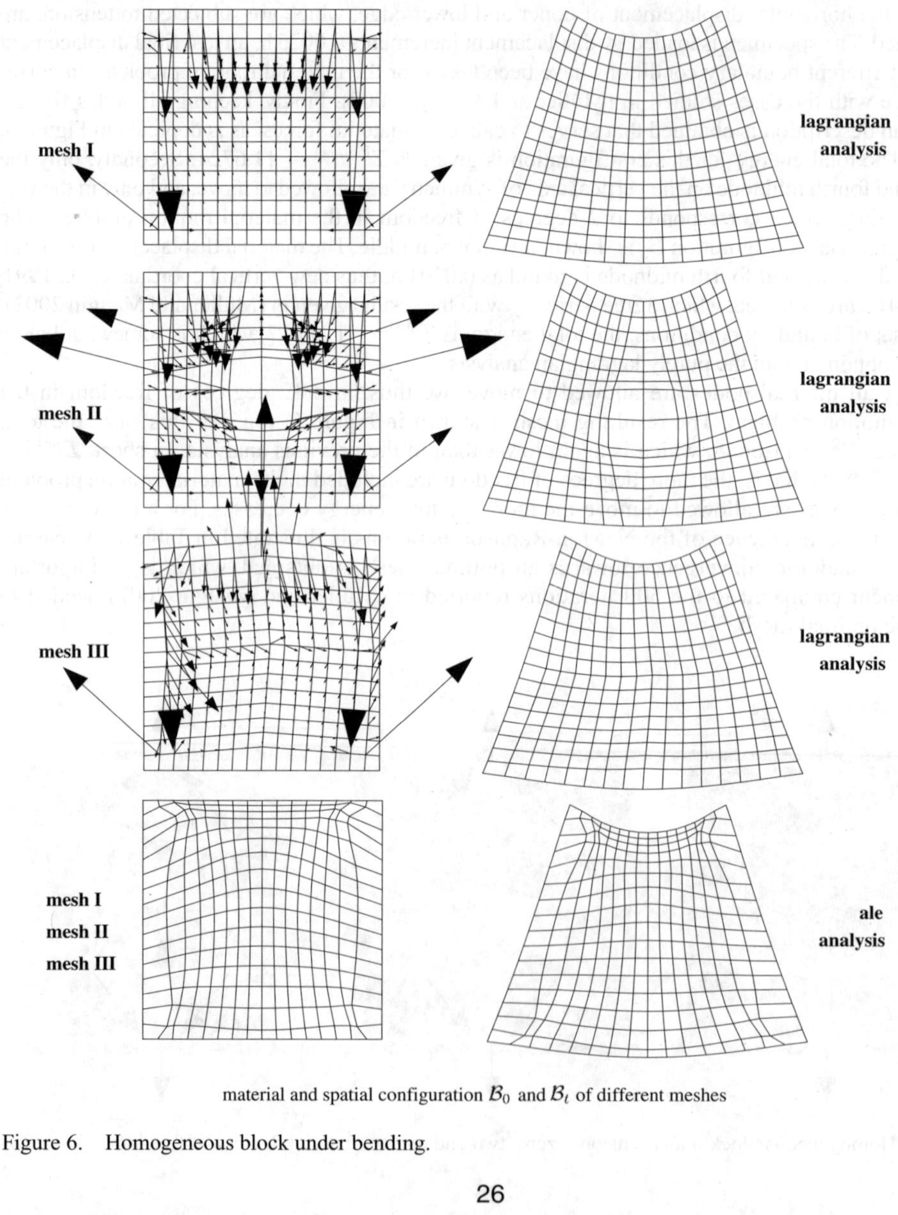

Figure 6. Homogeneous block under bending.

6.2 *Homogeneous block under bending*

Next, the same square, homogeneous block is subjected to bending. The material parameters are the same as in the previous section. For the discretisation, 8×8 elements are used. The boundary conditions of the spatial motion problem consist of prescribed displacements at vertical edges, whereas the left and right midhight node of each vertical edge are held fixed completely. For the material motion problem, nodes of all edges are constrained horizontally and vertically for the ALE analysis, and all nodes are fixed for the comparative lagrangian analysis. The specimen is loaded by applying horizontal displacements of 0.16 mm in 80 equal steps at the vertices of the specimen. In this example, inhomogeneities arise due to the nonlinearity, where it is noted that the structural behaviour differs in tension and compression, thus elements will tend to concentrate in the compressive regions.

Three different finite element discretisations have been used, one of which is regular. The other two are distorted versions of the regular mesh. Figure 6 shows the lagrangian results for the three meshes, obtained by keeping $\tilde{\boldsymbol{\Phi}}$ fixed and thus $\boldsymbol{\varphi} \equiv \bar{\boldsymbol{\varphi}}$. On the left of figure 6, the material domains $\mathcal{B}_0$ with corresponding discrete material forces are shown, while on the right the corresponding spatial domains $\mathcal{B}_t$ are plotted.

It is obvious, that different meshes lead to different discrete material force residuals. The total energy associated with these lagrangian solutions is $\mathcal{I}^{\text{lagrange}}_{\text{meshI}} = 3657.3$, $\mathcal{I}^{\text{lagrange}}_{\text{meshII}} = 3660.6$ and $\mathcal{I}^{\text{lagrange}}_{\text{meshIII}} = 3655.8$, respectively. Clearly, the potential energies differ for the three meshes. The bottom row of figure 6 illustrate the results of the ALE approach with $\tilde{\boldsymbol{\Phi}}$ free, thus nodes of $\mathcal{B}_0$ are allowed to move freely in space. Remarkable, identical results are obtained with all three starting meshes. Furthermore, all ALE analyses lead to a total potential energy $\mathcal{I}^{\text{ale}}_{\text{meshI}} = \mathcal{I}^{\text{ale}}_{\text{meshII}} = \mathcal{I}^{\text{ale}}_{\text{meshIII}} = 3650.7$, which is independent of the initial discretisation, and lower than the values obtained with either of the lagrangian meshes. On the bottom of figure 6, the material domain $\mathcal{B}_0$ is shown. It is emphasised that here vanishing discrete material are found, since material node point positions are determined such that discrete material forces vanish identically.

7 CONCLUSIONS

The central idea of the present contribution was the exploitation of the concept of configurational mechanics in the context of computational mechanics. To this end we derived an Arbitrary Lagrangian Eulerian formulation which is embedded in a consistent variational framework. Being essentially characterised through a spatial and a material mapping, the formulation is inherently related to the mechanics on the spatial and the material manifold. The governing equations follow from the evaluation of the corresponding ALE Dirichlet principle based on the total variation of the overall potential energy. By reformulating this total variation with respect to fixed reference coordinates as the sum of the variation with respect to the spatial coordinates at fixed material positions plus the variation with respect to the material coordinates at fixed spatial position, we obtained the Euler–Lagrange equations of the ALE formulation, i.e. the spatial and the material motion version of the balance of linear momentum. The solution of the former renders the spatial configuration while the latter defines the related material configuration and thus the optimal node point positions of the underlying mesh which minimises the overall potential energy.

The Euler–Lagrange equations can be interpreted as spatial and material force balance, respectively. In the continuous hyperelastic case, both statements are entirely equivalent for homogeneous problems. However, this equivalence is no longer valid upon discretisation, i.e. a vanishing discrete spatial forces do not necessarily have to coincide with vanishing discrete material forces and vice versa. Paralleling the traditional application, where continuous material forces are used to investigate material inhomogeneities, discrete material forces can be used to detect numerical inhomogeneities induced by the underlying finite element discretisation. An additional release of energy can be observed when moving nodes of the finite element mesh in the direction opposite to the material force acting on it. An optimal mesh thus corresponds to vanishing discrete material node point forces. In this sense, the remeshing strategy is no longer user-defined but objective with respect to energy minimisation.

Based on arguments of duality, our formulation is essentially characterised through the discretisation and simultaneous solution of both, the spatial and the material motion problem. In this sense, the variational ALE framework is particularly appealing since it renders symmetric system matrices upon discretisation. The presented generic framework is therefore considered as an elegant and powerful strategy to derive optimal meshes on a consistent variational basis.

REFERENCES

Armero, F. and E. Love (2003). An arbitrary Lagrangian–Eulerian finite element method for finite strain plasticity. *Int. J. Num. Meth. Eng. 35*, 737–766.

Askes, H. (2000). *Advanced spatial discretisation strategies for localised failure. Mesh adaptivity and meshless methods*. Ph.D. thesis, Technische Universiteit Delft, Delft, Nederlands.

Askes, H., E. Kuhl, and P. Steinmann (2003a). An ALE formulation based on spatial and material settings of continuum mechanics. Part 2: Classification and applications. *submitted for publication*.

Askes, H., E. Kuhl, and P. Steinmann (2003b). Arbitrary Lagrangian–Eulerian (ALE) mesh optimization by equilibration of discrete material forces. In E. Oñate and D.R.J. Owen (Eds.), *Proceedings of "COMPLAS VII", Barcelona, Spain, 07.-10. April 2003*. CIMNE, Barcelona, Spain.

Askes, H. and A. Rodriguez-Ferran (2001). A combined rh-adaptive scheme based on domain subdivision. Formulation and linear examples. *Int. J. Num. Meth. Eng. 51*, 253–273.

Askes, H., A. Rodriguez-Ferran, and A. Huerta (1999). Adaptive analysis of yield line patterns in plates with the arbitrary Lagrangian–Eulerian method. *Computers & Structures 70*, 257–271.

Askes, H. and L.J. Sluys (2000). Remeshing strategies for adaptive ALE analysis of strain localisation. *Eur. J. Mech./A: Solids 51*, 447–467.

Belytschko, T. and J.M. Kennedy (1978). Computer models for subassembly simulation. *Nuclear Engineering & Design 49*, 17–38.

Belytschko, T., W.K. Liu, and B. Moran (2000). *Nonlinear Finite Element Analysis for Continua and Structures*. John Wiley & Sons.

Braun, M. (1997). Configurational forces induced by finite-element discretization. *Proc. Estonian Acad. Sci. Phys. Math. 46*, 24–31.

Donéa, J. (1980). Finite element analysis of transient dynamic fluid-structure interaction. In J. Donéa (Ed.), *Advanced Structural Dynamics*, pp. 255–290. Applied Science Publishers, London.

Donéa, J. (1983). Arbitrary Lagrangian Eulerian finite element methods. In T. Belytschko and T.J.R. Hughes (Eds.), *Computer Methods for Transient Analysis*, pp. 473–516. Elsevier Science Publishers.

Donéa, J., P. Fasoli-Stella, and S. Giuliani (1977). Lagrangian and Eulerian finite element techniques for transient fluid-structure interaction problems. In T.A. Jaeger and B.A. Boley (Eds.), *Transactions of the 4th International Conference on Structural Mechanics in Reactor Technology – Volume B: Thermal and Fluid/Structure Dynamics Analysis*, pp. B 1/2: 1–12. North-Holland Publishing Company.

Donéa, J. and A. Huerta (2003). *Finite Element Methods for Flow Problems*. John Wiley & Sons, Chichester – New York.

Gurtin, M. E. (1995). On the nature of configurational forces. *Arch. Rat. Mech. Anal. 131*, 67–100.

Gurtin, M. E. (2000). *Configurational Forces as Basic Concepts of Continuum Physics*. Springer Verlag, New York – Berlin – Heidelberg.

Hughes, T.J.R., W.K. Liu, and T.K. Zimmermann (1981). Lagrangian–Eulerain finite element formulation for viscous flows. *Comp. Meth. Appl. Mech. Eng. 29*, 329–349.

Kienzler, R. and G. Herrmann (2000). *Mechanics in Material Space with Applications to Defect and Fracture Mechanics*. Springer Verlag, Berlin – Heidelberg – New York.

Kuhl, E., H. Askes, and P. Steinmann (2003). An ALE formulation based on spatial and material settings of continuum mechanics. Part 1: Generic hyperelastic formulation. *submitted for publication*.

Kuhl, E., S. Hulshoff, and R. de Borst (2003). An arbitrary lagrangian eulerian finite-element approach for fluid-structure interaction phenomena. *Int. J. Num. Meth. Eng. 57*, 117–142.

Kuhl, E. and P. Steinmann (2003). On spatial and material settings of thermo-hyperelastodynamics for open systems. *Acta Mechanica 160*, 179–217.

Maugin, G.A. (1993). *Material Inhomogeneities in Elasticity*. Chapman & Hall.

Maugin, G.A. (1995). Material forces: Concepts and applications. *Appl. Mech. Rev. 48*, 213–245.

Müller, R. and G. A. Maugin (2002). On material forces and finite element discretizations. *Comp. Mech. 29*, 52–60.

Rodriguez-Ferran, A., A. Pérez-Foguet, and A. Huerta (2002). Arbitrary Lagrangian–Eulerian (ALE) formulation for hyperelastoplasticity. *Int. J. Num. Meth. Eng. 53*, 1831–1851.

Shield, R.T. (1967). Inverse deformation results in finite elasticity. *ZAMP 18*, 490–500.

Steinmann, P. (2002a). On spatial and material settings of hyperelastodynamics. *Acta Mechanica 156*, 193–218.

Steinmann, P. (2002b). On spatial and material settings of thermo-hyperelastodynamics. *J. Elasticity 66*, 109–157.

Steinmann, P. (2003). On spatial and material settings of hyperelastostatic crystal defects. *J. Mech. Phys. Solids 50*, 1743–1766.

Thoutireddy, P. (2003). *Variational Arbitrary Lagrangian–Eulerian method.* Ph.D. thesis, California Institute of Technology, Passadena, California.

Yamada, T. and F. Kikuchi (1993). An arbitrary Lagrangian–Eulerian finite element method for incompressible hyperelasticity. *Comp. Meth. Appl. Mech. Eng. 102*, 149–177.

Configurational Mechanics, Kalpakides & Maugin (eds)
© 2004, Taylor & Francis Group, London, ISBN 90 5809 667 X

Topics in configurational mechanics of dielectrics

C. Trimarco
University of Pisa, Dipartimento di Matematica Applicata "U.Dini", Pisa, Italy

ABSTRACT: A first topic is concerned with some integral identities that are based on the material stress tensor. These identities, which hold true either in elastostatics or in electrostatics of elastic dielectrics, are interesting and useful as they appeal to classical boundary conditions of the electro-mechanical problem.

A second topic is connected with the first one and is concerned with the material Maxwell equations, which govern the material fields. These fields represent the classical electromagnetic fields that have been properly "convected" in the reference frame of the dielectric, through the deformation. As is known, the mathematical problem of electrostatics addresses to the Laplace equation for the electrostatic potential in the current configuration, provided that the dielectric is homogeneous and isotropic in this configuration. This may not be the case in the reference configuration. The discussion of this point addresses in turn towards a third topic that is concerned with the constitutive equations in electrostatics. These constitutive equations may hold true in one configuration and fail in a different configuration.

1 INTRODUCTION

The *material* or *configurational* stress tensor in dielectrics is an Eshelby-like tensor, which accounts for the force acting on material defects or on the tip of a crack. The general expression of the material tensor for a deformable electromagnetic body has been given in previous papers (Maugin 1993,Trimarco & Maugin 2001). This material tensor is here re-proposed in the framework of electrostatics in two slightly different forms. These two forms that are also found in the literature differ only apparently from one another. However, the difference plays a role in the boundary conditions. In fact, by adopting one of the two forms for the material tensor, the polarisation and the "elastic part of the stress" are the quantities that are involved at the boundary of the elastic dielectric. Differently, the electric displacement and the total stress naturally appear in the boundary conditions, if one refers to the second form. It is worth mentioning that the total stress includes the Maxwell stress tensor, which is also defined in a vacuum. Thus, the form of the material tensor that also includes the Maxwell tensor is expressed in terms of fields that are also defined in a vacuum. This result apparently contrasts with the current opinion, according to which the material stress tensor should express a peculiarity of the material behaviour. On the other hand, basing on this unexpected form of the material tensor, one can establish a useful integral identity, which generalises the Green's identity of elasticity (Green 1973, Trimarco, in press).

For a proper representation of the material stress tensor, the preliminary introduction of the *material* electromagnetic fields seems to be important and greatly helpful in some cases. These fields are related to the classical ones through the deformation, in such a way that the Maxwell equations preserve their form, passing from the current to the reference configuration of the dielectric. Based on this form-invariance of the Maxwell equations, one is encouraged to re-propose the classical procedures and results of electrostatics in the referential frame of the elastic dielectric.

In this view, one envisages the possibility of introducing a *material* electrostatics, as a self-consistent topic. In this material electrostatics, the material fields depend on the deformation gradient so that the electrical and the mechanical quantities are unavoidably coupled. In addition, the coupling of the mechanical and electrical equations is established in this framework in a natural way. Unfortunately, such a material electrostatics does not constitute a *replica* of the classical electrostatics. The main reason is due to an irrecoverable lack of invariance for the constitutive relations in the various configurations. A simple example is proposed in section 4, in which the constitutive relationship that holds true in an isotropic and homogeneous configuration of the dielectric, fails in the reference configuration. It is noticeable that the aforementioned comments on the constitutive relations extends to a vacuum. As a result, the electrostatic potential is still defined in the reference configuration, but it is not necessarily a harmonic function. In addition, the derivative of this potential with respect to the unit normal of the referential boundary of a conductor does not express the surface free charge, as in the classical electrostatics.

As a final remark, the following result of the *material* electrostatics is worthwhile to be noted. The right Cauchy-Green deformation tensor enters, as a rule, in the relationship between the *material* electric displacement and the *material* electric field and plays the role of a metric tensor. This remark shows in evidence the fact a metric tensor should ordinarily underlie this relationship.

2 MATERIAL FIELDS OF ELECTROSTATICS

In the following V and $\mathcal{V}$ denote the region occupied by the dielectric in its reference and in its current configuration, respectively. V and $\mathcal{V} \subseteq E_3$, E_3 representing the Euclidean space. $\mathcal{V}$ is the image of V through the deformation κ: $\mathbf{X} \rightarrow \mathbf{x}$, $\mathbf{X} \in$ V and $\mathbf{x} \in \mathcal{V}$. $\kappa \in C^3$ (V, $\mathcal{V}$). If the dielectric is of finite extent, ∂V and $\partial\mathcal{V}$ represent the boundary of V and $\mathcal{V}$, respectively. **N** and **n** denote the outward unit normal to these boundaries, respectively. The deformation gradient is denoted by **F** and its deteminant J is assumed to be strictly positive, as usual (Truesdell & Noll 1965). **E**(**x**) and **D**(**x**) represent the classical electric field and the electric displacement, respectively. As is known from electrostatics (Becker 1964, Jackson 1962), **E** is an irrotational field, whereas **D** is a solenoidal one in the absence of free charges. Along with these classical fields, one can introduce the following *material* fields (Nelson 1979, Maugin 1993, Trimarco & Maugin 2001):

$$\mathcal{E} = \mathbf{F}^{\mathrm{T}}\,\mathbf{E}, \tag{1}$$

$$\mathcal{D} = \mathrm{J}\,\mathbf{F}^{-1}\,\mathbf{D}, \tag{2}$$

$$\mathcal{P} = \mathrm{J}\,\mathbf{F}^{-1}\,\mathbf{P}. \tag{3}$$

These material fields satisfy the Maxwell equations in the reference configuration of the elastic dielectric (Maugin 1993, Trimarco, in press).

$$\mathrm{Div}\,\mathcal{D} = \mathrm{J}\,\rho, \tag{4}$$

$$\mathrm{Curl}\,\mathcal{E} = 0 \qquad \text{in} \qquad E_3 - \partial\mathrm{V}; \tag{5}$$

$$[\mathcal{D}] \bullet \mathbf{N} = \Sigma \tag{6}$$

$$[\mathcal{E}] \times \mathbf{N} = 0 \qquad \text{across } \partial\mathrm{V}. \tag{7}$$

Jρ and Σ represent the free electric charge per unit volume and per unit surface of the reference configuration, respectively. Square brackets denote the finite discontinuity of the quantities included. Based on the equation (4), the electric potential Φ can be introduced in the reference configuration as

$$\mathcal{E} = -\,\mathrm{Grad}\,\Phi. \tag{8}$$

It is worth remarking that the Div, Curl and Grad differential operators act in the reference configuration, whereas div, curl and grad will denote hereon the corresponding differential operators in the current configuration. The relationships among the above-introduced material electrostatic fields deserve a special attention and will be discussed in a subsequent section. It is reported here only the classical link that relates **E**, **D** and the polarisation **P**, in S.I. units.

$$\mathbf{D} = \varepsilon_o \mathbf{E} \qquad \text{in } E_3 - \mathcal{V}, \tag{9}$$

$$\mathbf{D} = \varepsilon_o \mathbf{E} + \mathbf{P} \qquad \text{in } \mathcal{V}. \tag{10}$$

ε_o represents the electric permeability of a vacuum.

As **P** is a feature of the material, it has to be understood as defined only inside the material in the present context.

3 MATERIAL OR CONFIGURATIONAL TENSORS IN DIELECTRICS

The general form of the electromagnetic stress tensor can be found in (Maugin, 1993, Trimarco & Maugin, 2001). The specific form to which this tensor reduces in the case of elastic dielectrics is reported below

$$\mathbf{T} = \mathbf{T}_M + \mathbf{E} \otimes \mathbf{P} + J^{-1} (\partial W(\mathbf{F}, \Pi)/\partial \mathbf{F})\, \mathbf{F}^T, \tag{11}$$

where $\mathbf{T}_M$ is the Maxwell stress tensor, which explicitly reads

$$\mathbf{T}_M = \varepsilon_o (\mathbf{E} \otimes \mathbf{E} - \tfrac{1}{2} E^2 \mathbf{I}), \tag{12}$$

I denoting the second order identity tensor. In the formula (11), W is the stored energy of the hyperelastic dielectric and Π represents the polarisation per unit volume of the reference configuration.

The *material* or *configurational* stress for a dielectric, in the absence of all magnetic quantities (Trimarco & Maugin 2001, Trimarco, in press), reads

$$\mathbf{b} = (W - \boldsymbol{E} \bullet \boldsymbol{P})\, \mathbf{I} + \boldsymbol{E} \otimes \boldsymbol{P} - \mathbf{F}^T (\partial \hat{W}(\mathbf{F}, \mathbf{F}\boldsymbol{P})/\partial \mathbf{F}). \tag{13}$$

With the help of formula (3), one can note that $\Pi \equiv J\mathbf{P} = \mathbf{F}\boldsymbol{P}$.

The second order tensor **b** is an Eshelby-like tensor and is naturally defined in a referential frame. By contrast, **T** and $\mathbf{T}_M$ are Cauchy-like stress tensors, which are naturally defined in the current frame of reference and are related to the traction. An alternate form of the configurational stress is the following:

$$\boldsymbol{b} = L\, \mathbf{I} - \boldsymbol{E} \otimes \boldsymbol{D} + \mathbf{F}^T (\partial L/\partial \mathbf{F}). \tag{14}$$

where L represents a Lagrangian density per unit volume of the reference configuration. From this Lagrangian is possible to derive the whole set of equations, which include the Maxwell equations in material form, (Maugin 1993, Trimarco & Maugin 2001). In dielectrics $L = L(\mathbf{F}, \boldsymbol{E}, \boldsymbol{P}, \mathbf{X})$ and more specifically

$$L = \tfrac{1}{2} \varepsilon_o J \boldsymbol{E} \bullet \mathbf{C}^{-1} \boldsymbol{E} + \boldsymbol{P} \bullet \boldsymbol{E} - \hat{W}(\mathbf{F}, \mathbf{F}\boldsymbol{P}, \mathbf{X}). \tag{15}$$

where $\mathbf{C} = \mathbf{F}^T\mathbf{F}$ is the right Cauchy-Green deformation tensor.

The Lagrangian (15) can be also written in the following more familiar form:

$$L = \tfrac{1}{2} \varepsilon_o J\, \mathbf{E}^2 + J\, \mathbf{E} \bullet \mathbf{P} - \hat{W}(\mathbf{F}, \mathbf{P}, \mathbf{X}). \tag{16}$$

The expressions (15) and (16) coincide, provided that the fields **E** and **P** are understood as $\mathbf{E} = \boldsymbol{E}(\mathbf{F}, \boldsymbol{E})$ and $\mathbf{P} = \boldsymbol{P}(\mathbf{F}, \boldsymbol{P})$. The Lagrangian L, such as expressed by the formula (16), is also known as the electric enthalpy that is associated with the dielectric of interest (Mindlin 1972).

By differentiating L with respect to the deformation gradient, one is able to write the total stress in the Piola-Kirchhoff form

$$\partial L / \partial \mathbf{F} = -(J\mathbf{T}_M \mathbf{F}^{-T} + \partial \hat{W} / \partial \mathbf{F}), \qquad \text{in V.} \tag{17}$$

Note that L is also defined in $E_3 - V$. In this domain, **P** and W identically vanish and L reduces to the electrostatic energy density in a vacuum. Accordingly, the formula (17) naturally extends to a vacuum and reads

$$J(\partial L / \partial \mathbf{F})\mathbf{F}^T = -\mathbf{T}_M . \tag{18}$$

This formula is interesting, as it provides the Maxwell stress tensor due to an electric field that pervades a vacuum. The procedure can be considered rather artificial, as an unlike deformation of a vacuum underlies the treatment. However, nothing prevents one to conceive a vacuum as a deformable material, whose mass density tends to zero.

If the dielectric is homogeneous, the following identities hold true:

$$\text{Div}\, \mathbf{b} = 0, \tag{19}$$

and

$$\text{Div}\, \boldsymbol{b} = 0. \tag{20}$$

The latter equality, though written in different notation, can be found in (Herrmann & Pack 1984). The tensors **b** and ***b*** are each other equivalent (apart from a minus sign), if some additional identities are taken into account (Maugin & Epstein 1991). These identities, that are not discussed hereby, are also found and commented in other works (Maugin 1993, Trimarco & Maugin 2001, Trimarco, in press). With reference to the equalities (13), (14), (19) and (20), one foresees the possibility of prescribing different set of conditions at the boundary. In fact, despite the equivalence between **b** and ***b***, different physical quantities are involved in their respective forms. In order to point out the role of these different forms, a few identities are reported below. A first identity is trivially achieved by integrating equation (19) and by appealing to the Gauss theorem.

$$\Xi_{\partial V}\ \mathbf{b}\,\mathbf{N}\, dV = \Xi_{\partial V}\ \{(W - \boldsymbol{E} \bullet \boldsymbol{P})\mathbf{N} + \boldsymbol{E}\,(\boldsymbol{P} \bullet \mathbf{N}) - \mathbf{F}^T(\partial \hat{W} / \partial \mathbf{F})\mathbf{N}\} dS . \tag{21}$$

Notice that in this equation the "elastic part of the stress" $(\partial \hat{W}/\partial \mathbf{F})$, which is part of the total stress, is involved at the boundary. Also notice that the polarisation density at the surface-boundary of the reference configuration $(\boldsymbol{P} \bullet \mathbf{N})$ appears at r.h.s. of the equality (21). It can be readily shown that $(\boldsymbol{P} \bullet \mathbf{N})dS = (\mathbf{P} \bullet \mathbf{N})ds$, ds representing the elementary surface area of $\partial \mathcal{V}$. One can note that $(\mathbf{P} \bullet \mathbf{N})$ is the classical surface polarisation and is related to the jump of the normal component of the material electric field across $\partial \mathcal{V}$, (Becker 1964, Jackson 1962). Following a similar procedure, one can perform the integration of equation (20) and write a second integral identity

$$\Xi_{\partial V}\ \boldsymbol{b}\,\mathbf{N}\, dV = -\Xi_{\partial V}\ \{[L]\,\mathbf{N} - [\boldsymbol{E}]\,(\boldsymbol{D} \bullet \mathbf{N}) + [\mathbf{F}^T(\partial L / \partial \mathbf{F})]\mathbf{N}\} dS . \tag{22}$$

The integrand at l.h.s. of the equality (21) is written by taking into account the fact that ***b*** vanishes outside V. This identity will be discussed below, after establishing another integral identity. We only note here that the quantity $(\boldsymbol{D} \bullet \mathbf{N})$ is continuous across ∂V.

It is not difficult to show that the formula (18) entails the following identity:

$$\text{Div}\,(\boldsymbol{b}\,\mathbf{X}) = \text{tr}\, \boldsymbol{b}. \tag{23}$$

where tr $\boldsymbol{b}$ denotes the spur of $\boldsymbol{b}$. Integration in V of the equality (23) leads to an additional remarkable integral equality. This reads

$$3\Xi_V \, L \, dV$$

$$=\Xi_{\partial V}\{(L)(\mathbf{N}\bullet\mathbf{X}) - ((\phi - (\nabla_{\mathbf{R}}\phi).\mathbf{X})(\boldsymbol{D}\bullet\mathbf{N}) - (\partial L/\partial\mathbf{F})\mathbf{N}\bullet(\mathbf{x}-\mathbf{F}\mathbf{X})\}dS. \quad (24)$$

The integration can be extended to E_3, provided that the electric field **E** suitably decays to infinity. In this case, the jump of the integrand across ∂V would appear at r.h.s. of the equality (24). It is worth mentioning that this equality generalises the Green's identity for elastic bodies (Green 1973, Trimarco, in press).

From the physical standpoint, it seems feasible to consider **b** as the proper representative of the material stress tensor, as it is expressed in terms of fields that are solely defined in the dielectric. However, $\boldsymbol{b}$ is very useful. In fact, the quantity $(\boldsymbol{D}\bullet\mathbf{N})$ that appears at r.h.s. of the integral formulas (22) and (24) represents the natural electrostatic condition either across the boundary of the dielectric or at the boundary of a conductor. More specifically, either $(\boldsymbol{D}\bullet\mathbf{N})$ is continuous across ∂V, as in the present case, or its jump $[\boldsymbol{D}\bullet\mathbf{N}]$ is equal to the surface free charge Σ (if present), in accordance with the formula (6).

It is worth noting that the identity (24) represents a suitable premise for establishing unique results in elastic dielectrics (Knops et al., in press).

4 MATERIAL CONSTITUTIVE RELATIONSHIPS

The link that relates the *material* electrostatic fields stems straightforwardly from the formulas (1), (2), (3), (9) and (10) and reads

$$\boldsymbol{D} = \varepsilon_o \mathbf{C}^{-1} \boldsymbol{E} + \boldsymbol{P}. \quad (25)$$

One can note the presence of the inverse of the right Cauchy-Green deformation tensor in the formula (25). This tensor acts like a metric tensor on the material electric field and this role of metric tensor survives also in the absence of polarisation. This remark warns against a straightforward identification of the electric displacement with the electric field in a vacuum.

Consider now the simplest classical constitutive relationships for a homogeneous and isotropic dielectric in the classical context.

$$\mathbf{P} = \varepsilon_0\chi\,\mathbf{E}, \quad (26)$$

$$\mathbf{D} = \varepsilon_0\varepsilon\,\mathbf{E} \quad (27)$$

where $\varepsilon = 1 + \chi$. ε and χ represent the electric permeability and the electric susceptibility of the dielectric, respectively. (Becker 1960, Jackson 1962). It is worth recalling that χ identically vanish and that ε reduces to unity in a vacuum, accordingly. The mathematical problem that is associated with the equations (26) and (27) reduces to the following one for the electrostatic potential φ, in the absence of free charges:

$$A\,\varphi = 0 \qquad \text{in } E_3 - \partial\mathcal{V}, \quad (28)$$

$$[\varphi] = 0 \qquad \text{across } \partial\mathcal{V}, \quad (29)$$

$$\varepsilon_0(\partial\varphi/\partial\mathbf{n})^+ = \varepsilon_0\varepsilon\,(\partial\varphi/\partial\mathbf{n})^- \qquad \text{across } \partial\mathcal{V}, \quad (30)$$

φ is given at infinity.

Δ represents the Laplace operator. Superscripts + and − in the formula (30) denote the limits of the related quantities, as the boundary is approached from the exterior and from the interior domain, respectively. If the dielectric is elastic, these equations should be satisfied along with the equilibrium equations, which involve the total stress such as given by the formula (17).

A similar electrostatic problem can be framed in the reference configuration of the elastic dielectric. If one wishes to preserve the constitutive features of the material behaviour such as expressed by the equations (26) and (27), the following constitutive equations should be written in terms of the material electrostatic fields, in accordance with all relationships of section 2:

$$\boldsymbol{P} = \varepsilon_0 \chi\, \mathbf{C}^{-1} \boldsymbol{E}, \tag{31}$$

$$\boldsymbol{D} = \varepsilon_0 \varepsilon\, \mathbf{C}^{-1} \boldsymbol{E}. \tag{32}$$

These equations show in evidence the fact that the homogeneity and the isotropy of the dielectric in the current configuration are not preserved in the reference configuration. If the equations (4), (8) and (32) are taken into account, the equation that governs the electrostatic potential Φ in this configuration is

$$\mathrm{Div}\,(\mathrm{J}\,\mathbf{C}^{-1}\,\mathrm{Grad}\,\Phi) = 0 \qquad \text{in } E_3 - \partial \mathrm{V}. \tag{33}$$

Clearly Φ is not a harmonic function as φ is.

Alternatively, one can consider the case of a dielectric, in which the following constitutive equations hold true:

$$\boldsymbol{D} = \varepsilon_0 \varepsilon\, \boldsymbol{E} \tag{34}$$

and

$$\boldsymbol{P} = \varepsilon_0 \chi\, \boldsymbol{E}, \tag{35}$$

in a rather formal analogy with the equations (26) and (27).

Differently with respect to the previous case, the dielectric that is described by the equations (34) and (35) is homogeneous and isotropic only in the reference configuration. This dielectric is a different material with respect to the previous one. Basing on the equations (34) and (35), the electrostatic potential $\Phi(\mathbf{X})$ turns out to be a harmonic function for this dielectric, contrary to $\varphi(\mathbf{x})$, which has to satisfy a more involved field equation in this case.

Note, however, that Φ and φ are related by the equality (1), which in electrostatics also reads

$$\mathrm{Grad}\,\Phi = \mathbf{F}^{\mathrm{T}}\,\mathrm{grad}\,\varphi. \tag{36}$$

As the deformation is smooth apart from the boundary, the equation (36) entails the equality

$$\Phi(\mathbf{X}) = \varphi\,(\mathbf{x}(\mathbf{X})), \tag{37}$$

according to which, the electrostatic potentials Φ and φ have the same value in each point of E_3, but different forms: one addresses towards a *material* or Lagrangian description, whereas the other accounts for an Eulerian description.

The formula (37) may be misleading in some circumstances or become a source of misunderstandings, if out of the appropriate context. For instance, the electric field at the boundary of the current configuration of a conductor is given by $-(\partial\varphi/\partial\mathbf{n})$, as this boundary is an equipotential surface. Accordingly, the related surface free charge density on the conductor that is surrounded by the dielectric is given by the quantity $-\varepsilon_0\varepsilon\,(\partial\varphi/\partial\mathbf{n})$ (Becker 1964, Jackson 1962).

One can easily check that this result does not hold true in the reference configuration. The reference configuration of the boundary is still an equipotential surface, but the free charge surface density on the undeformed conductor does not correspond to $-\varepsilon_0\varepsilon\ (\partial\Phi/\partial\mathbf{N})$ in this configuration.

5 CONCLUSION

The form-invariance of the Maxwell equations in the various configuration of a deformable electromagnetic material is almost completely determined by general physical principles, such as the conservation of the electric charge. This invariance holds true provided that the fields of interest are properly transformed in *material* fields. This may not be the case of the constitutive relationships that relate the electrostatic fields. As is shown in section 4, a lack of form-invariance can be pointed out even in the simplest cases. One can also envisage in section 4 interesting results that emerge in the electrostatics of the *material* fields, in the presence of a conductor.

An additional remark is that the *material* fields couple in a natural way with the classical mechanical quantities, which in turn address to interesting identities that are expounded in section 3.

As a final remark, it is worth mentioning the role of metric tensor played by the right Cauchy-Green deformation tensor in the relationship between the *material* electric displacement and the *material* electric field, even in a vacuum. This remark suggests that the identification between the electric displacement and the electric field in a vacuum is possible only in the Euclidean space.

REFERENCES

Becker, R. 1964. *Electromagnetic Fields and Interactions,* Vol. 1. London: Blackie & Son.

Green, A.E. 1973. On some general formulae in finite elasticity. *Arch. Rat. Mech. Anal.*, 50: 73–80.

Jackson, J.D. 1962. *Classical Electrodynamics*. New York: J. Wiley & Sons.

Kienzler, R. & Herrmann, G. 2000. *Mechanics in Material Space*. Berlin: Springer-Verlag.

Knops, R.J., Trimarco, C. & Williams, H.T. 2003. Uniqueness and complementary energy in nonlinear elastostatics (Forthcoming).

Maugin, G.A. 1993. *Material Inhomogeneities in Elasticity*. London: Chapman and Hall. Series "Applied Mathematics and Mathematical Computation", Vol. 3.

Maugin, G.A. & Epstein, M. 1991. The electroelastic energy-momentum tensor, *Proc. Roy. Soc. Lond.*, A433: 299–312.

Maugin, G.A. & Trimarco, C. 2001. Elements of field theory in inhomogeneous and defective materials, in *CISM courses and lectures n. 427 on "Configurational Mechanics of Materials"*, Wien: Springer-Verlag, 55–128.

Mindlin, R.D. 1972. Elasticity, Piezoelasticity and Crystal Lattice Dynamics, *J. of Elasticity*, 2: 17–282.

Nelson, D.F. 1979. Electric, Optic and Acoustic Interactions in Dielectrics. New York: John Wiley.

Pack, Y.E. & Herrmann, G. 1986. Conservation laws and the material momentum tensor for the elastic dielectric, *Int. J. Engng. Sci.*, 24: 1365–1374.

Toupin, R.A. 1956. The Elastic dielectrics, *J. Rational Mech. An.*, 5: 849–915.

Trimarco, C. 2002. A Lagrangian approach to electromagnetic bodies, *Technische Mechanik*, 22(3): 175–180.

Trimarco, C. 2002. Stresses and momenta in electromagnetic materials, *Mechanics Research Communications*, 29(6): 485–492.

Trimarco, C. 2003. The Green's and the Eshelby's identities in generalised continua and in dielectrics (Forthcoming).

Trimarco, C. & Maugin, G.A. 2001. Material mechanics of electromagnetic bodies, in *CISM courses and lectures n. 427 on "Configurational Mechanics of Materials"*, Wien: Springer-Verlag, 129–172.

Truesdell, C.A. & Toupin R.A. 1960. The Classical Theory of Fields, in *Handbuch der Physik*, Bd. III/1. S. Flügge (ed.). Berlin: Springer-Verlag.

Configurational Mechanics, Kalpakides & Maugin (eds)
© 2004, Taylor & Francis Group, London, ISBN 90 5809 667 X

Configurational forces and couples in micropolar continua

Jovo P. Jarić
University of Belgrade, Faculty of Mathematics, Belgrade, Serbia and Montenegro

ABSTRACT: Following the approach of Gurtin and Podio-Guidugli (1996), (1998), the problem of crack propagation based on the notion of configurational forces and couples in micropolar continua is considered. The derivation of the Eshelby relation was accomplished without recourse to constitutive equations or to a variational principle; the derivation was based on a version of the second law appropriate to referential control volumes whose boundaries evolve with time.

1 INTRODUCTION

Very recently Gurtin and Podio-Guidugli (1996), developed a framework for dynamical fracture, concerning on the derivation of balance equations and constitutive equations that describe the motion of the crack tip in two-space dimensions. They worked within the nonlinear theory because the basic ideas are most easily explained within a framework that distinguishes between reference and deformed configurations; moreover, instead of laying down specific assumptions regarding the strength of the crack-tip singularities, they consider hypotheses motivated by the requirement that the underlying physical laws make sense. The theory is based on a configurational force balance and a mechanical version of the second law of thermodynamics.

In classical continuum mechanics the response of a body to deformation is described by standard deformational forces consistent with balance laws for linear and angular momentum. But, as they stated, configurational forces are less intuitive: they are related to the intrinsic coherency of a body's material structure and perform work in the addition and removal of material and in the evolution of structural defects. Following Gurtin and Struthers (1990) and Gurtin (1995), they viewed configurational forces as basic primitive objects consistent with their own force balance. Gurtin's argument, with particular reference to the requirement of invariance under re-parameterization of the boundary of the reference control volume, is utilized by Mariano (2000) to deduce the expression of configurational forces in the case of continua with microstructure. Configurational forces defined via the calculus of variations as derivatives of an energy have been introduced earlier, e.g. in the classic works of Eshelby (1951), (1975) on lattice defects.

The role of configurational forces, however, seems more pervasive and fundamental than problems susceptible to a variational formulation can indicate, a view they demonstrated within the context of fracture dynamics.

We refer a reader also to the recent monograph by Maugin (1993) and (1998), for a discussion of configurational forces within an Eshelbian framework, and for related references.

In this communication we strictly follow the approach of Gurtin and Podio-Guidugli (1996), (1998), in a case of (two-dimensional) micropolar continua. For particular reviews of version sectors of this field see, first of all, Eringen (1976), Stojanović (1969), Kafadar and Eringen (1971). Our final results are not different in form results well known (see Jarić (2000)). What is different is a derivation and a interpretation in terms of configurational forces, which we believe most accurately describe the underlying physics.

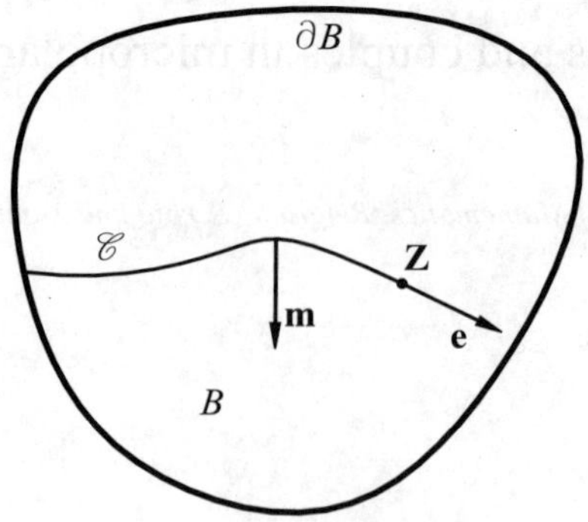

Figure 1.

The scope of the paper is the following: at the beginning we review some basic formulae of the geometry and the kinematics of a cracked body given by Gurtin and Podio-Guidugli (1996), we need in our investigation; for the same reason we described the motion of the crack bodies; next we state the balances laws for deformational and configurational forces for micropolar continua; the last part belongs to the derivation of Eshelby tensor as a consequence of invariance under reparametrization.

2 CRACKS; TIME-DEPENDENT CONTROL VOLUMES

Let B denote a closed region of $\mathbb{R}^2$ with boundary ∂B and, for each t in some open time interval, let $\mathcal{C}(t)$ be a smooth, connected, oriented curve in B with one end, $\mathbf{Z}_0$, fixed at the boundary ∂B, with the remainder of $\mathcal{C}(t)$ – including the other end point $\mathbf{Z}(t)$ – contained in the interior of B, and with

$$\mathcal{C}(\tau) \subset \mathcal{C}(t) \quad \text{for all} \quad t \geq \tau \tag{2.1}$$

Also

$$\mathcal{B}(t) = B \backslash \mathcal{C}(t) \tag{2.2}$$

is viewed as a referential neighborhood of a growing crack $\mathcal{C}(t)$ with $\mathbf{Z}(t)$ the crack tip. Note that B contains the points of $\mathcal{C}(t)$ while $\mathcal{B}(t)$ does not; hence $\mathcal{B}(t)$ is cracked, while B is not. Note also that the assumed regularity of $\mathcal{C}(t)$ precludes singularities such as kinks and bifurcations. We let $\mathbf{e}(t)$ denote the unit tangent to $\mathcal{C}(t)$ at $\mathbf{Z}(t)$ in the direction of (possible) propagation (Fig. 1).

Then the tip velocity

$$\mathbf{v}(t) = \mathrm{d}\mathbf{Z}(t)/\mathrm{d}t \tag{2.3}$$

may be written in the form

$$\mathbf{v}(t) = V(t)\mathbf{e}(t), \qquad V(t) \geq 0, \tag{2.4}$$

with V the speed. Finally, a continuous unit normal field of $\mathcal{C}(t)$ is denoted by $\mathbf{m}(\mathbf{X}, t)$.

A control volume is defined as closed subregion $R(t)$ of B for which $\partial R(t)$ evolves smoothly with t, and for which

$$\mathcal{C}_R(t) = \mathcal{C}(t) \cap R(t). \tag{2.5}$$

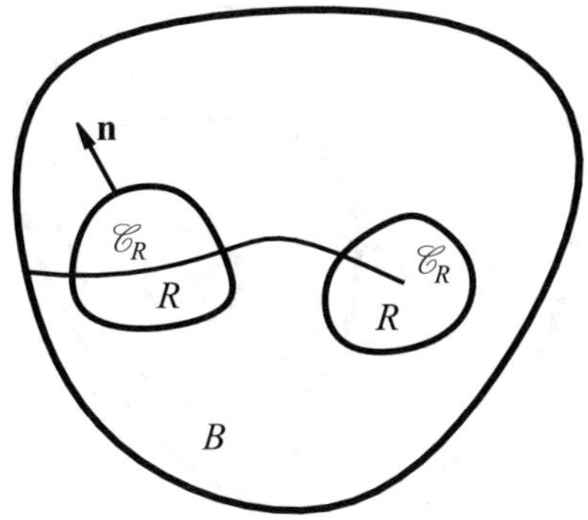

Figure 2.

It is assumed that the portion of the crack in $R(t)$, does not intersect $\partial R(t)$ at more than two points (Fig. 2). For convenience, two classes of control volumes are discussed: those that do not intersect the tip and those that contain the tip in their interior. The dependence of $R(t)$ on t is viewed as resulting from the addition and removal of material points. So defined a control volume does not preclude control volumes R that are independent of time.

For $R(t)$ a control volume, $\mathbf{n}(\mathbf{X}, t)$ designates the outward unit normal to $\partial R(t)$, and $U_{\partial R}$ the (scalar) normal velocity of the boundary curve in the direction $\mathbf{n}$. A useful example of a time-dependent control volume is the tip disc

$$D_\delta(t) = \{\mathbf{X} \in B : |\mathbf{X} - \mathbf{Z}(t)| \leq \delta\}, \tag{2.6}$$

For convenience, we write

$$\mathcal{C}_\delta(t) = \mathcal{C}_{D_\delta}(t) = \mathcal{C}(t) \cap D_\delta(t). \tag{2.7}$$

2.1 *Derivatives following the crack tip; tip integrals; transport theorems*

According to Gurtin and Podio-Guidugli (1996), a field $\Phi(\mathbf{X}, t)$ is referred as smooth away from the tip if $\Phi(\mathbf{X}, t)$ is defined for all $\mathbf{X} \in \mathcal{B}(t)$ and all t, and if, away from the tip, $\Phi(\mathbf{X}, t)$ and its derivatives have limits up to the crack from either side. Then we write, for $\mathbf{X} \in \mathcal{C}(t)$, $\mathbf{X} \neq \mathbf{Z}(t)$

$$\Phi^{\pm}(\mathbf{X}, t) = \lim_{\varepsilon \to 0} \Phi(\mathbf{X} \pm \varepsilon \mathbf{m}(\mathbf{X}, t), t), \quad [\Phi] = \Phi^{+} - \Phi^{-}. \tag{2.8}$$

Given such a field $\Phi(\mathbf{X}, t)$, consider the corresponding field $\hat{\Phi}(\mathbf{Y}, t)$ in which $\mathbf{Y}$ represents the position of the material point $\mathbf{X}$ relative to the tip $\mathbf{Z}(t)$

$$\hat{\Phi}(\mathbf{Y}, t) = \Phi(\mathbf{X}, t), \quad \mathbf{Y} = \mathbf{X} - \mathbf{Z}(t). \tag{2.9}$$

The partial derivative

$$\Phi^{0}(\mathbf{X}, t) = \frac{\partial \hat{\Phi}(\mathbf{Y}, t)}{\partial t}$$

with respect to t holding $\mathbf{Y}$ fixed represents the time derivative of $\Phi(\mathbf{X}, t)$ following the tip $\mathbf{Z}(t)$; by the chain rule,

$$\Phi^{0} = \Phi^{\bullet} + \nabla\Phi \cdot \mathbf{v} \tag{2.10}$$

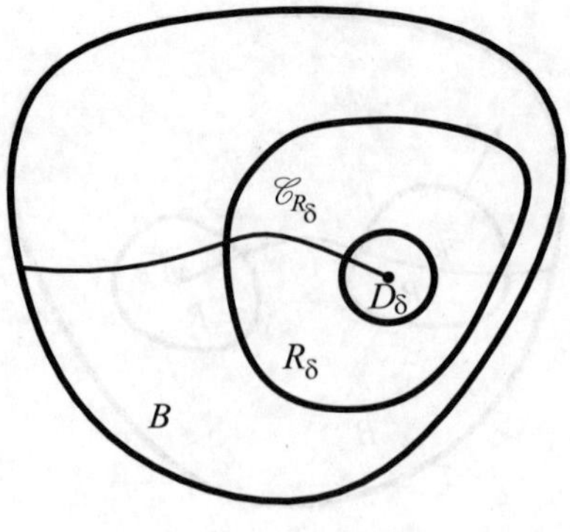

Figure 3.

away from the tip, where

$$\Phi^{\bullet}(\mathbf{X},t) = \frac{\partial\Phi(\mathbf{X},t)}{\partial t}.$$

We will repeatedly take limits, as $\delta \to 0$, of integrals of fields over $\partial D_\delta(t)$; such limits, when meaningful, are called tip integrals. Thus,

$$\oint_{tip} \varphi\{\mathbf{n}\} = \lim_{\delta\to 0} \int_{\partial D_\delta(t)} \varphi\{\mathbf{n}\}, \tag{2.11}$$

for any tensor field φ and any meaningful product { } of φ and the outward unite normal vector $\mathbf{n}$ on $D_\delta(t)$.

Let $R(t)$ be a control volume that includes the tip and consider the region

$$R_\delta(t) = R(t)\backslash D_\delta(t), \tag{2.12}$$

with $\delta > 0$ sufficiently small that $\partial R_\delta(t) = \partial R(t) \cup \partial D_\delta(t)$. Then, using the same letter $\mathbf{n}$ for the outward unit normal on both ∂R and ∂D_δ, and bearing in mind that the outward unit normal to ∂R_δ on ∂D_δ is $\mathbf{n}$, we may use the gradient theorem in the usual manner – with C_{R_δ} considered as a "slit in R_δ" giving rise to an additional pair of boundary segments (Fig. 3), and with $\int_{R_\delta} \nabla\Phi$ interpreted accordingly – to conclude that, for Φ smooth away from the tip,

$$\int_{R_\delta} \nabla\Phi = \int_{\partial R} \Phi\mathbf{n} - \int_{C_{R_\delta}} [\Phi]\mathbf{m} - \int_{\partial D_\delta} \Phi\mathbf{n}. \tag{2.13}$$

Here, for convenience, the argument t has been suppressed. Thus, if $\oint_{tip} \Phi\mathbf{n}$ exists, and if $[\Phi]$ is integrable on C then $\int_R \nabla\Phi$ exists as the limit

$$\lim_{\delta\to 0} \int_{R_\delta} \nabla\Phi$$

and we have the generalized gradient theorem

$$\int_R \nabla\Phi = \int_{\partial R} \Phi\mathbf{n} - \int_{C_R} [\Phi]\mathbf{m} - \oint_{tip} \Phi\mathbf{n}. \tag{2.14}$$

The next definition allows one to state succinctly the hypotheses concerning momenta and energies. We will refer to Φ as regular if, in addition to being smooth away from the tip,

(R1) Φ is integrable on B; given any control volume $R(t)$, the mapping $t \to \int_{R(t)}\Phi$ is differentiable;
(R2) Φ^o is integrable on B and $[\Phi]\mathbf{m} \cdot \mathbf{v}$ is integrable on $\mathcal{C}(t)$, both uniformly in t;
(R3) $\oint_{tip} \Phi\mathbf{n}$ exists.

(The phrase "uniformly in t" signifies "uniformly for t in any compact interval".) By (R2) and (2.7),

$$\int\limits_{\mathcal{D}_\delta(t)} [\Phi]\mathbf{m}{\cdot}\mathbf{v} \quad \text{approaches zero as } \delta \to 0. \tag{2.15}$$

The following well-known transport theorem is valid when $\Phi(\mathbf{X}, t)$ is smooth away from the tip and $R(t)$ does not contain the tip

$$\frac{\mathrm{d}}{\mathrm{d}t}\left\{\int\limits_{R(t)} \Phi\right\} = \int\limits_{R(t)} \Phi^{\bullet} + \int\limits_{\partial R(t)} \Phi U_{\partial R}. \tag{2.16}$$

Then Gurtin and Podio-Guidugli (1996) gave two generalizations of (2.16) that account for the crack tip by the following *Transport theorem.* For $R(t)$ a control volume that includes the tip, if $\Phi(\mathbf{X}, t)$ is regular, then

$$\frac{\mathrm{d}}{\mathrm{d}t}\left\{\int\limits_{R(t)} \Phi\right\} = \int\limits_{R(t)} \Phi^{\bullet} + \int\limits_{\partial R(t)} \Phi U_{\partial R} - \oint\limits_{tip} \Phi(\mathbf{v}{\cdot}\mathbf{n}), \tag{2.17}$$

$$\frac{\mathrm{d}}{\mathrm{d}t}\left\{\int\limits_{R(t)} \Phi\right\} = \int\limits_{R(t)} \Phi^{o} + \int\limits_{\partial R(t)} \Phi(U_{\partial R} - \mathbf{v}{\cdot}\mathbf{n}) + \int\limits_{\mathcal{C}_{R(t)}} [\Phi]\mathbf{m}{\cdot}\mathbf{v} \tag{2.18}$$

(with $\int_{R(t)} \Phi^{\bullet}$ defined as $\lim_{\delta\to 0} \int_{R_\delta} \Phi^{\bullet}$, which exists).

The normal velocity $U_{\partial R}$ is defined in the next section (see (2.21)).

This transport theorem is considered as a central result of their theory. We shall come to this theorem in the case of a micropolar continua.

2.2 *Motions of cracked micropolar bodies*

The motion of micropolar body $\mathcal{B}(t)$ is described completely by $\mathbf{y}(\mathbf{X}, t)$ and $\boldsymbol{\chi}(\mathbf{X}, t)$, where orthogonal $\boldsymbol{\chi}$ is called microrotation (see Eringen (1976)). Let $\mathbf{y}(\mathbf{X}, t)$ and $\boldsymbol{\chi}(\mathbf{X}, t)$ be smooth away from the tip and let $\mathbf{y}(\mathbf{X}, t)$ be one-to-one in $\mathbf{X}$ on $\mathcal{B}(t)$ for each t. Then the deformation gradient

$$\mathbf{F} = \nabla\mathbf{y} \tag{2.19}$$

and the material velocity $\mathbf{y}^{\bullet}$ is then smooth away from the tip.

Let $R(t)$ be a control volume. The boundary curve $\partial R(t)$ may be parameterized in a sufficiently small time interval and in a neighborhood of any of its points by a function of the form $\mathbf{X} = \hat{\mathbf{X}}(\sigma, t)$; the field

$$\mathbf{u}(\mathbf{X}, t) = \frac{\partial\hat{\mathbf{X}}(\sigma, t)}{\partial t} \tag{2.20}$$

then represents a velocity field for $\partial R(t)$ in that neighborhood. It is possible to use such parametrizations to construct a velocity field for $\partial R(t)$; that is, a smooth field $\mathbf{u}(\mathbf{X}, t)$ defined for all $\mathbf{X}$ on

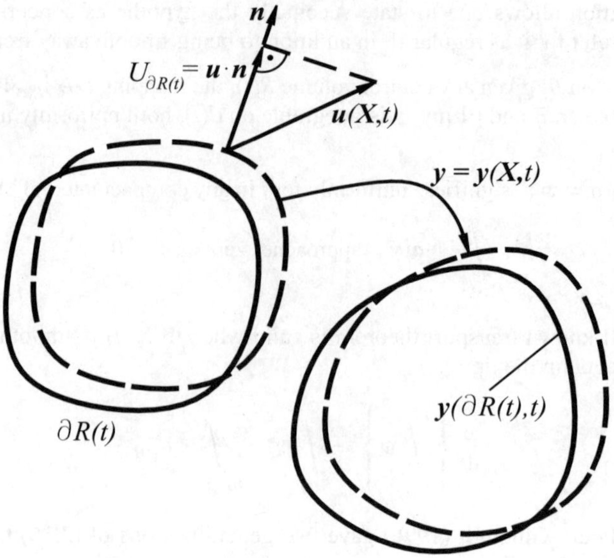

Figure 4.

$\partial R(t)$ and all t in any (sufficiently small) time interval. A field $\mathbf{u}$ so constructed depends on the choice of local parametrizations, but its normal component is intrinsic (see Fig. 4)

$$\mathbf{u}{\cdot}\mathbf{n} = U_{\partial R}. \tag{2.21}$$

Each local parametrization $\mathbf{X} = \hat{\mathbf{X}}(\sigma, t)$ induces a corresponding local parametrization $\mathbf{x} = \hat{\mathbf{x}}(\sigma, t) = \mathbf{y}(\hat{\mathbf{X}}(\sigma, t), t)$ for the deformed boundary curve $\mathbf{y}(\partial R(t), t)$; the corresponding induced velocity field

$$\bar{\mathbf{u}}(\mathbf{X}, t) = \frac{\partial \hat{\mathbf{x}}(\sigma, t)}{\partial t} \tag{2.22}$$

for the deformed boundary $\mathbf{y}(\partial R(t), t)$ is related to $\mathbf{u}$ by the formula

$$\bar{\mathbf{u}} = \mathbf{y}^{\bullet} + \mathbf{F}\mathbf{u}. \tag{2.23}$$

In Gurtin and Podio-Guidugli's (1996) approach the tip velocity $\mathbf{v}(t)$ has been considered as a velocity field for the boundary of the disc $D_\delta(t)$ using as a parametrization

$$\mathbf{X} = \hat{\mathbf{X}}_\delta(\sigma, t) = \mathbf{Z}(t) + \delta\mathbf{n}(\sigma) \tag{2.24}$$

with $\mathbf{n}(\sigma)$ a unit vector at an angle σ from a fixed axis. Then

$$\mathbf{y}^{o} = \mathbf{y}^{\bullet} + \mathbf{F}\mathbf{v} \tag{2.25}$$

the time derivative following $\mathbf{Z}(t)$, represents the corresponding induced velocity field for $\mathbf{y}(\partial D_\delta(t), t)$.

Here we consider the case when σ is a function of time t, i.e. when $\sigma = \sigma(t)$. Then $D_\delta(t)$ represents a disk which rotates with the motion of crack tip $\mathbf{Z}(t)$. This assumption is in accordance with mathematical model of micropolar continua. Since $\dot{\mathbf{n}} = \boldsymbol{\omega} \times \mathbf{n}$, where $\boldsymbol{\omega}(t) = \dot{\sigma}(t)\mathbf{k}$ is an angular velocity, and $\mathbf{k}$ – the unite vector orthogonal to the region $\mathbb{R}^2$, then from (2.24) we have (see Fig. 5)

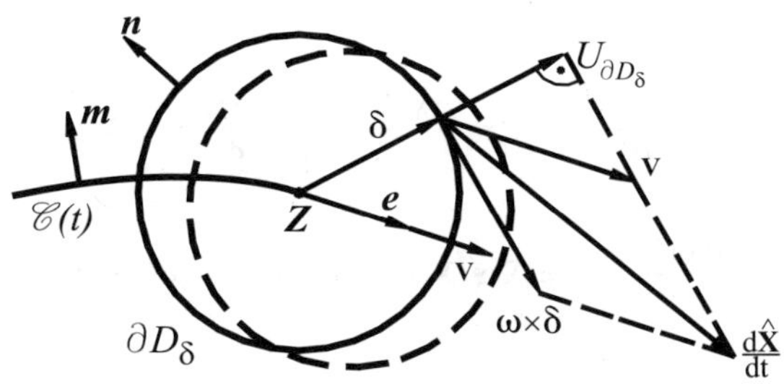

Figure 5.

$$\mathbf{v}_\delta = \frac{d\hat{\mathbf{X}}_\delta}{dt} = \mathbf{v}(t) + \delta\boldsymbol{\omega} \times \mathbf{n} = \mathbf{v}(t) + \boldsymbol{\omega} \times \boldsymbol{\delta}. \tag{2.26}$$

Remark. Note that

$$U_{\partial D_\delta} = \frac{d\hat{\mathbf{X}}_\delta}{dt}\cdot \mathbf{n} = \mathbf{v}(t)\cdot \mathbf{n} \tag{2.27}$$

does not depend on the parameterization of $\partial D_\delta(t)$. Then (2.17) and (2.18) holds true for a micropilar continua with the crack as well.

Accordingly, from $\mathbf{y}(\partial D_\delta(t), t) = \mathbf{y}(\hat{\mathbf{X}}(\sigma(t), t))$ we have

$$\mathbf{y}^\square = \frac{d\mathbf{y}}{dt} = \mathbf{y}^\bullet + \mathbf{F}(\mathbf{v} + \boldsymbol{\omega} \times \boldsymbol{\delta}) = \mathbf{y}^o + \mathbf{F}(\boldsymbol{\omega} \times \boldsymbol{\delta}). \tag{2.28}$$

This formula represents generalization of (2.25).

We assume that:

(Al) there is a function $\bar{\mathbf{v}}(t)$ such that

$$\mathbf{y}^\square(\mathbf{X}, t) \to \bar{\mathbf{v}}(t) \quad \text{as } \mathbf{X} \to \mathbf{Z}(t) \text{ uniformly in } t, \tag{2.29}$$

where $\bar{\mathbf{v}}(t)$ represents the velocity of the deformed crack tip. This is indeed the case since $\mathbf{y}^\square \to \mathbf{y}^o$ as $\delta \to 0$. The rest of the proof is identical as in Gurtin and Podio-Guidugli (1998).

Moreover,

$$\bar{\mathbf{v}}(t) = \frac{d\mathbf{y}(\mathbf{Z}(t), t)}{dt}. \tag{2.30}$$

We recall that for micropolar continua, the skew-symmetric gyration tensor is defined by

$$\mathbf{N} = \boldsymbol{\chi}^\bullet \boldsymbol{\chi}^T, \quad \mathbf{N} = -\mathbf{N}^T. \tag{2.31}$$

The corresponding angular velocity $\boldsymbol{\nu}$ is defined by

$$\boldsymbol{\nu} = -\frac{1}{2}\boldsymbol{\varepsilon}\cdot\cdot\,\mathbf{N}, \quad \text{or} \quad \nu^k = -\frac{1}{2}\varepsilon^{klm} N_{lm}. \tag{2.32}$$

Then, on $\partial R(t)$: $\chi = \chi(\hat{\mathbf{X}}(\sigma, t), t)$, we defined gyration tensor $\boldsymbol{M}$ by

$$\boldsymbol{M} = \left.\frac{\partial \boldsymbol{\chi}(\hat{\mathbf{X}}(\sigma, t), t)}{\partial t}\right|_\sigma \boldsymbol{\chi}(\hat{\mathbf{X}}(\sigma, t), t)^T.$$

But

$$\left.\frac{\partial\chi(\hat{\mathbf{X}}(\sigma,t),t)}{\partial t}\right|_{\sigma}=\left.\frac{\partial\chi(\hat{\mathbf{X}},t)}{\partial t}\right|_{\hat{\mathbf{X}}}+\frac{\mathrm{d}\hat{\mathbf{X}}}{\mathrm{d}t}\cdot\nabla\chi(\hat{\mathbf{X}},t)$$

$$=\frac{\partial\chi(\hat{\mathbf{X}},t)}{\partial t}+\mathbf{u}\cdot\nabla\chi(\hat{\mathbf{X}},t)=\chi^{\bullet}(\hat{\mathbf{X}},t)+\mathbf{u}\cdot\nabla\chi(\hat{\mathbf{X}},t)$$

so that

$$\boldsymbol{M}=\mathbf{N}+(\mathbf{u}\cdot\nabla\chi)\,\chi^{T}. \tag{2.33}$$

Thus, the corresponding gyration vector $\mu = -1/2\ \varepsilon\cdot\cdot\ \boldsymbol{M}$ is given by

$$\boldsymbol{\mu}=\boldsymbol{\nu}-\frac{1}{2}\varepsilon\cdot\cdot(\mathbf{u}\cdot\nabla\chi)\,\chi^{T}\quad\text{or}\quad\mu_k=\nu_k-\frac{1}{2}\varepsilon_{klm}\chi^{l}_{K;L}u^{L}\chi^{mK}. \tag{2.34}$$

Next we investigate the skew-symmetric tensor $\boldsymbol{M}_\delta$ on $\partial D_{\delta(t)}$. Then, $\boldsymbol{\chi}_\delta(\sigma(t),t)=\boldsymbol{\chi}(\hat{\mathbf{X}}_\delta(\sigma(t),t),t)$ having in mind that $\sigma=\sigma(t)$. Hence,

$$\chi_\delta^{\square}=\frac{\mathrm{d}\chi_\delta}{\mathrm{d}t}=\frac{\partial\chi_\delta}{\partial t}+\frac{\mathrm{d}\mathbf{X}_\delta}{\mathrm{d}t}\cdot\nabla\chi_\delta=\frac{\partial\chi_\delta}{\partial t}+(\mathbf{v}+\boldsymbol{\omega}\times\boldsymbol{\delta})\cdot\nabla\chi_\delta=\chi_\delta^{\bullet}+(\boldsymbol{\omega}\times\boldsymbol{\delta})\cdot\nabla\chi_\delta, \tag{2.35}$$

where

$$\chi_\delta^{\bullet}=\frac{\partial\chi_\delta}{\partial t}+\mathbf{v}\cdot\nabla\chi_\delta. \tag{2.36}$$

But

$$\boldsymbol{M}_\delta=\chi_\delta^{\square}\chi_\delta^{T}. \tag{2.37}$$

Thus,

$$\boldsymbol{M}_\delta=(\chi_\delta^{\bullet}+(\boldsymbol{\omega}\times\boldsymbol{\delta})\cdot\nabla\chi_\delta)\chi_\delta^{T}=\mathbf{N}_\delta+((\boldsymbol{\omega}\times\boldsymbol{\delta})\cdot\nabla\chi_\delta)\chi_\delta^{T} \tag{2.38}$$

where

$$\mathbf{N}_\delta=\chi_\delta^{\bullet}\chi_\delta^{T}=\left(\frac{\partial\chi_\delta}{\partial t}+\mathbf{v}\cdot\nabla\chi_\delta\right)\chi_\delta^{T}. \tag{2.39}$$

The corresponding gyration vector $\boldsymbol{\mu}_\delta$ is given by

$$\boldsymbol{\mu}_\delta=-\frac{1}{2}\varepsilon\cdot\cdot\,\mathbf{M}_\delta=\boldsymbol{\nu}_\delta-\frac{1}{2}\varepsilon\cdot\cdot((\boldsymbol{\omega}\times\boldsymbol{\delta})\cdot\nabla\chi_\delta)\,\chi_\delta^{T}, \tag{2.40}$$

where

$$\boldsymbol{\nu}_\delta=-\frac{1}{2}\varepsilon\cdot\cdot\mathbf{N}_\delta=\bar{\boldsymbol{\nu}}_\delta-\frac{1}{2}\varepsilon\cdot\cdot(\mathbf{v}\cdot\nabla\chi_\delta)\chi_\delta^{T}, \tag{2.41a}$$

$$\bar{\boldsymbol{\nu}}_\delta=-\frac{1}{2}\varepsilon\cdot\cdot\frac{\partial\chi_\delta}{\partial t}\chi_\delta^{T}. \tag{2.41b}$$

Note that at the tip of the crack

$$\boldsymbol{\mu}_\delta\to\boldsymbol{\mu}_0=\boldsymbol{\nu}_0=\bar{\boldsymbol{\nu}}_0(\mathbf{Z}(t),t)-\frac{1}{2}\varepsilon\cdot\cdot[\mathbf{v}\cdot\nabla\chi(\mathbf{Z}(t),t)]\chi^{T}(\mathbf{Z}(t),t) \tag{2.42}$$

or

$$\mu_{0k} = \nu_{0k} = \bar{\nu}_k(\mathbf{Z}(t),t) - \frac{1}{2}\varepsilon_{klm}\chi^l_{K;L}(\mathbf{Z}(t),t)v^L\chi^{mK}(\mathbf{Z}(t),t) \tag{2.43}$$

as $\delta \to 0$, where

$$\bar{\boldsymbol{\nu}}_0(t) = -\frac{1}{2}\boldsymbol{\varepsilon}\cdot\cdot\frac{\partial \chi_\delta(\mathbf{Z}(t),t)}{\partial t}\chi_\delta^T(\mathbf{Z}(t),t) = \boldsymbol{\nu}(\mathbf{Z}(t),t). \tag{2.44}$$

3 BASIC LAWS

3.1 *Balance laws for deformational and configurational forces*

We let ρ denote the reference mass density, write

$$\mathbf{p} = \rho\mathbf{y}^{\bullet}, \qquad \rho\boldsymbol{\sigma} = \boldsymbol{\Pi} \tag{3.1}$$

for the momentum, and intrinsic angular momentum density due to rotation; $\boldsymbol{\sigma}$ is a spin density. Let $\mathbf{S}$ denote the Piola-Kirchhoff stress and $\mathbf{M}$ couple stress that arise in response to deformation. We neglect external body forces and body couples, and assume that the crack faces are traction-free, i.e

$$\mathbf{S}^{\pm}\mathbf{m} = 0, \qquad \mathbf{M}^{\pm}\mathbf{m} = 0, \quad \text{on } \mathcal{C}(t). \tag{3.2}$$

The balance laws for linear and angular momentum then take the form

$$\frac{\mathrm{d}}{\mathrm{d}t}\left\{\int_{R(t)} \mathbf{p}\right\} = \int_{\partial R(t)} (\mathbf{p}\,U_{\partial R} + \mathbf{Sn}) \tag{3.3a}$$

$$\frac{\mathrm{d}}{\mathrm{d}t}\left\{\int_{R(t)} (\mathbf{y}\times\mathbf{p}+\boldsymbol{\Pi})\right\} = \int_{\partial R(t)} \left\{(\mathbf{y}\times\mathbf{p}+\boldsymbol{\Pi})\,U_{\partial R} + \mathbf{y}\times\mathbf{Sn} + \mathbf{Mn}\right\}, \tag{3.3b}$$

for each control volume $R(t)$.

We consider, in addition, a configurational stress $\mathbf{C}$, a configurational couple stress $\boldsymbol{\Lambda}$, a configurational force $\mathbf{f}$ and a configurational moment Δ distributed over $\mathcal{B}(t)$; a configurational force $\mathbf{g}$, and a configurational moment $\mathfrak{M}(t)$ concentrated at the tip; these are presumed consistent with the configurational force balance and moment balance, if $R(t)$ does not contain the tip

$$\int_{\partial R(t)} \mathbf{Cn} + \int_{\mathcal{C}_{R(t)}} [\![\mathbf{C}]\!]\,\mathbf{m} + \int_{R(t)} \mathbf{f} = 0, \tag{3.4a}$$

$$\int_{\partial R(t)} (\mathbf{X}\times\mathbf{Cn} + \boldsymbol{\Lambda}\mathbf{n}) + \int_{\mathcal{C}_{R(t)}} (\mathbf{X}\times[\![\mathbf{C}]\!]\,\mathbf{m} + [\![\boldsymbol{\Lambda}]\!]\,\mathbf{m}) + \int_{R(t)} (\mathbf{X}\times\mathbf{f} + \boldsymbol{\Delta}) = 0, \tag{3.4b}$$

and, if $R(t)$ contain the tip

$$\int_{\partial R(t)} \mathbf{Cn} + \int_{\mathcal{C}_{R(t)}} [\![\mathbf{C}]\!]\,\mathbf{m} + \int_{R(t)} \mathbf{f} + \mathbf{g}(t) = 0, \tag{3.5a}$$

$$\int_{\partial R(t)} (\mathbf{X} \times \mathbf{Cn} + \mathbf{\Lambda n}) + \int_{\mathcal{C}_{R(t)}} (\mathbf{X} \times [\![\mathbf{C}]\!]\, \mathbf{m} + [\![\mathbf{\Lambda}]\!]\, \mathbf{m}) +$$

$$\int_{R(t)} (\mathbf{X} \times \mathbf{f} + \mathbf{\Delta}) + \mathbf{Z}(t) \times \mathbf{g}(t) + \mathfrak{M}(t) = 0. \tag{3.5b}$$

We assume that each of **f** and **g** consists of internal and inertial portions. The decomposition of **f** is irrelevant to most of the discussion, determining the inertial portion of **g** will form a major part of this analysis. The same conclusion hold for $\mathbf{\Delta}$ and $\mathfrak{M}$.

To ensure that the balances (3.4a)–(3.5b) are well defined and that their localization to the crack tip (in the next section) is meaningful, we assume that:

(A2) ρ is continuous; **p** and $\mathbf{y} \times \mathbf{p} + \mathbf{\Pi}$ are regular; **S**,**M**, $\mathbf{\Lambda}$ and **C** are smooth away from the tip; **f** and $\mathbf{\Delta}$ are integrable over B; $\int_{\partial D_\delta} |\mathbf{Sn}|$ and $\int_{\partial D_\delta} |\mathbf{Mn}|$ remains bounded as $\delta \to 0$; $[\![\mathbf{C}]\!]\, \mathbf{m}$ and $[\![\mathbf{\Lambda}]\!]\, \mathbf{m}$ are integrable on $\mathcal{C}(t)$.

Then the following local relations away from the crack hold:

$$\mathrm{div}\mathbf{S} = \mathbf{p}^{\bullet}, \tag{3.6a}$$

$$\mathbf{\Pi}^{\bullet} = \mathrm{div}\mathbf{M} + \mathbf{s}^{A}, \tag{3.6b}$$

where

$$\mathbf{s}^A = \varepsilon \cdot\cdot \mathbf{FS}^T \quad \text{or} \quad s_i^A = \varepsilon_{ijk} y^j_{;K} S^{kK}, \qquad (\mathbf{FS}^T)^A = \frac{1}{2}\varepsilon \cdot \mathbf{s}^A.$$

From (3.4b) and (3.5b) it follows

$$\mathrm{div}\mathbf{C} + \mathbf{f} = 0 \tag{3.7a}$$

$$-\mathbf{c}^A + \mathrm{div}\mathbf{\Lambda} + \mathbf{\Delta} = 0 \tag{3.7b}$$

where

$$\mathbf{c}^A = \varepsilon \cdot\cdot \mathbf{C} \quad \text{or} \quad c_K^A = \varepsilon_{KLM} C^{LM}.$$

From (3.7b) it follows that the configurational stress **C** is not, in general, symmetric. Of cause, in the absence of the couple stress $\mathbf{\Lambda}$ and the configurational moment $\mathbf{\Delta}$ the configurational stress **C** is symmetric.

3.2 *Mechanical version of the second law*

In the absence of defects (such as cracks), of external body forces, and of thermal and compositional effects, micropolar continuum mechanics may be based on a "second law" that utilizes stationary control volumes R and has the form

$$\frac{\mathrm{d}}{\mathrm{d}t}\left\{\int_R \psi\right\} + \mathcal{K}(R) \le \mathcal{W}(R), \tag{3.8}$$

where ψ is the free energy density, and

$$\mathcal{K}(R) = \frac{\mathrm{d}}{\mathrm{d}t}\left\{\int_R (k + \kappa)\right\} \tag{3.9}$$

Thus, $\mathcal{K}(R)$ is defined by

$$k + \kappa, \tag{3.10c}$$

with
– the kinetic energy density due to the translation motion,

$$k = \frac{1}{2}\rho|\mathbf{y}^{\bullet}|^2 \tag{3.10a}$$

– the kinetic energy density due to the microrotation of the particle,

$$\kappa = \frac{1}{2}\rho\boldsymbol{\sigma}\cdot\boldsymbol{\nu}. \tag{3.10b}$$

Also

$$\mathcal{W}(R) = \int\limits_{\partial R} (\mathbf{Sn}\cdot\mathbf{y}^{\bullet} + \mathbf{Mn}\cdot\boldsymbol{\nu}) \tag{3.11}$$

is the boundary working (Gurtin and Struthers (1990); Gurtin (1995)).

For an evolving control volume $R(t)$ generalization of (3.8)–(3.11) is necessary. We consider the dependence of $R(t)$ on t as representing the addition of material to (or the removal of material from) the boundary $\partial R(t)$, and we write the second law in a manner reflecting this view. To begin with we take

$$\frac{\mathrm{d}}{\mathrm{d}t}\left\{\int\limits_{R(t)} \psi\right\}$$

as the sole term involving free energy; we do not include the outflow term

$$\int\limits_{\partial R(t)} \psi U_{\partial R}$$

as we view noninertial interactions with the material exterior to $R(\mathrm{t})$ in terms of working, rather than transport.

This leads to the main issue: generalization of the expression (3.11) to account for the work performed in the addition and removal of material at the boundary. According to Gurtin and Podio-Guidugli (1996) $\mathbf{Cn}\cdot\mathbf{u}$ represents the boundary working of the configurational stress $\mathbf{C}$, where $\mathbf{u}$ is the velocity field computed via a particular choice of local parametrizations $\mathbf{X} = \hat{\mathbf{X}}(\sigma, t)$ for $\partial R(t)$. The working of the deformational stress $\mathbf{S}$ must also be taken into account. When the control volume depends on time there is no intrinsic material description of its deformed boundary $\mathbf{y}(\partial R(t), t)$ as material is continually being added and removed, and it would seem appropriate to use, as a velocity for $\mathbf{y}(\partial R(t), t)$, the derivative $\overline{\mathbf{u}}(\mathbf{X}, t)$ of $\mathbf{y}(\hat{\mathbf{X}}(\sigma, t), t)$ with respect to t holding the surface parameter σ fixed; therefore, the boundary working of $\mathbf{S}$ is given by $\mathbf{Sn}\cdot\overline{\mathbf{u}}$. The same reasoning holds for $\mathbf{M}$. Particulary, $\mathbf{Mn}\cdot\boldsymbol{\mu}$ is the boundary working of $\mathbf{M}$.

A very different case is with the configurational couple stress $\boldsymbol{\Lambda}$: contrary to $\mathrm{D}_\delta(t)$ we do not expect that $R(t)$ perform any rotation. Therefore, $\boldsymbol{\Lambda}$ do not perform any working on $\partial R(t)$.

Finally, as we view the kinetic energy as "independent" of the internal structure of the material, we generalize $\mathcal{K}(R)$ in the standard manner, viz.,

$$\mathcal{K}(R(t)) = \frac{\mathrm{d}}{\mathrm{d}t}\left\{\int\limits_{R} (k+\kappa)\right\} - \int\limits_{\partial R(t)} (k+\kappa)U_{\partial R}. \tag{3.12}$$

In conclusion, we write the second law for an evolving control volume $R(t)$ that do not contain the crack tip in the form

$$\frac{\mathrm{d}}{\mathrm{d}t}\left\{\int\limits_{R(t)} \psi\right\} + \mathcal{K}(R(t)) \leq \int\limits_{\partial R(t)} (\mathbf{Sn}\cdot\bar{\mathbf{u}} + \mathbf{Cn}\cdot\mathbf{u} + \mathbf{Mn}\cdot\boldsymbol{\mu}), \tag{3.13}$$

with $\mathbf{u}$ a velocity field for $\partial R(t)$ $\bar{\mathbf{u}}$ and $\boldsymbol{\mu}$ the corresponding induced velocity field for $\mathbf{y}(\partial R(t), t)$ and $\boldsymbol{\chi}(\partial R(t), t)$. (The configurational forces $\mathbf{f}$ and $\mathbf{g}$ perform no work, as their inertial components are accounted for by $\mathcal{K}(R(t))$, while their noninertial components are internal; the same conclusion holds for configurational moments $\boldsymbol{\Delta}$ and $\boldsymbol{\mathcal{M}}$; moreover, there is no contribution from $C(t)$ because of (3.2) and since only the tip of $C(t)$ evolves.) Note that, by (2.23), the deformational working $\mathbf{Sn}\cdot\bar{\mathbf{u}}$ consists of a classical term $\mathbf{Sn}\cdot\mathbf{y}^{\bullet}$ plus a term $\mathbf{Sn}\cdot\mathbf{Fu}$; also $\mathbf{Mn}\cdot\boldsymbol{\mu}$ consist a classical term $\mathbf{Mn}\cdot\boldsymbol{\nu}$ and $-1/2\,\mathbf{Mn}\cdot(\boldsymbol{\epsilon}\cdot\cdot(\mathbf{u}\cdot\nabla\boldsymbol{\chi})\boldsymbol{\chi}^T)$. They account for the addition of strained material to ∂R. Note also that for R independent of time (3.13) reduces to the standard inequality (3.8), so there is no conflict with classical continuum mechanics.

To ensure that this version of the second law be meaningful, and to allow for its localization, we assume that:

(A3) ψ, k and κ are regular.

3.3 *The Eshelby tensor as a consequence of invariance under reparameterization (Gurtin, 1995)*

We require that our theory be independent of the choice of parametrization for $\partial R(t)$. This requirement of invariance under reparametrization has important consequences. In particular, the invariance of (3.13) is equivalent to invariance of boundary working, which, by (2.23), can be given the form

$$\mathcal{W}(R(t)) = \int\limits_{\partial R(t)} (\mathbf{Sn}\cdot\bar{\mathbf{u}} + \mathbf{Cn}\cdot\mathbf{u} + \mathbf{Mn}\cdot\boldsymbol{\mu})$$

$$= \int\limits_{\partial R(t)} \left\{\mathbf{Sn}\cdot\mathbf{y}^{\bullet} + (\mathbf{F}^T\mathbf{Sn} + \mathbf{Cn})\cdot\mathbf{u} + \mathbf{Mn}\cdot\boldsymbol{\nu} - \frac{1}{2}\mathbf{Mn}\cdot(\boldsymbol{\epsilon}\cdot\cdot(\mathbf{u}\cdot\nabla\boldsymbol{\chi})\boldsymbol{\chi}^T)\right. \tag{3.14}$$

where,

$$\frac{1}{2}\mathbf{Mn}\cdot(\boldsymbol{\epsilon}\cdot\cdot(\mathbf{u}\cdot\nabla\boldsymbol{\chi})\boldsymbol{\chi}^T) = \frac{1}{2}M_{kK}n^K\epsilon^{klm}\chi_{lL,M}\chi_m{}^L u^M.$$

By definition,

$$\frac{1}{2}\epsilon^{klm}M_{lK}\chi_{mL} \equiv \mathcal{M}^k_{\cdot LK}$$

so that

$$\frac{1}{2}\mathbf{Mn}\cdot(\boldsymbol{\epsilon}\cdot\cdot[\mathbf{u}\cdot\nabla\boldsymbol{\chi})\boldsymbol{\chi}^T] = -u^M\chi^l_{L,M}\mathcal{M}_l^{\cdot LK}n^K = -\mathrm{tr}(\nabla\boldsymbol{\chi}\cdot\cdot\boldsymbol{\mathcal{M}})\mathbf{n}\otimes\mathbf{n}.$$

Also

$$(\mathbf{F}^T\mathbf{Sn} + \mathbf{Cn})\cdot\mathbf{u} = \mathrm{tr}(\mathbf{F}^T\mathbf{S} + \mathbf{C})\mathbf{n}\otimes\mathbf{n}$$

Then

$$(\mathbf{F}^T\mathbf{S}+\mathbf{Cn})\cdot\mathbf{u}-\frac{1}{2}\mathbf{Mn}\cdot(\epsilon\cdot\cdot[\mathbf{u}\cdot\nabla\chi)\chi^T]$$

$$=\operatorname{tr}(\mathbf{C}+\mathbf{F}^T\mathbf{S}+\nabla\chi\cdot\cdot\boldsymbol{\mathcal{M}})\mathbf{n}\otimes\mathbf{u}.$$

Changes in parametrization affect the tangential component of $\mathbf{u}$, but leave the normal component unaltered. In fact, invariance of (3.14) under reparametrization is equivalent to the requirement that $[(\mathbf{C}+\mathbf{F}^T\mathbf{S}+\nabla\boldsymbol{\chi}\cdot\cdot\boldsymbol{\mathcal{M}})\mathbf{n}]\cdot\mathbf{a}_\alpha=0$ on $\partial R(t)$ for all tangential vector fields $\mathbf{a}_\alpha$ on $\partial R(t)$; thus, since $R(t)$ is arbitrary, $(\mathbf{C}+\mathbf{F}^T\mathbf{S}+\nabla\boldsymbol{\chi}\cdot\cdot\boldsymbol{\mathcal{M}})\mathbf{n}$ must be parallel to $\mathbf{n}$ for all $\mathbf{n}$, so that

$$\Rightarrow\ \mathbf{C}+\mathbf{F}^T\mathbf{S}+\nabla\chi\cdot\cdot\boldsymbol{\mathcal{M}}=\pi\mathbf{I} \tag{3.15}$$

and, by (2.21), the working has the intrinsic form

$$\mathcal{W}(R(t),t)=\int\limits_{R(t)}(\mathbf{Sn}\cdot\mathbf{y}^\bullet+\mathbf{Mn}\cdot\boldsymbol{\nu})+\int\limits_{\partial R(t)}\pi U_{\partial R} \tag{3.16}$$

The scalar field π is a configurational tension that works to increase the volume of $R(t)$ through the addition of material at its boundary. Referring to the final term in (3.16) as the configurational working, (3.16) may be stated more suggestively as boundary working equals deformational working plus configurational working. Note that the configurational working $\pi U_{\partial R(t)}$ is not due solely to the action of the configurational stress $\mathbf{C}$; the deformational stress and couple stress contribute also through the term $\mathbf{n}\cdot(\mathbf{F}^T\mathbf{S}+\nabla\boldsymbol{\chi}\cdot\cdot\boldsymbol{\mathcal{M}})\mathbf{n}U_{\partial R}$.

Next, having in mind that $R(t)$ does not contain the crack, and using (2.17) and (3.16), the inequality (3.13) may be rewritten as

$$\int\limits_{R(t)}(\psi+k+\kappa)^\bullet\le\int\limits_{\partial R(t)}\{\mathbf{Sn}\cdot\mathbf{y}^\bullet+\mathbf{Mn}\cdot\boldsymbol{\nu}+(\pi-\psi)U_{\partial R}\cdot\} \tag{3.17}$$

Given a time τ, it is possible to find a second referential control volume $R'(t)$ with $R'(t)=R(t)$, but with $U_{\partial R'}(\mathbf{X},\tau)$, the normal velocity of $\partial R'(\tau)$, an arbitrary scalar field on $\partial R'(\tau)$; satisfaction of (3.17) for all such $U_{\partial R'}$ implies

$$\pi=\psi. \tag{3.18}$$

Therefore, configurational tension coincides with free energy, a result analogous to the coincidence of surface tension and surface free-energy; what is more important, (3.16) and (3.18) yield the Eshelby relation

$$\mathbf{C}=\psi\mathbf{I}-\mathbf{F}^T\mathbf{S}-\nabla\chi\cdot\cdot\boldsymbol{\mathcal{M}} \tag{3.19}$$

for the configurational stress $\mathbf{C}$.

4 THE CRACK TIP

4.1 *Basic laws*

We now localize the balance laws (3.3a,b) and (3.5a,b) to the crack tip. Then, by (*A*2), (*A*3), (2.11) and (2.27) the momentum balances have the limiting forms

$$\oint_{tip} \mathbf{Sn} = -\oint_{tip} \mathbf{p}(\mathbf{v}\cdot\mathbf{n}), \tag{4.1}$$

$$\oint_{tip} \mathbf{Mn} = -\oint_{tip} \mathbf{\Pi}(\mathbf{v}\cdot\mathbf{n}), \tag{4.2}$$

while the limiting forms of the balances for configurational force and moment are

$$\oint_{tip} \mathbf{Cn} = -\mathbf{g}, \tag{4.3}$$

$$\oint_{tip} \mathbf{\Lambda n} = -\mathfrak{M}. \tag{4.4}$$

4.2 *Tip inertia; resistance to the breaking of bonds*

At the tip of the crack the deformational inertial force $\mathbf{b}_d$ and moment $\mathfrak{N}_d$ are defined through the balances

$$\oint_{tip} \mathbf{Sn} + \mathbf{b}_d = 0, \tag{4.5}$$

$$\oint_{tip} \mathbf{Mn} + \mathfrak{N}_d = 0. \tag{4.6}$$

Hence, $\mathbf{b}_d$ and $\mathfrak{N}_d$ are identified as the limiting values of the inflow of momentum and angular momentum, respectively, across ∂D_δ.

Further, the configurational force $\mathbf{g}$ and moment $\mathfrak{M}$ concentrated at the tip admit the decompositions

$$\mathbf{g} = \mathbf{g}_b + \mathbf{g}_d, \tag{4.7}$$

$$\mathfrak{M} = \mathfrak{M}_b + \mathfrak{M}_d, \tag{4.8}$$

with $\mathbf{g}_b$ – internal force, $\mathbf{g}_d$ – inertial force, $\mathfrak{M}_b$ – internal moment, $\mathfrak{M}_d$ – inertial moment. Here $\mathbf{g}_b$ and $\mathfrak{M}_b$ represent the internal force and moment respectively, that resist the breaking of bonds during crack propagation.

In order to identify $\mathbf{g}_d$ and $\mathfrak{M}_d$ we generalize the procedure of Podio-Guidugli, P. (1997), through the equivalence of inertial working and temporal changes in kinetic energy

$$\lim_{\delta\to 0}\left(\int_{\partial D_\delta(t)} (k+\kappa)(\mathbf{v}\cdot\mathbf{n}) - \frac{\mathrm{d}}{\mathrm{d}t}\left\{\int_{D_\delta(t)} (k+\kappa)\right\}\right) = \mathbf{b}_d\cdot\bar{\mathbf{v}} + \mathbf{g}_d\cdot\mathbf{v} + \mathfrak{N}_d\cdot\boldsymbol{\mu}_0 + \mathfrak{M}_d\cdot\boldsymbol{\omega}. \tag{4.9}$$

But, from (2.15), (2.18) and (2.27) it follows that

$$\frac{\mathrm{d}}{\mathrm{d}t}\int_{D_\delta(t)} \phi \to 0 \tag{4.10}$$

for any regular ϕ. Thus, for $\phi = k + \kappa$, from (4.9) we have

$$\oint_{tip} (k+\kappa)(\mathbf{v}\cdot\mathbf{n}) = \mathbf{b}_d\cdot\bar{\mathbf{v}} + \mathbf{g}_d\cdot\mathbf{v} + \mathfrak{N}_d\cdot\boldsymbol{\mu}_0 + \mathfrak{M}_d\cdot\boldsymbol{\omega} \tag{4.11}$$

Now, according to (4.5), (4.6), (4.1) and (4.2)

$$\mathbf{b}_d\cdot\bar{\mathbf{v}} + \mathfrak{N}_d\cdot\boldsymbol{\mu}_0 = \oint_{tip}(\mathbf{p}\cdot\bar{\mathbf{v}} + \boldsymbol{\Pi}\cdot\boldsymbol{\mu}_0)\mathbf{v}\cdot\mathbf{n}.$$

Hence,

$$\oint_{tip}(k+\kappa)(\mathbf{v}\cdot\mathbf{n}) - \mathbf{b}_d\cdot\bar{\mathbf{v}} - \mathfrak{N}_d\cdot\boldsymbol{\mu}_0 = \oint_{tip}(k+\kappa-\mathbf{p}\cdot\bar{\mathbf{v}} - \boldsymbol{\Pi}\cdot\boldsymbol{\mu}_0)(\mathbf{v}\cdot\mathbf{n}). \tag{4.12}$$

It is easy to show that

$$k - \mathbf{p}\cdot\bar{\mathbf{v}} = k_{rel} - \frac{1}{2}\rho|\bar{\mathbf{v}}|^2,$$

where

$$k_{rel} = \frac{1}{2}\rho|\mathbf{y}^\bullet - \bar{\mathbf{v}}|^2,$$

is the kinetic energy measured relative to the tip. The integral of $\rho|\bar{\mathbf{v}}|^2$ over ∂D_δ tend to zero; therefore

$$\oint_{tip}(k - \mathbf{p}\cdot\bar{\mathbf{v}})\mathbf{n} = \oint_{tip} k_{rel}\mathbf{n}. \tag{4.13}$$

Also, from (3.1) it follows that

$$\kappa - \boldsymbol{\Pi}\cdot\boldsymbol{\mu}_0 = \frac{1}{2}\rho\boldsymbol{\sigma}\cdot\boldsymbol{\nu} - \rho\boldsymbol{\sigma}\cdot\boldsymbol{\mu}_0 = \frac{1}{2}\rho\boldsymbol{\sigma}\cdot(\boldsymbol{\nu} - 2\boldsymbol{\mu}_0).$$

Next, according to Eringen (1976), $\boldsymbol{\sigma} = \mathbf{J}\boldsymbol{\nu}$, $\mathbf{J} = \mathbf{J}^T$, so that

$$\boldsymbol{\sigma}\cdot(\boldsymbol{\nu} - 2\boldsymbol{\mu}_0) = \mathbf{J}\boldsymbol{\nu}\cdot(\boldsymbol{\nu} - 2\boldsymbol{\mu}_0) = \mathbf{J}(\boldsymbol{\nu} - \boldsymbol{\mu}_0)\cdot(\boldsymbol{\nu} - \boldsymbol{\mu}_0) - \mathbf{J}\boldsymbol{\mu}_0\cdot\boldsymbol{\mu}_0.$$

Thus

$$\frac{1}{2}\rho\boldsymbol{\sigma}\cdot(\boldsymbol{\nu} - 2\boldsymbol{\mu}_0) = \frac{1}{2}\rho\mathbf{J}(\boldsymbol{\nu} - \boldsymbol{\mu}_0)\cdot(\boldsymbol{\nu} - \boldsymbol{\mu}_0) - \frac{1}{2}\rho\mathbf{J}\boldsymbol{\mu}_0\cdot\boldsymbol{\mu}_0.$$

Hence

$$\frac{1}{2}\rho\boldsymbol{\sigma}\cdot(\boldsymbol{\nu} - 2\boldsymbol{\mu}_0) = \kappa_{rel} - \frac{1}{2}\rho\mathbf{J}\boldsymbol{\mu}_0\cdot\boldsymbol{\mu}_0,$$

where

$$\kappa_{rel} = \frac{1}{2}\rho\mathbf{J}(\boldsymbol{\nu} - \boldsymbol{\mu}_0)\cdot(\boldsymbol{\nu} - \boldsymbol{\mu}_0)$$

is the kinetic energy of "rotation" measured relative to the tip. Since ρ and $\mathbf{J}$ are continuous, the integral of $\rho\mathbf{J}\boldsymbol{\mu}_0\cdot\boldsymbol{\mu}_0$ over $D_\delta(t)$ tends to zero as $\delta \to 0$.

Therefore,

$$\oint_{tip} (\kappa - \mathbf{\Pi} \cdot \boldsymbol{\mu}_0)\mathbf{n} = \oint_{tip} \kappa_{rel}\mathbf{n}. \tag{4.14}$$

Finally, from (4.11)–(4.14) we obtain

$$\mathbf{g}_d \cdot \mathbf{v} + \mathfrak{M}_d \cdot \omega = \mathbf{v} \cdot \oint_{tip} (k_{rel} + \kappa_{rel})\mathbf{n}. \tag{4.15}$$

Note that $\mathbf{v}$ and ω are linearly independent vectors. Thus, under the assumption that

$$\mathfrak{M}_d = 0, \tag{4.16}$$

we have

$$\mathbf{g}_d \cdot \mathbf{v} = \mathbf{v} \cdot \oint_{tip} (k_{rel} + \kappa_{rel})\mathbf{n}.$$

This motivates us to identify $\mathbf{g}_d$ with the tip integral of the sum of kinetic energy of relative translation and rotation, i.e.

$$\mathbf{g}_d = \oint_{tip} (k_{rel} + \kappa_{rel})\mathbf{n}. \tag{4.17}$$

Further, making use of (4.7), (4.3), (4.17) and (3.19) we have that

$$-\mathbf{g}_b = \oint_{tip} \left[(\psi + k_{rel} + \kappa_{rel})\mathbf{I} - \mathbf{F}^T\mathbf{S} - \nabla\chi \cdot\cdot\, \boldsymbol{\mathcal{M}} \right] \mathbf{n}. \tag{4.18}$$

In addition to this, from (4.16), (4.8) and (4.4) we obtain

$$\mathfrak{M}_b = -\oint_{tip} \mathbf{\Lambda n}. \tag{4.19}$$

Remark. If $\omega = 0$ then (4.16) does not necessary hold, and (4.19) becomes (4.4). In this case kinematics of disc $D_\delta(t)$, given in section 2, significantly, simplifies.

Obviously, in a case of classical continua with the crack, i.e. when $\mathbf{\Lambda}, \boldsymbol{\mathcal{M}}, \kappa = 0$, (4.18) becomes

$$-\mathbf{g}_b = \oint_{tip} \left[(\psi + k_{rel} + \kappa_{rel})\mathbf{I} - \mathbf{F}^T\mathbf{S} \right] \mathbf{n}.$$

which is the same as (4.13) in Gurtin, M.E, and Podio-Guidugli, P. (1996). Then, in addition to (4.16), we have that $\mathfrak{M}_b = 0$.

The expressions (4.17) and (4.18) are basic relations for further investigation.

4.3 *Internal dissipation inequality*

We now turn to the local form of second laws (3.13) to the crack tip. Then $R(t) = D_\delta(t)$, $\mathbf{u} = \mathbf{v}$, $\bar{\mathbf{u}} = \mathbf{y}^o$ and $\boldsymbol{\mu} = \boldsymbol{\mu}_0$. Taking into account (3.12), (4.10) and (4.16) we write

$$\int_{\partial D_\delta(t)} [(k+\kappa)U_{\partial D_\delta(t)} + \mathbf{Sn}\cdot\mathbf{y} \quad + \mathbf{Cn}\cdot\mathbf{v} + \mathbf{Mn}\cdot\boldsymbol{\mu}_0] + O(1) \geq 0, \tag{4.20}$$

or, according to (4.11) and (4.16),

$$\int_{\partial D_\delta(t)} (\mathbf{Sn}\cdot\mathbf{y} \quad + \mathbf{Cn}\cdot\mathbf{v} + \mathbf{Mn}\cdot\boldsymbol{\mu}_0) + \mathbf{b}_d\cdot\bar{\mathbf{v}} + \mathbf{g}_d\cdot\mathbf{v} + \mathfrak{M}_d\cdot\boldsymbol{\mu}_0 + O(1) \geq 0. \tag{4.21}$$

But, making use of (2.29) and (4.5), we have

$$\int_{\partial D_\delta(t)} \mathbf{Sn}\cdot\mathbf{y} \quad = \int_{\partial D_\delta(t)} \mathbf{Sn}\cdot\bar{\mathbf{v}} + O(1) = -\mathbf{b}_d\cdot\bar{\mathbf{v}} + O(1). \tag{4.22}$$

Also from (4.6) we have that

$$\int_{\partial D_\delta(t)} \mathbf{Mn} + \mathfrak{M}_d = O(1). \tag{4.23}$$

Thus (4.21) reduces to

$$\left(\int_{\partial D_\delta(t)} \mathbf{Cn} + \mathbf{g}_d\right)\cdot\mathbf{v} + O(1) \geq 0. \tag{4.24}$$

Now, from (4.3) we have

$$\int_{\partial D_\delta(t)} \mathbf{Cn} + \mathbf{g} = O(1), \tag{4.25}$$

and by (4.7), the above inequality, in the limit, reduces to

$$\mathbf{g}_b\cdot\mathbf{v} \leq 0. \tag{4.26}$$

This inequality represents an internal-dissipation inequality for the crack tip. From physical point of view it means that the internal configurational force oppose motion of crack tip.

4.4 *Entropy production; energy release-rate*

By means of (2.16) and (3.18), we write (3.17) as

$$\mathcal{E}(R(t)) = \frac{\mathrm{d}}{\mathrm{d}t}\int_{R(t)} (\psi + k + \kappa) - \int_{\partial R(t)} \mathbf{Sn}\cdot\mathbf{y}^\bullet + \mathbf{Mn}\cdot\boldsymbol{\nu} - \int_{\partial R(t)} (\psi + k + \kappa)U_{\partial R} \leq 0. \tag{4.27}$$

So defined quantity $\mathcal{E}(R(t))$ represents the production of energy in $R(t)$, with $-\mathcal{E}(R(t))$ the corresponding energy dissipated. Applying this to (3.13) for the crack tip, by the used of (4.10), we have

$$-\lim_{\delta\to 0}\mathcal{E}(D_\delta) = \oint_{tip} [\mathbf{Sn}\cdot\mathbf{y}^{\bullet\prime} + \mathbf{Mn}\cdot\boldsymbol{\nu} + (\psi + k + \kappa)(\mathbf{v}\cdot\mathbf{n})] \tag{4.28}$$

provided that the tip integral exist. It is easy to show that the right integral, which represent the working on and energy flow out of an infinitesimal neighborhood of the tip, exists. Moreover, from (2.28) and (3.19), we have that

$$\begin{aligned}\mathbf{Sn}\cdot\mathbf{y}^{\bullet} &= \mathbf{Sn}\cdot[\mathbf{y}^{\square} - \mathbf{F}(\mathbf{v}+\boldsymbol{\omega}\times\boldsymbol{\delta})]\\ &= \mathbf{Sn}\cdot\mathbf{y}^{\square} - \mathbf{v}\cdot\mathbf{F}^{T}\mathbf{Sn} - \mathbf{Sn}\cdot\mathbf{F}(\boldsymbol{\omega}\times\boldsymbol{\delta})\\ &= \mathbf{Sn}\cdot\mathbf{y}^{\square} - \mathbf{v}\cdot(\psi\mathbf{I} - \mathbb{C} - \nabla\chi\cdot\cdot\boldsymbol{\mathcal{M}})\mathbf{n} - \mathbf{Sn}\cdot\mathbf{F}(\boldsymbol{\omega}\times\boldsymbol{\delta}).\end{aligned}$$

It is more convenient to write it in the following form

$$\mathbf{Sn}\cdot\mathbf{y}^{\bullet} + \psi(\mathbf{v}\cdot\mathbf{n}) = \mathbf{Sn}\cdot\mathbf{y}^{\square} + \mathbf{v}\cdot\mathbb{C}\mathbf{n} + (\mathbf{v}\cdot\nabla\chi)\cdot\cdot\boldsymbol{\mathcal{M}}\mathbf{n} - \mathbf{Sn}\cdot\mathbf{F}(\boldsymbol{\omega}\times\boldsymbol{\delta}). \tag{4.29}$$

Then, by virtue of (4.22) and (4.3) we obtain

$$\oint_{tip} \mathbf{Sn}\cdot\mathbf{y}^{\bullet} + \psi(\mathbf{v}\cdot\mathbf{n}) = -\mathbf{b}_d\cdot\bar{\mathbf{v}} - \mathbf{g}\cdot\mathbf{v} + \oint_{tip}(\mathbf{v}\cdot\nabla\chi)\cdot\cdot\boldsymbol{\mathcal{M}}\mathbf{n} \tag{4.30}$$

Further, by (2.40),

$$\int_{\partial D_\delta} \mathbf{Mn}\cdot\boldsymbol{\nu}_\delta = \int_{\partial D_\delta} \mathbf{Mn}\cdot\boldsymbol{\mu}_\delta + \frac{1}{2}\int_{\partial D_\delta} \mathbf{Mn}\cdot[\varepsilon\cdot\cdot(\mathbf{v}\cdot\nabla\chi_\delta)\chi^{T}] + O(1).$$

Hence, making use of (2.42), (2.44), (4.22) and (3.14a), it becomes

$$\begin{aligned}\oint_{tip} \mathbf{Mn}\cdot\boldsymbol{\nu} &= \oint_{tip}\left\{\mathbf{Mn}\cdot\boldsymbol{\mu}_0 + \frac{1}{2}\mathbf{Mn}\cdot[\varepsilon\cdot\cdot(\mathbf{v}\cdot\nabla\chi)\chi^{T}]\right\}\\ &= -\mathfrak{M}_d\boldsymbol{\mu}_0 - \oint_{tip}(\mathbf{v}\cdot\nabla\chi)\cdot\cdot\boldsymbol{\mathcal{M}}\mathbf{n}.\end{aligned} \tag{4.31}$$

Also, (4.11) and (4.16) gives

$$\oint_{tip}(k+\kappa)(\mathbf{v}\cdot\mathbf{n}) = \mathbf{b}_d\cdot\bar{\mathbf{v}} + \mathbf{g}_d\cdot\mathbf{v} + \mathfrak{M}_d\cdot\boldsymbol{\mu}_0. \tag{4.32}$$

Substituting (4.30)–(4.32) into (4.28), with the help of (4.7), we obtain (4.33)

$$\mathbf{g}_b\cdot\mathbf{v} = \lim_{\delta\to 0}\mathcal{E}(D_\delta) \tag{4.33}$$

Thus, $-\mathbf{g}_b\cdot\mathbf{v}$ is the sole source of dissipation at the tip.

4.5 *Energy release-rate*

As a consequence of (4.28) and (4.33) it the tip balance

$$\oint_{tip}[\mathbf{Sn}\cdot\mathbf{y}^{\bullet} + \mathbf{Mn}\cdot\boldsymbol{\nu} + (\psi + k + \kappa)(\mathbf{v}\cdot\mathbf{n})] = -\mathbf{g}_b\cdot\mathbf{v} \tag{4.34}$$

Having in mind that $\mathbf{v} = V\mathbf{e}$, $V \neq 0$, we write the left side of (4.34) as $\mathcal{S}(t)V$.

Then, according to (4.18),

$$\mathcal{S} = -\mathbf{g}_b \cdot \mathbf{e} = \mathbf{e} \cdot \oint_{tip} \left[(\psi + k_{rel} + \kappa_{rel})\mathbf{I} - \mathbf{F}^T\mathbf{S} - \nabla\chi \cdot\cdot \boldsymbol{\mathcal{M}} \right] \mathbf{n}. \tag{4.35}$$

So defined quantity $\mathcal{S}(t)$ is called the energy release-rate. Obviously, the energy release-rate coincides with the component of the internal configurational force opposing crack propagation.

5 CONCLUSION

The derivation of the Eshelby relation was accomplished without recourse to constitutive equations or to a variational principle; the derivation was based on a version of the second law appropriate to referential control volumes whose boundaries evolve with time. The result (3.18) is a consequence of the invariance of $\mathcal{W}(R)$ under re-parametrization; it is independent of the particular form chosen for the second law and is hence more basic than (3.19).

ACKNOWLEDGMENT

This paper is a part of scientific projects No.1793 which is supported by MNTR of Republic Serbia. The author is very grateful to P. Podio-Guidguli and G.A. Maugin for very helpful discussions and comments on the topics considered in the paper.

REFERENCES

Eringen, C.A. (1976). *Continuum physics*, Vol. IV, New York: Academic Press.

Eshelby, J.D. (1951). The force on an elastic singularity, *Phil. Trans. Roy. Soc. Lond.* A244: 87–112.

Eshelby, J.D. (1975). The elastic energy-momentum tensor, *J. Elasticity*, 5: 321–335.

Gurtin, M.E. and Struthers, A. (1990). Multiphase thermomechanics with interfascial structure, *Arch. Rat. Mech. Anal.*, 112: 97–160.

Gurtin, M.E. (1995). The nature of configurational forces, *Arch. Rat. Mech. Anal.*, 131: 67–100.

Gurtin, M.E. and Podio-Guidugli, P. (1996). Configurational forces and the basic laws for crack propagations. *J. Mech. Phys. Solids*, 44(6): 905–927.

Gurtin, M.E. and Podio-Guidugli, P. (1998). Configurational forces and a constitutive theory for crack propagation that allows for kinking and curving. *J. Mech. Phys. Solids*, 46(8): 1343–1378.

Jarić, J.P. (2000). On Noether's theorem in nonlinear continua, *Journal of the Mechanical Behavior of materials*, 11(1–3): 73–79.

Kafadar, C.B. and Eringen, A.C. (1971). Micropolar media, I and II, *Int. J. Eng. Sci.* 9: 271–307.

Mariano, P.M. (2000). Configurational forces in continua with microstructure, *Z. angew. Math. Phys.* 51: 752–791.

Maugin, G.A. (1993). *Material inhomogenities in elasticity*, London: Chapman and Hall.

Maugin, G.A. (1998). On the structure of the theory of polar elasticity, *Phil. Trans.R.Soc.Lond.* A 356: 1367–1395.

Podio-Guidugli, P. (1997). Inertia and invariance, *Ann. Mat. Pura Appl.* (IV) 171: 103.

Stojanović, R. (1969). *Mechanics of polar continua*, (lecture notes), Int. Center Mech. Sci., Udine, Italy.

Configurational Mechanics, Kalpakides & Maugin (eds)

On the configurational thermomechanics of nonchemically reacting solid–fluid mixtures

S. Quiligotti & G.A. Maugin
Université Pierre et Marie Curie, Paris, France

ABSTRACT: Focusing on geometric and kinematic aspects of thermal stresses, this work aims at proposing a continuum theory of nonlinear thermoelasticity of a porous solid, infused with a compressible fluid. Within the framework of the Eshelbian thermomechanics of multiphase continua, such a material is naturally regarded as a binary solid–fluid mixture, consisting of different body manifolds embedded into the three-dimensional Euclidean space, so as to share a smooth region of the physical environment while undertaking independent motions. Coupled thermal phenomena are coarsely taken into account by following the time evolution of the solid stress-free configuration, which is, in the most general case, geometrically incompatible. Some relevant constitutive implications are finally investigated and discussed.

1 INTRODUCTION

A multiphase body is here regarded as a set of material manifolds, embedded into the three-dimensional physical environment while undertaking independent motions (see e.g. Atkin and Craine 1976; Bowen 1976; Truesdell 1957; Nunziato and Walsh 1980; Rajagopal and Tao 1995; Krishnaswamy and Batra 1997). By virtue of a customary fundamental assumption, any place in the current configuration of the mixture is simultaneously occupied by a set of material points, one per each constituent. For this reason, under suitable smoothness assumptions, any fluid particle is naturally associated with a 1-parameter family of reference places (Wilmanski 1996, 1998, 2000). Accordingly, Eulerian fluid and solid velocity fields are pulled back to the linear vector space associated with the reference configuration of the solid constituent, and the dual notion of generalized material forces (Eshelby 1951, 1956, 1975; Maugin 1993,1995) is extended to any rational theory of solid–fluid mixtures. In particular, two partial Eshelby-like stress tensors (Quiligotti 2000, Quiligotti, Maugin, and dell'Isola 2003) are introduced, by duality, through the definition of the internal power that they expend on their conjugate kinematical descriptors (Germain 1972; Germain 1973; Maugin 1980; DiCarlo 1996).

When properly enriched by the concept of volume fractions (figure 1), a mathematical theory of binary solid–fluid mixtures provides a suitable framework for the development of a consistent macroscopic theory of porous solids, partially or totally saturated with fluids (see e.g. deBoer 1996, 2000a, 2000b; Bowen 1982, 1976; Wilmanski 1996, 1998, 2000). If the saturation constraint is satisfied (Klisch 1999), the stress response is constitutively determined except for a pressure reaction, which arises in the material to maintain each constituent in contact with the other one (Quiligotti, Maugin, and dell'Isola 2002, 2003).

Focusing on kinematic aspects of thermal stresses (Vujošević and Lubarda 2002), in this paper we investigate a proposal for a thermomechanical theory of nonchemically reacting solid–fluid mixtures, based on the multiplicative decomposition of the gradient of the solid motion, $\mathbf{F}_s = \mathbf{F}\mathbf{F}_\vartheta$ (figure 3).

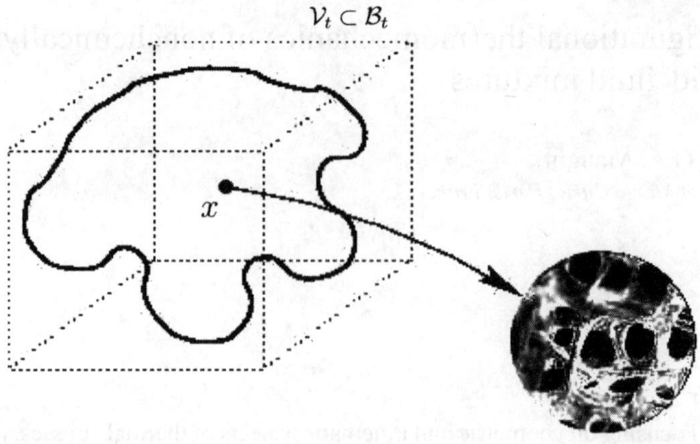

Figure 1. A fluid-saturated porous solid can be regarded as an internally constrained solid–fluid mixture, introducing the notion of volume fraction.

Thermal expansions of the solid constituent are described by following the time evolution of corresponding stress-free configurations. The geometrical incompatibility, associated with the additional descriptor introduced (denoted by $\mathbf{F}_\vartheta$ in figure 3), is mostly due to the fact that whenever a solid particle is allowed to expand, independently of neighbouring ones, after its thermal expansion it may be necessary to deform it, in order to make particles fit together again (see also Lee 1969; Taber 1995; Epstein and Maugin 2000; DiCarlo and Quiligotti 2002). As in the most general case this phenomenon gives rise to a residual stress field, the kinematic tensor $\mathbf{F}_\vartheta$, closely related to the standard notion of stress-free configuration, contributes relevant constitutive information, while enriching the overall kinematic description. Unfortunately, the decomposition of the gradient of the solid motion is intrinsically nonunique, and the thorny issue of invariance requirements needs to be carefully addressed (Casey and Naghdi 1980, 1981, Casey 1987).

Some remarks on inherent constitutive implications and perspectives applications are finally sketched, to complete the analysis.

2 KINEMATICS

When excluding the existence of fluid drops and solid slivers, a binary solid–fluid mixture is customarily regarded as a multiphase continuum, consisting of two smooth three-dimensional material manifolds, respectively denoted by $\mathcal{B}_s$ and $\mathcal{B}_f$ in figure 2.

In order to avoid any confusion between particles which belong to different constituents of the mixture, we refer to material points $\mathcal{X}_\alpha \in \mathcal{B}_\alpha$ as *α-points*, where $\alpha \in \{\mathrm{S}, \mathrm{F}\}$.

By assumption, there exists a smooth embedding of the solid body $\mathcal{B}_s$ into the three-dimensional Euclidean space ε,

$$\mathcal{K}_s : \mathcal{B}_s \rightarrow \mathcal{E}\,, \tag{1}$$

$$\mathcal{X}_s \mapsto X\,, \tag{2}$$

which associates any material s-point with a time-independent reference place. Hence, a smooth motion of $\mathcal{B}_s$ may be regarded as a time sequence of mappings,

$$\chi_s(\cdot, t) : \mathcal{B} \rightarrow \mathcal{E}\,, \tag{3}$$

$$X \mapsto x\,, \tag{4}$$

which carry any given s-point $\mathcal{X}_s$ from its reference place X to its current place x.

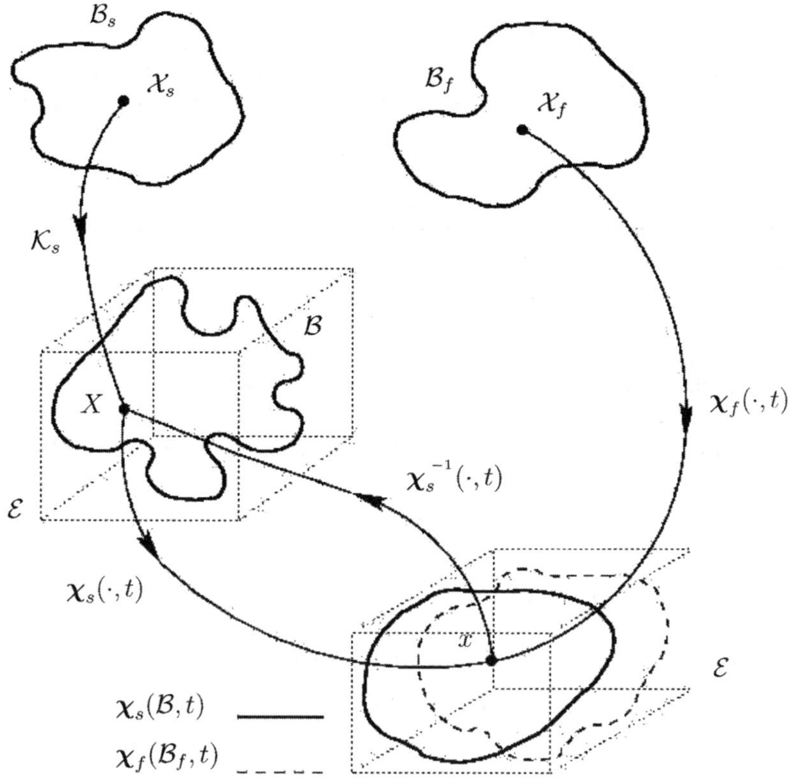

Figure 2. Standard kinematics of a binary solid–fluid mixture.

In a similar way, a smooth motion of the fluid body $\mathcal{B}_f$ may be regarded as a time sequence of embeddings,

$$\chi_f(\cdot,t) : \mathcal{B}_f \rightarrow \mathcal{E}, \tag{5}$$

$$\mathcal{X}_f \mapsto x, \tag{6}$$

which map, at any given time t, the body manifold $\mathcal{B}_f$ onto its current shape $\chi_f(\mathcal{B}_f, t)$.

According to a fundamental kinematic assumption in mixture theories (Atkin and Craine 1976; Bowen 1980; Nunziato andWalsh 1980; Truesdell 1957; Rajagopal and Tao 1995), any place x in the current configuration $\mathcal{B}_t := \{\chi_s(\mathcal{K}_s(\mathcal{B}_s), t)\} \cap \{\chi_f(\mathcal{B}_f, t)\}$ is simultaneously occupied by a pair of material points, $\mathcal{X}_s \in \mathcal{B}_s$ and $\mathcal{X}_f \in \mathcal{B}_f$, such that

$$x = \chi_s(\mathcal{K}_s(\mathcal{X}_s), t) = \chi_f(\mathcal{X}_f, t). \tag{7}$$

As the reference shape $\mathcal{B}$ does not depend on time, the solid velocity field is obtained by taking the partial derivative of the s-motion χ_s with respect to time,

$$\mathbf{v}_s(x,t) := \frac{\mathcal{D}^S}{\mathcal{D}t}[\chi_s(\cdot,t)\circ\mathcal{K}_s](\mathcal{X}_s) = \frac{\partial\chi_s}{\partial t}(X,t), \tag{8}$$

where, as usual, $X = \mathcal{K}_s(\mathcal{X}_s)$. Moreover, excluding a priori the possibility that a three-dimensional neighbourhood of any given reference place collapses under the s-motion,

$$\det \mathbf{F}_s > 0, \quad \mathbf{F}_s := \operatorname{Grad} \chi_s = \frac{\partial \chi_s}{\partial X}, \tag{9}$$

we assume that the inverse smooth mapping χ_s^{-1} exists as well, and satisfies the identity

$$X = \chi_s^{-1}(\chi_s(X,t),t), \quad \text{for any} \quad X \in \mathcal{B}. \tag{10}$$

Denoting by $\mathbf{V}_s$ the referential description of the partial derivative of χ_s^{-1} with respect to time (Maugin 1993, 1995),

$$\mathbf{V}_s(\cdot,t) \; : \; X \mapsto \mathbf{V}_s(X,t) := \frac{\partial \chi_s^{-1}}{\partial t}(x,t), \tag{11}$$

it can be shown, taking partial derivatives of the identity (10) with respect to X and t, that

$$\mathbf{I} = \left(\frac{\partial \chi_s^{-1}}{\partial x} \circ \chi_s \right) \frac{\partial \chi_s}{\partial X} \implies \operatorname{grad}\left(\chi_s^{-1}\right)\Big|_{x,t} = \left(\operatorname{Grad} \chi_s\right)^{-1}\Big|_{X,t}, \tag{12}$$

and

$$\mathbf{F}_s(X,t)\, \mathbf{V}_s(X,t) + \frac{\partial \chi_s}{\partial t}(x,t) \;=\; \mathbf{0}. \tag{13}$$

Bearing in mind the identity (13) and the following definition,

$$\operatorname{Grad} \mathbf{F}_s \;:=\; \operatorname{Grad}\left(\operatorname{Grad} \chi_s\right), \tag{14}$$

it is furthermore possible to deduce that

$$\operatorname{Grad} \frac{\partial \chi_s}{\partial t} + \left(\operatorname{Grad} \mathbf{F}_s\right) \mathbf{V}_s + \mathbf{F}_s \left(\operatorname{Grad} \mathbf{V}_s\right) = \mathbf{O}. \tag{15}$$

In order to describe the relative motion of solid and fluid overlapped particles through the reference shape of the solid constituent (Bowen 1980, 1982; Quiligotti, Maugin, and dell'Isola 2003; Wilmanski 1996, 2000), we notice that any f-point belonging to the mixture at time t (namely, any $\mathcal{X}_f \in \mathcal{B}_f$ such that $\chi_f(\mathcal{X}_f, t) \in \mathcal{B}_t$), interacts with a 1-parameter family of s-points, moving along the smooth curve

$$\chi_s^{-1}\left(\chi_f(\mathcal{X}_f,\cdot),\cdot\right) : t \mapsto X, \tag{16}$$

at the velocity $\mathbf{w}_f(X, t)$, defined by the expression

$$\mathbf{v}_f(x,t) \;:=\; \mathbf{F}_s(X,t)\; \mathbf{w}_f(X,t) + \mathbf{v}_s(x,t), \tag{17}$$

where $x = \chi_f(\mathcal{X}_f, t) = \chi_s(X, t) \in \mathcal{B}_t$.

Finally, we introduce the velocity field $\mathbf{V}_f$, which identically meets the requirement

$$\mathbf{v}_f(x,t) + \mathbf{F}_s(X,t)\, \mathbf{V}_f(X,t) := \mathbf{0}, \tag{18}$$

such that

$$\mathbf{V}_f - \mathbf{V}_s = -\,\mathbf{w}_f. \tag{19}$$

To outline the role played in the theory by the velocity field $\mathbf{w}_f$, let us consider a moving deformable surface which envelops, at the given time t, a smooth region $\gamma(t)$ of the current configuration of the mixture.

If the surface $\partial\gamma(t)$ evolves independently of the s-motion, the time derivative of the integral over $\gamma(t)$ of any smooth scalar field φ, following the motion of such a deformable surface, is given by the expression

$$\left\{\frac{d}{d\tau}\int_{\gamma(\tau)}\varphi\right\}_{\tau=t} = \left\{\frac{d}{d\tau}\int_{\gamma_s(\tau)}\varphi\right\}_{\tau=t} + \int_{\partial\gamma(t)}\varphi\,(\mathbf{v}-\mathbf{v}_s)\cdot\mathbf{n}_t\,, \tag{20}$$

where $\mathbf{v}$ represents the independent velocity of the moving boundary $\partial\gamma(t)$, $\mathbf{n}_t$ is the outward normal to the migrating surface, and $\gamma_s(\tau)$ is, at time τ, the shape of the solid subbody that occupies, at time t, the smooth region $\gamma(t)$ of the current configuration.

As the inverse s-motion $\boldsymbol{\chi}_s^{-1}(\cdot,\tau)$ maps, respectively, $\gamma(\tau)$ onto $\gamma^*(\tau)$, and $\gamma_s(\tau)$ onto $\gamma_s^*(\tau)$, denoting by $\mathbf{w}$ the velocity at which the boundary $\partial\gamma^*(\tau)$ moves through the reference time-independent configuration $\mathcal{B}$,

$$\mathbf{w}(X,t) = \mathbf{F}_s^{-1}(X,t)\left\{\mathbf{v}(x,t)-\mathbf{v}_s(x,t)\right\}, \tag{21}$$

and by φ^* the referential description of the Eulerian scalar field φ,

$$\varphi^*(X,t) = \det\mathbf{F}_s(X,t)\;\varphi(x,t)\,, \tag{22}$$

we may also consider the alternative expression

$$\left\{\frac{d}{d\tau}\int_{\gamma^*(\tau)}\varphi^*\right\}_{\tau=t} = \left\{\frac{d}{d\tau}\int_{\gamma_s^*(\tau)}\varphi^*\right\}_{\tau=t} + \int_{\partial\gamma^*(t)}\varphi^*\,\mathbf{w}\cdot\mathbf{n}\,, \tag{23}$$

where, by assumption, $\partial\gamma^*$ coincides with $\partial\gamma_s^*$ at time $\tau = t$, and $\mathbf{n}$ is the outward normal to the moving surface. If we finally introduce the velocity field $\mathbf{V}$, such that

$$\mathbf{v}(x,t) + \mathbf{F}_s(X,t)\,\mathbf{V}(X,t) := \mathbf{0}\,, \tag{24}$$

we can also deduce that

$$\int_{\partial\gamma(t)}\varphi\,(\mathbf{v}-\mathbf{v}_s)\cdot\mathbf{n}_t = -\int_{\partial\gamma^*(t)}\varphi^*\,(\mathbf{V}-\mathbf{V}_s)\cdot\mathbf{n} = \int_{\partial\gamma^*(t)}\varphi^*\,\mathbf{w}\cdot\mathbf{n}\,. \tag{25}$$

3 NONLINEAR THERMOELASTICITY

With the aim of describing incompatible thermal deformations of the solid constituent, we follow the time evolution of its stress-free configuration, described by a smooth (but geometrically non-compatible) tensor field on $\mathcal{B}$, denoted by $\mathbf{F}_\vartheta$ in figure 3.

In order to extend the action of the group of change in observers on the manifold of admissible solid and fluid motions, we consider the customary relations

$$\tilde{\chi}_s(X,t) = \tilde{x}_o(t) + \tilde{\mathbf{Q}}(t)\left(\chi_s(X,t) - x_o(t)\right), \quad \forall X\in\mathcal{B} \tag{26}$$

$$\tilde{\chi}_f(\mathcal{X}_f,t) = \tilde{x}_o(t) + \tilde{\mathbf{Q}}(t)\left(\chi_f(\mathcal{X}_f,t) - x_o(t)\right), \quad \forall \mathcal{X}_f\in\mathcal{B}_f \tag{27}$$

and their partial time derivatives

$$\tilde{\mathbf{v}}_s(\tilde{x},t) = \boldsymbol{\omega}(t) + \tilde{\boldsymbol{\Omega}}(t)\left(\tilde{x}-\tilde{x}_o(t)\right) + \tilde{\mathbf{Q}}(t)\,\mathbf{v}_s(x,t) \tag{28}$$

$$\tilde{\mathbf{v}}_f(\tilde{x},t) = \boldsymbol{\omega}(t) + \tilde{\boldsymbol{\Omega}}(t)\left(\tilde{x}-\tilde{x}_o(t)\right) + \tilde{\mathbf{Q}}(t)\,\mathbf{v}_f(x,t)\,, \tag{29}$$

where

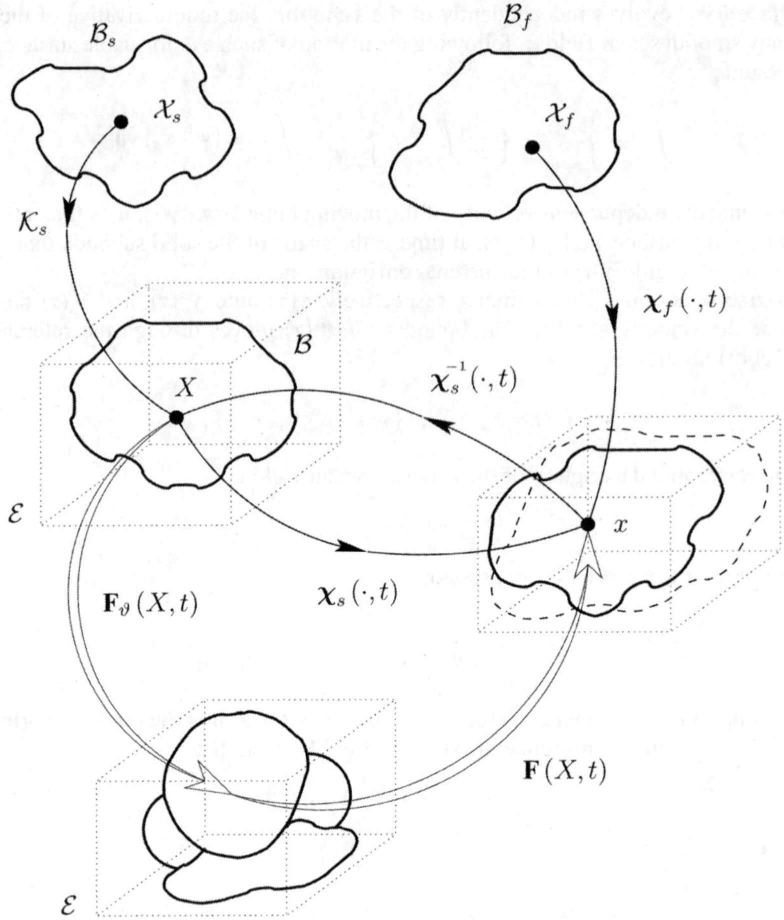

Figure 3. The multiplicative decomposition, $\mathbf{F}_s = \mathbf{F}\mathbf{F}_\vartheta$, where $\mathbf{F}_s$ = Grad χ_s.

$$\tilde{x} := \tilde{\chi}_s(X,t) = \tilde{\chi}_f(\mathcal{X}_f,t)\ , \quad x := \chi_s(X,t) = \chi_f(\mathcal{X}_f,t)$$

$$\omega(t) := \dot{\tilde{x}}_o(t) - \tilde{\mathbf{Q}}(t)\,\dot{x}_o(t)\ , \quad \tilde{\boldsymbol{\Omega}}(t) := \dot{\tilde{\mathbf{Q}}}(t)\,\tilde{\mathbf{Q}}^T(t)\ \in \text{Skw},$$

while bearing in mind (Casey and Naghdi 1980, 1981, Casey 1987, Quiligotti 2002) that the nonunique decomposition $\mathbf{F}_s = \mathbf{F}\mathbf{F}_\vartheta$ should be independent of the observer,

$$\tilde{\mathbf{F}}_s(X,t) = \tilde{\mathbf{Q}}(t)\,\mathbf{F}_s(X,t) \tag{30}$$

$$\tilde{\mathbf{F}}(X,t) = \tilde{\mathbf{Q}}(t)\,\mathbf{F}(X,t)\,\bar{\mathbf{Q}}^T(t) \tag{31}$$

$$\tilde{\mathbf{F}}_\vartheta(X,t) = \bar{\mathbf{Q}}(t)\,\mathbf{F}_\vartheta(X,t)\ , \tag{32}$$

for all orthogonal tensor-valued functions of time $\tilde{\mathbf{Q}}(t), \bar{\mathbf{Q}}(t) \in$ Orth, such that $\tilde{\mathbf{F}}_s = \tilde{\mathbf{F}}\tilde{\mathbf{F}}_\vartheta$.

These general invariance requirements, early observed in the context of elastic-plastic deformation at finite strain (Green and Naghdi 1971), formalize the idea that two different observers have to agree on the choice of the stress-free configuration only up to an arbitrary rotation, as no physical argument can be invoked to lead such a choice.

Finally, partial time derivatives of expressions (30)–(32) yield

$$\dot{\tilde{\mathbf{F}}}_s \tilde{\mathbf{F}}_s^{-1} = \tilde{\boldsymbol{\Omega}} + \tilde{\mathbf{Q}}\left\{\dot{\mathbf{F}}_s \mathbf{F}_s^{-1}\right\}\tilde{\mathbf{Q}}^T, \quad \tilde{\boldsymbol{\Omega}} := \dot{\tilde{\mathbf{Q}}}\tilde{\mathbf{Q}}^T \in \mathrm{Skw} \tag{33}$$

$$\dot{\bar{\mathbf{F}}}_\theta \bar{\mathbf{F}}_\vartheta^{-1} = \bar{\boldsymbol{\Omega}} + \bar{\mathbf{Q}}\left\{\dot{\mathbf{F}}_\vartheta \mathbf{F}_\vartheta^{-1}\right\}\bar{\mathbf{Q}}^T, \quad \bar{\boldsymbol{\Omega}} := \dot{\bar{\mathbf{Q}}}\bar{\mathbf{Q}}^T \in \mathrm{Skw} \tag{34}$$

$$\dot{\tilde{\mathbf{F}}}\tilde{\mathbf{F}}^{-1} = \tilde{\boldsymbol{\Omega}} + \tilde{\mathbf{Q}}\left\{\dot{\mathbf{F}}\mathbf{F}^{-1}\right\}\tilde{\mathbf{Q}}^T - \tilde{\mathbf{F}}\bar{\boldsymbol{\Omega}}\tilde{\mathbf{F}}^{-1}, \tag{35}$$

while

$$\left.\frac{\partial \tilde{\mathbf{v}}_\alpha}{\partial \tilde{x}}\right|_t (\tilde{x}, t) = \tilde{\boldsymbol{\Omega}}(t) + \tilde{\mathbf{Q}}(t)\left\{\left.\frac{\partial \mathbf{v}_\alpha}{\partial x}\right|_t (x,t)\right\}\tilde{\mathbf{Q}}^T(t). \tag{36}$$

4 MASS CONSERVATION

Focusing attention on the solid constituent, we can introduce two different mass densities per unit volume of the mixture, ϱ_s and ϱ_s^*, defined in such a way that the solid-mass content of both three-dimensional smooth regions $\mathcal{V}_s^* \subset \mathcal{B}$ and $\mathcal{V}_s(t) = \boldsymbol{\chi}_s(\mathcal{V}_s^*, t) \subset \mathcal{B}_t$ equals the measure

$$\mathcal{M}_s(\mathcal{V}_t) = \int_{\mathcal{V}_s(t)} \varrho_s = \int_{\mathcal{V}_s^\star} \varrho_s^\star, \quad \varrho_s^\star(X,t) = J_s(x,t)\,\varrho_s(x,t), \tag{37}$$

where $J_s(x, t) = \det \mathbf{F}_s(X, t)$ and $\mathcal{V}_t = \mathcal{V}_s(t)$.

If the mixture does not undergo any chemical reaction, at any given time $\tau = t$, the value of the material s-derivative of $\mathcal{M}_s(\mathcal{V}_\tau)$ vanishes for any $\mathcal{V}_t \subset \mathcal{B}_t$,

$$\left\{\frac{d}{d\tau}\int_{\mathcal{V}_s(\tau)} \varrho_s\right\}_{\tau=t} = \int_{\mathcal{V}_s(t)} \left(\frac{\partial \varrho_s}{\partial t} + \mathrm{div}\,(\varrho_s \mathbf{v}_s)\right) = 0, \tag{38}$$

leading to the following expression of the local conservation law,

$$\frac{\partial \varrho_s}{\partial t} + \mathrm{div}\,(\varrho_s \mathbf{v}_s) = 0. \tag{39}$$

Recalling the definition of the referential mass density (37), we may also notice that the smooth scalar map ϱ_s^* does not depend on time,

$$\left\{\frac{d}{d\tau}\int_{\mathcal{V}_s(\tau)} \varrho_s\right\}_{\tau=t} = \left\{\frac{d}{d\tau}\int_{\mathcal{V}_s^\star} \varrho_s^\star\right\}_{\tau=t} = \int_{\mathcal{V}_s^\star}\left\{\frac{\partial \varrho_s^\star}{\partial \tau}\right\}_{\tau=t} = 0, \tag{40}$$

and therefore the local conservation law (39) can be rewritten in the alternative form

$$\frac{\partial \varrho_s^\star}{\partial t} = 0. \tag{41}$$

Because of the coexistence of different constituents, the smooth region $\mathcal{V}_t$ of the current configuration of the mixture can also be associated with a fluid subbody. As a consequence, we can state the integral conservation law of the fluid mass,

$$\left\{\frac{d}{d\tau}\int_{\mathcal{V}_f(\tau)} \varrho_f\right\}_{\tau=t} = \int_{\mathcal{V}_f(t)} \left(\frac{\partial \varrho_f}{\partial t} + \text{div}\,(\varrho_f \mathbf{v}_f)\right) = 0\,, \tag{42}$$

which yields the local expression

$$\frac{\partial \varrho_f}{\partial t} + \text{div}\,(\varrho_f \mathbf{v}_f) = 0\,. \tag{43}$$

As the fluid-mass content of the smooth current region $\mathcal{V}_t = \mathcal{V}_f(t) \subset \mathcal{B}_t$ equals the measure (Wilmanski 1996)

$$\mathcal{M}_f(\mathcal{V}_t) = \int_{\mathcal{V}_f(t)} \varrho_f = \int_{\mathcal{V}_f^\star(t)} \varrho_f^\star\,, \tag{44}$$

where ϱ_f^* is the fluid-mass density per unit reference volume,

$$\varrho_f^\star(X,t) = J_s(x,t)\,\varrho_f(x,t)\,, \tag{45}$$

and

$$\mathcal{V}_f(t) = \chi_s\left(\mathcal{V}_f^\star(t), t\right) \subset \mathcal{B}_t\,, \tag{46}$$

at any given time t, the value of the material f-derivative of $\mathcal{M}_f(\mathcal{V}_\tau)$ vanishes for any choice of $\mathcal{V}_t = \mathcal{V}_f(t)$, namely

$$\left\{\frac{d}{d\tau}\int_{\mathcal{V}_f(\tau)} \varrho_f\right\}_{\tau=t} = \left\{\frac{d}{d\tau}\int_{\mathcal{V}_f^\star(\tau)} \varrho_f^\star\right\}_{\tau=t} = \int_{\mathcal{V}_f^\star(t)} \left(\frac{\partial \varrho_f^\star}{\partial t} + \text{Div}\,\left(\varrho_f^\star \mathbf{w}_f\right)\right) = 0\,, \tag{47}$$

leading to the following expression of the local fluid-mass conservation law,

$$\frac{\partial \varrho_f^\star}{\partial t} + \text{Div}\,\left(\varrho_f^\star \mathbf{w}_f\right) = 0\,. \tag{48}$$

Finally, let us consider both integral mass conservation laws (38) and (42) from a slightly different point of view, referring to the motion of the mixture as a single body (Truesdell 1969). In particular, let us assume that any material point of the mixture as a whole (figure 1) moves at the given velocity

$$\mathbf{v} := \xi_s \mathbf{v}_s + \xi_f \mathbf{v}_f\,, \tag{49}$$

where ξ_α represents the mass fraction of the α-th constituent,

$$\xi_\alpha := \frac{\varrho_\alpha}{\varrho}\,, \quad \alpha \in \{\text{S}, \text{F}\}\,, \tag{50}$$

i.e. the dimensionless ratio between the current mass density of the α-th constituent and the current mass density of the mixture per unit volume,

$$\varrho := \varrho_s + \varrho_f\,, \tag{51}$$

such that, by definition,

$$\xi_s + \xi_f = 1\,. \tag{52}$$

Taking the sum of left-hand sides of (38) and (42), an integral conservation law of global mass can be stated for any smooth region of the current shape of the mixture,

$$\left\{\frac{d}{d\tau}\left(\int_{\mathcal{V}_s(\tau)}\varrho_s+\int_{\mathcal{V}_f(\tau)}\varrho_f\right)\right\}_{\tau=t}=\int_{\mathcal{V}_t}\left(\frac{\partial\varrho}{\partial t}+\operatorname{div}\left(\varrho\mathbf{v}\right)\right)=0\,, \tag{53}$$

where, at time $\tau = t$, $\mathcal{V}_t = \mathcal{V}_s(t) = \mathcal{V}_f(t)$.

Similarly, referring to the reference shape of the solid constituent,

$$\left\{\frac{d}{d\tau}\left(\int_{\mathcal{V}_s^\star}\varrho_s^\star+\int_{\mathcal{V}_f^\star(\tau)}\varrho_f^\star\right)\right\}_{\tau=t}=\int_{\mathcal{V}_t^\star}\left(\frac{\partial\varrho^\star}{\partial t}+\operatorname{Div}\left(\varrho^\star\mathbf{w}\right)\right)=0\,, \tag{54}$$

where ϱ^* is the overall mass density of the mixture per unit reference volume,

$$\varrho^\star(X,t):=\varrho_s^\star(X,t)+\varrho_f^\star(X,t)=J_s(x,t)\,\varrho(x,t)\,, \tag{55}$$

and

$$\mathbf{w}:=\xi_f\,\mathbf{w}_f\,. \tag{56}$$

As a general rule, we notice that the time derivative of the α-mass content of any smooth region $\mathcal{V}_\tau \subset \mathcal{B}_\tau$, enveloped by a migrating surface which follows the motion of the mixture as a whole, does not vanish at any time $\tau = t$,

$$\left\{\frac{d}{d\tau}\int_{\mathcal{V}_\tau}\varrho_\alpha\right\}_{\tau=t}=\int_{\mathcal{V}_t}\varrho\left\{\frac{\mathcal{D}\xi_\alpha}{\mathcal{D}\tau}\right\}_{\tau=t}=-\int_{\partial\mathcal{V}_t}\varrho_\alpha\left(\mathbf{v}_\alpha-\mathbf{v}\right)\cdot\mathbf{n}_t\neq 0\,, \tag{57}$$

where the relative velocity field $\mathbf{v}_\alpha - \mathbf{v}$ represents the macroscopic velocity of diffusion of α-th particles through the mixture.

5 THE CONFIGURATIONAL NATURE OF LOCAL INTERACTIONS

Within the framework of a first-order gradient theory, the internal power expended by local interactions, exchanged by overlapped constituents on any pair of smooth velocity fields ($\mathbf{v}_s$, $\mathbf{v}_f$), is given by the sum (dell'Isola, Guarascio, and Hutter 2000; Quiligotti, Maugin, and dell'Isola 2003)

$$\sum_{\alpha\in\{\mathrm{S,F}\}}\left(\boldsymbol{\pi}_\alpha\cdot\mathbf{v}_\alpha+\boldsymbol{\sigma}_\alpha\cdot\operatorname{grad}\mathbf{v}_\alpha\right). \tag{58}$$

By virtue of the principle of material frame-indifference, the internal power expended on any rigid-body velocity field,

$$\mathbf{v}_s(x,t)=\mathbf{v}_f(x,t)=\mathbf{w}_o(t)+\mathbf{W}(t)\,(x-x_o)\,,\quad \mathbf{W}(t)\in\mathrm{Skw}, \tag{59}$$

has to vanish for any choice of spatially uniform $\mathbf{w}_o(t)$ and $\mathbf{W}(t)$,

$$\mathbf{w}_o\cdot\sum_{\alpha\in\{\mathrm{S,F}\}}\boldsymbol{\pi}_\alpha+\mathbf{W}\cdot\sum_{\alpha\in\{\mathrm{S,F}\}}\left\{\boldsymbol{\sigma}_\alpha+\boldsymbol{\pi}_\alpha\otimes(x-x_o)\right\}=0\,. \tag{60}$$

As a consequence, this assumption leads straightforwardly to the deduction of some preliminary requirements on admissible constitutive assumptions,

$$\operatorname{skw}\left(\boldsymbol{\sigma}_S+\boldsymbol{\sigma}_F\right)=\mathbf{O}\,, \tag{61}$$

$$\pi_S + \pi_F = \mathbf{0}, \tag{62}$$

which state, respectively, that the sum of peculiar Cauchy-like stress tensors has to be symmetric, and zero-th order interactions, exerted on any material α-point by the overlapped one, are opposite.

Let us now introduce a partial Piola-Kirchhoff stress tensor $\mathbf{T}_\alpha$ for each constituent of the mixture (Wilmanski 1996, cf. Bowen 1980, 1982), such that, for any pair of corresponding smooth regions, $\mathcal{V} \subset \mathcal{B}$ and $\mathcal{V}_t = \chi_s(\mathcal{V}, t) \subset \mathcal{B}_t$, we have

$$\int_{\mathcal{V}} \mathbf{T}_\alpha \cdot \text{Grad } \mathbf{v}_\alpha := \int_{\mathcal{V}_t} \boldsymbol{\sigma}_\alpha \cdot \text{grad } \mathbf{v}_\alpha \,, \tag{63}$$

namely, under suitable smoothness assumptions,

$$\mathbf{T}_\alpha := J_s \boldsymbol{\sigma}_\alpha \mathbf{F}_s^{-T}, \quad \forall\, \alpha \in \{\text{S, F}\}\,. \tag{64}$$

The expression of the internal energy density (58) may be alternatively rewritten, by taking into account the relation (17) and the preliminary requirement (62), as

$$\mathbf{T} \cdot \text{Grad } \mathbf{v}_s + \boldsymbol{\tau} \cdot \mathbf{w}_f + \boldsymbol{\mathcal{T}} \cdot \text{Grad } \mathbf{w}_f \,, \tag{65}$$

where

$$\mathbf{T} := \mathbf{T}_S + \mathbf{T}_F \tag{66}$$

$$\boldsymbol{\tau} := (\text{Grad } \mathbf{F}_S)^T \mathbf{T}_F + J_S \mathbf{F}_S^T \boldsymbol{\pi}_F \tag{67}$$

$$\boldsymbol{\mathcal{T}} := \mathbf{F}_S^T \mathbf{T}_F \,. \tag{68}$$

Finally, we may consider the further possibility of defining a new set of dynamical descriptors $(\mathbf{b}_\alpha, \boldsymbol{\tau}_\alpha)$, which satisfy the identity

$$\sum_{\alpha \in \{\text{S,F}\}} \int_{\mathcal{B}} (J_s \boldsymbol{\pi}_\alpha \cdot \mathbf{v}_\alpha + \mathbf{T}_\alpha \cdot \text{Grad } \mathbf{v}_\alpha) + \int_{\partial\mathcal{B}} \varrho^\star \psi\, \mathbf{V} \cdot \mathbf{n}$$
$$= \sum_{\alpha \in \{\text{S,F}\}} \int_{\mathcal{B}} (\boldsymbol{\tau}_\alpha \cdot \mathbf{V}_\alpha + \mathbf{b}_\alpha \cdot \text{Grad } \mathbf{V}_\alpha)\,, \tag{69}$$

for any choice of smooth velocity fields $\mathbf{V}$, $\mathbf{V}_\alpha$ and $\mathbf{v}_\alpha$, such that

$$\mathbf{v}_\alpha + \mathbf{F}_s \mathbf{V}_\alpha = \mathbf{0}\,, \quad \alpha \in \{\text{S,F}\} \tag{70}$$

$$\mathbf{V} = \xi_s \mathbf{V}_s + \xi_f \mathbf{V}_f \,, \tag{71}$$

denoting by ψ in (69) the Helmholtz free-energy per unit mass of the mixture (see also sections 6 and 7). As a result, we obtain that

$$\mathbf{b}_\alpha = \varrho_\alpha^\star \psi\, \mathbf{I} - \mathbf{F}_s^T \mathbf{T}_\alpha \,, \tag{72}$$

$$\boldsymbol{\tau}_\alpha = \text{Grad}\,(\varrho_\alpha^\star \psi) - (\text{Grad } \mathbf{F}_s)^T \mathbf{T}_\alpha - J_s \mathbf{F}_s^T \boldsymbol{\pi}_\alpha \,, \tag{73}$$

where $\mathbf{b}_\alpha$ is the *peculiar Eshelby stress tensor* (Eshelby 1951, 1956, 1975; Maugin 1993, 1995), associated with the α-th constituent of the mixture (Quiligotti 2000; Quiligotti, Maugin, and dell'Isola 2003). In order to meet the requirement (61), the sum of peculiar stress tensors (72) needs to be symmetric with respect to the right Cauchy-Green tensor,

$$\mathrm{skw}\,[(\mathbf{b}_S + \mathbf{b}_F)\,\mathbf{C}_S] = \mathbf{O}\,, \qquad \mathbf{C}_s = \mathbf{F}_s^T \mathbf{F}_s\,. \tag{74}$$

6 ENERGY BALANCE

Referring to a solid–fluid mixture as a single continuum, and following its mean motion while taking material derivatives of overall thermodynamic potentials, we assume the balance law of energy for any referential smooth region $\mathcal{V} \subset \mathcal{B}$ to be given by the following integral equation,

$$\left\{\frac{d}{d\tau}\int_{\mathcal{V}} \varrho^\star \mathcal{E}\right\}_{\tau=t} = \mathcal{P}_i\,(\mathcal{V},t) + \int_{\mathcal{V}} \varrho^\star h - \int_{\mathcal{V}} J_s \,\mathrm{div}\,\mathbf{q}\,, \tag{75}$$

where h and ε represent, respectively, the external heat supply and the internal energy per unit mass of the mixture, whereas $\mathbf{q}$ is the inward heat flux,

$$\int_{\partial\mathcal{B}_t} \mathbf{q}\cdot\mathbf{n}_t = \int_{\mathcal{B}_t} \mathrm{div}\,\mathbf{q} = \int_{\mathcal{B}} J_s\,\mathrm{div}\,\mathbf{q}\,, \tag{76}$$

and $\mathcal{P}_i\,(\mathcal{V},\,t)$ is the internal power expended by local interactions,

$$\mathcal{P}_i\,(\mathcal{B},t) := \int_{\mathcal{B}} (\mathbf{T}\cdot \mathrm{Grad}\;\mathbf{v}_s + \boldsymbol{\tau}\cdot\mathbf{w}_f + \boldsymbol{\mathcal{T}}\cdot \mathrm{Grad}\;\mathbf{w}_f)\,. \tag{77}$$

Bearing in mind that the overall mass is conserved along the mean motion, under suitable smoothness assumptions, the integral equation (75) can be also rewritten in the local form

$$\varrho^\star\frac{\mathcal{D}\mathcal{E}}{\mathcal{D}t} = \mathbf{T}\cdot \mathrm{Grad}\;\mathbf{v}_s + \boldsymbol{\tau}\cdot\mathbf{w}_f + \boldsymbol{\mathcal{T}}\cdot \mathrm{Grad}\;\mathbf{w}_f + \varrho^\star h - J_s\,\mathrm{div}\,\mathbf{q}\,. \tag{78}$$

7 DISSIPATION INEQUALITY

When considering a thermomechanical process in which both the entropy influx and the entropy supply per unit mass are due to heat inputs (Eringen 1967), the second principle of thermodynamics can be stated, for the overall mixture, in the customary integral form

$$\left\{\frac{d}{d\tau}\int_{\mathcal{B}_t} \varrho\eta\right\}_{\tau=t} \geq \int_{\mathcal{B}_t} \varrho\frac{h}{\vartheta} - \int_{\partial\mathcal{B}_t} \frac{1}{\vartheta}\,\mathbf{q}\cdot\mathbf{n}_t\,, \tag{79}$$

denoting by η and ϑ, respectively, the overall entropy per unit mass of the mixture and the mean temperature field. By usual arguments, we prefer dealing with the local form of this inequality with respect to the reference configuration of the solid constituent,

$$\vartheta\varrho^\star\frac{\mathcal{D}\eta}{\mathcal{D}t} - \varrho^\star h + \vartheta J_s\,\mathrm{div}\,\frac{\mathbf{q}}{\vartheta} \geq 0\,. \tag{80}$$

Moreover, an overall Helmholtz free-energy per unit mass of the mixture (Atkin and Craine 1976; Bowen 1976), can be defined

$$\psi := \mathcal{E} - \vartheta\eta\,, \tag{81}$$

in order to obtain, by substitution into (80), the local inequality

$$\varrho^\star \frac{\mathcal{D}\psi}{\mathcal{D}t} - \varrho^\star \frac{\mathcal{D}\mathcal{E}}{\mathcal{D}t} + \varrho^\star \eta \frac{\mathcal{D}\vartheta}{\mathcal{D}t} + \varrho^\star h - \vartheta J_s \operatorname{div} \frac{\mathbf{q}}{\vartheta} \leq 0 . \tag{82}$$

With the aim of describing macroscopic local interactions exchanged by nonchemically reacting constituent, we consider, at any given time t, a dependence of the Helmholtz free-energy density per unit mass of the mixture, $\psi(X, \mathbf{F}_s, \varrho_f^*, \vartheta)$, on

- the reference place $X = \mathcal{K}_s(\mathcal{X}_s) = \boldsymbol{\chi}_s^{-1}(\boldsymbol{\chi}_f(\mathcal{X}_f, t), t) \in \mathcal{B}$, associated with the pair of solid and fluid particles, $\mathcal{X}_s \in \mathcal{B}_s$ and $\mathcal{X}_f \in \mathcal{B}_f$, interacting at the given time t,
- the gradient of the solid motion $\mathbf{F}_s (X, t) = \mathbf{F}(X, t)\ \mathbf{F}_\vartheta\ (X, t)$ (see figure 3), where $X = \mathcal{K}_s(\mathcal{X}_s) = \boldsymbol{\chi}_s^{-1}(\boldsymbol{\chi}_f(\mathcal{X}_f, t), t) \in \mathcal{B}$,
- the fluid-mass density per unit reference volume of the mixture, $\varrho_f^* (X, t)$, associated with the fluid particle $\mathcal{X}_f \in \mathcal{B}_f$, such that $X = \boldsymbol{\chi}_s^{-1}(\boldsymbol{\chi}_f(\mathcal{X}_f, t), t) \in \mathcal{B}$,
- the mean temperature $\vartheta\ (X, t)$, associated with the pair of solid and fluid particles, $\mathcal{X}_s \in \mathcal{B}_s$ and $\mathcal{X}_f \in \mathcal{B}_f$, interacting at the given place $X \in \mathcal{B}$, at time t.

By virtue of the local balance of energy (78), we have

$$\varrho^\star \frac{\mathcal{D}\psi}{\mathcal{D}t} - \mathbf{T} \cdot \operatorname{Grad} \mathbf{v}_s - \boldsymbol{\tau} \cdot \mathbf{w}_f - \boldsymbol{\mathcal{T}} \cdot \operatorname{Grad} \mathbf{w}_f + \varrho^\star \eta \frac{\mathcal{D}\vartheta}{\mathcal{D}t} + \frac{J_s}{\vartheta} \mathbf{q} \cdot \operatorname{grad} \vartheta \leq 0 , \tag{83}$$

where

$$\begin{aligned} \frac{\mathcal{D}\psi}{\mathcal{D}t} &= \left(\xi_f \frac{\partial \psi}{\partial X} + \xi_f (\operatorname{Grad} \mathbf{F}_s)^T \frac{\partial \psi}{\partial \mathbf{F}_s} - \xi_s \frac{\partial \psi}{\partial \varrho_f^\star} \operatorname{Grad} \varrho_f^\star \right) \cdot \mathbf{w}_f \\ &+ \frac{\partial \psi}{\partial \mathbf{F}_s} \cdot \operatorname{Grad} \mathbf{v}_s + \frac{\partial \psi}{\partial \vartheta} \frac{\mathcal{D}\vartheta}{\mathcal{D}t} - \varrho_f^\star \frac{\partial \psi}{\partial \varrho_f^\star} \mathbf{I} \cdot \operatorname{Grad} \mathbf{w}_f . \end{aligned} \tag{84}$$

If zero-th and first order local interactions $\boldsymbol{\tau}$, $\mathbf{T}$, and $\boldsymbol{\mathcal{T}}$, expending power on macroscopic solid and fluid motions, can be reasonably decomposed (Eringen 1967; Maugin 1998) into, respectively, a dissipative part ($\hat{\boldsymbol{\tau}}$, $\hat{\mathbf{T}}$, and $\hat{\boldsymbol{\mathcal{T}}}$) and a recoverable hyperelastic part,

$$\mathbf{T} := \hat{\mathbf{T}} + \varrho^\star \frac{\partial \psi}{\partial \mathbf{F}_s} , \tag{85}$$

$$\boldsymbol{\tau} := \hat{\boldsymbol{\tau}} + \left(\varrho_f^\star \frac{\partial \psi}{\partial X} + \varrho_f^\star (\operatorname{Grad} \mathbf{F}_s)^T \frac{\partial \psi}{\partial \mathbf{F}_s} - \varrho_s^\star \frac{\partial \psi}{\partial \varrho_f^\star} \operatorname{Grad} \varrho_f^\star \right) , \tag{86}$$

$$\boldsymbol{\mathcal{T}} := \hat{\boldsymbol{\mathcal{T}}} - \varrho^\star \left(\varrho_f^\star \frac{\partial \psi}{\partial \varrho_f^\star} \right) \mathbf{I} , \tag{87}$$

it is possible to derive straightforwardly, from the inequality (83) and the expression (84), an alternative form of the local dissipation inequality,

$$\hat{\mathbf{T}} \cdot \operatorname{Grad} \mathbf{v}_s + \hat{\boldsymbol{\tau}} \cdot \mathbf{w}_f + \hat{\boldsymbol{\mathcal{T}}} \cdot \operatorname{Grad} \mathbf{w}_f - \varrho^\star \left(\eta + \frac{\partial \psi}{\partial \vartheta} \right) \frac{\mathcal{D}\vartheta}{\mathcal{D}t} - \frac{J_s}{\vartheta} \mathbf{q} \cdot \operatorname{grad} \vartheta \geq 0 . \tag{88}$$

8 CONCLUSION AND PERSPECTIVES

As a further step (Quiligotti and Maugin 2003), we can reconsider, within the framework of multiphase continua, some constitutive proposals available for nonlinear continuum theories of single bodies, based on the multiplicative decomposition of the gradient of the solid motion (Vujošević and Lubarda 2002; Epstein and Maugin 2000; DiCarlo and Quiligotti 2002). In order to formalize the idea that the intermediate configuration can be obtained, by means of a thought experiment, by relaxing isothermally the current configuration of the mixture, we may define an overall Helmholtz free-energy per unit volume of the relaxed configuration,

$$\varrho_\theta\,\psi_\vartheta(X, \mathbf{F}, \varrho_{f\vartheta}, \vartheta) = \varrho^\star \psi(X, \mathbf{F}_s, \varrho_f^\star, \vartheta)\,, \tag{89}$$

where

- ϱ_θ is the global mass density of the mixture per unit relaxed volume ($\varrho^* = \varrho_\theta \det \mathbf{F}_\vartheta$),
- $\varrho_{f\theta}$ is the fluid-mass density per unit relaxed volume ($\varrho_f^* = \varrho_{f\theta} \det \mathbf{F}_\vartheta$),
- $X \in \mathcal{B}$ represents, as usual, the reference place associated with a pair of solid and fluid particles, $\mathcal{X}_s \in \mathcal{B}_s$ and $\mathcal{X}_f \in \mathcal{B}_f$, which share, at the given time t, the current place $x = \boldsymbol{\chi}_s(X, t) = \boldsymbol{\chi}_f(\mathcal{X}_f, t), \in \mathcal{B}_t$,
- $\mathbf{F}$ is the tensor field given by the expression $\mathbf{F}(X, t) = \mathbf{F}_s(X, t)\,\mathbf{F}_\vartheta^{-1}(X, t)$, where $X = \mathcal{K}_s(\mathcal{X}_s) = \boldsymbol{\chi}_s^{-1}(\boldsymbol{\chi}_f(\mathcal{X}_f, t), t) \in \mathcal{B}$,
- $\vartheta(X, t)$ is the mean temperature associated with the pair of solid and fluid particles, $\mathcal{X}_s \in \mathcal{B}_s$ and $\mathcal{X}_f \in \mathcal{B}_f$, interacting at the given place $X \in \mathcal{B}$, at time t.

Consistently, the internal power density (65) could be properly rewritten, bearing in mind that, by virtue of the thermoelastic multiplicative decomposition,

$$\operatorname{grad} \mathbf{v}_s = \dot{\mathbf{F}}_s \mathbf{F}_s^{-1} = \dot{\mathbf{F}} \mathbf{F}^{-1} + \mathbf{F}\left(\dot{\mathbf{F}}_\vartheta \mathbf{F}_\vartheta^{-1}\right)\mathbf{F}^{-1} \tag{90}$$

$$\mathbf{v}_f = \mathbf{v}_s + \mathbf{F}_s \mathbf{w}_f = \mathbf{v}_s + \mathbf{F}\left(\mathbf{F}_\vartheta \mathbf{w}_f\right). \tag{91}$$

Moreover, focusing on constitutive aspects of thermal stresses in the more general case of phase changes and chemically reacting solid–fluid mixtures, a two-temperature continuum theory of nonlinear thermoelasticity of a porous solid, infused with a compressible fluid, could be established (cf. Lhuillier 2003), with the aim of describing coupled thermal phenomena by following the time evolution of the solid stress-free configuration.

REFERENCES

Atkin, R.J. and R.E. Craine (1976). Continuum Theories of Mixtures: Basic Theory and Historical Development. *Quart. J. Mech. Appl. Math.* 29: 209–244.

Bowen, R.M. (1976). Theory of Mixtures. In A.C. Eringen (Ed.), Continuum Physics, Volume III: 1–127. New York: Academic Press.

Bowen, R.M. (1980). Incompressible porous media models by use of the theory of mixtures. *Int. J. Engng. Sci.* 18: 1129–1148.

Bowen, R.M. (1982). Compressible porous media models by use of the theory of mixtures. *Int. J. Engng. Sci.* 20: 697–735.

Casey, J. (1987). Invariance considerations in large strain elasto-plasticity. *ASME J. Appl. Mech.* 54: 247. (Discussion).

Casey, J. and P.M. Naghdi (1980). A remark on the use of the decomposition $\mathbf{F} = \mathbf{F}_e\,\mathbf{F}_p$ in plasticity. *ASME J. Appl. Mech.* 47: 672–675.

Casey, J. and P.M. Naghdi (1981). A correct definition of elastic and plastic deformation and its computational significance. *ASME J. Appl. Mech.* 48: 983–984. (Discussion).

deBoer, R. (1996). Highlights in the historical development of the porous media theory: Toward a consistent macroscopic theory. *ASME Appl. Mech. Rev.* 49: 201–262.

deBoer, R. (2000a). Contemporary progress in porous media theory. *ASME Appl. Mech. Rev.* 51: 323–370.

deBoer, R. (2000b). *Theory of porous media. Highlights in the Historical Development and current state.* Springer–Verlag.

dell'Isola, F., M. Guarascio, and K. Hutter (2000). A variational approach for the deformation of a saturated porous solid. A second gradient theory extending Terzaghi's effective stress principle. *Arch. Appl. Mech.* 70: 323–337.

DiCarlo, A. (1996). A Non-Standard Format for Continuum Mechanics. In R.C. Batra and M.F. Beatty (Eds.), *Contemporary Research in the Mechanics and Mathematics of Materials*, CIMNE, Barcelona: 92–104.

DiCarlo, A. and S. Quiligotti (2002). Growth & Balance. *Mech. Res. Comm.* 29: 449– 456.

Epstein, M. and G.A. Maugin (2000). Thermomechanics of volumetric growth in uniform bodies. *Int. J. Plasticity* 16: 951–978.

Eringen, A.C. (1967). *Mechanics of continua*. John Wiley & Sons.

Eshelby, J.D. (1951). The force on an elastic singularity. *Phil. Trans. Roy. Soc.* A 244: 87–112.

Eshelby, J.D. (1956). The continuum theory of lattice defects. In F. Seitz and D. Turnbull (Eds.), *Solid State Physics*, Volume 3: 79–144. New York: Academic Press.

Eshelby, J.D. (1975). The elastic energy-momentum tensor. *J. Elasticity* 5: 321–335.

Germain, P. (1972). La méthode des puissances virtuelles en mécanique des milieux continus. Première partie: Théorie du second gradient. *Journal de Mécanique* 12: 235–264.

Germain, P. (1973). The method of virtual power in Continuum Mechanics. Part II: Microstructure. *SIAM J. Appl. Math* 25: 556–575.

Green, A.E. and P.M. Naghdi (1971). Some remarks on elastic-plastic deformation at finite strain. *Int. J. Engng. Sci.* 9: 1219–1229.

Klisch, S.M. (1999). Internally constrained mixtures of elastic continua. *Math. Mech. Solids* 4: 481–498.

Krishnaswamy, S. and R.C. Batra (1997). A thermomechanical theory of solid–fluid mixtures. *Math. Mech. Solids* 2: 143–151.

Lee, E.H. (1969). Elastic-plastic deformation at finite strains. *ASME J. Appl. Mech.* 36: 1–6.

Lhuillier, D. (2003). A mean-field description of two-phase flows with phase changes. *Int. J. Multiphase Flow* 29: 511–525.

Maugin, G.A. (1980). The Method of Virtual Power in Continuum Mechanics: Application to Coupled Fields. *Acta Mech.* 35: 1–70.

Maugin, G.A. (1993). *Material Inhomogeneities in Elasticity*. London: Chapman & Hall.

Maugin, G.A. (1995). Material Forces: Concepts and applications. *ASME Appl. Mech. Rev.* 48: 213–245.

Maugin, G.A. (1998). *The thermomechanics of nonlinear irreversible behaviours: An introduction*, Volume 27: World Scientific.

Nunziato, J.W. and E. Walsh (1980). On ideal multiphase mixtures with chemical reactions and diffusion. *Arch. Rational Mech. Anal.* 73: 287–311.

Quiligotti, S. (2000). *Un Approccio Configurazionale alla Teoria delle Miscele Binarie*. Ph.D. thesis, Università degli Studi "La Sapienza", Rome.

Quiligotti, S. (2002). On bulk growth mechanics of solid–fluid mixtures: kinematics and invariance requirements. *TAM* 28–29: 277–288.

Quiligotti, S. and G.A. Maugin (2003). On the thermomechanics of nonchemically reacting solid–fluid mixtures. *Preprint*.

Quiligotti, S., G.A. Maugin, and F. dell'Isola (2002). Wave motions in unbounded poroelastic solids infused with compressible fluids. *ZAMP* 53: 1110–1138.

Quiligotti, S., G.A. Maugin, and F. dell'Isola (2003). An Eshelbian approach to the nonlinear mechanics of constrained solid–fluid mixtures. *Acta Mech.* 160: 45–60.

Rajagopal, K.R. and L. Tao (1995). *Mechanics of Mixtures*, Volume 35 of *Advances in Mathematics for Applied Sciences*. Singapore: World Scientific.

Taber, L.A. (1995). Biomechanics of growth, remodeling, and morphogenesis. *Appl. Mech. Rev.* 48: 487–545.

Truesdell, C.A. (1957). Sulle basi della termomeccanica. *Lincei – Rend. Sc. fis. mat. nat.* 22: 33–38, 158–166.

Truesdell, C.A. (1969). *Rational Thermodynamics*. New York: McGraw–Hill.

Vujošević, L. and V.A. Lubarda (2002). Finite-strain thermoelasticity based on multiplicative decomposition of deformation gradient. *TAM* 28–29: 379–399.

Wilmanski, K. (1996). Dynamics of porous materials under large deformations and changing porosity. In R.C. Batra and M.F. Beatty (Eds.), *Contemporary Research in the Mechanics and Mathematics of Materials*, CIMNE, Barcelona: 343–356.

Wilmanski, K. (1998). A Thermodynamic Model of Compressible Porous Materials with the Balance Equation of Porosity. *Transport Porous Media* 32: 21–47.

Wilmanski, K. (2000). Mathematical Theory of Porous Media. Ravello (Italy). Lecture notes of the XXV Summer School on Mathematical Physics.

Configurational Mechanics, Kalpakides & Maugin (eds)
© 2004, Taylor & Francis Group, London, ISBN 90 5809 667 X

On consistent weakly-nonlocal modeling of granular media

J. Makowski, H. Stumpf, K. Hackl
Lehrstuhl für Allgemeine Mechanik, Ruhr-Universität Bochum, Germany

T. Triantafyllidis
Lehrstuhl für Grundbau und Bodenmechanik, Ruhr-Universität Bochum, Germany

ABSTRACT: The aim of the paper is the presentation of a thermodynamically consistent weakly-nonlocal (gradient type) modeling of materials characterized by translational and rotational degrees of freedom of material particles in the physical space and dissipative changes of the internal structure caused by material micro-forces and micro-couples. These properties can be observed for a wide range of media including concrete, rocks, sands and polycrystalline metals. Since these materials under external actions localize in bands of intensive shearing or narrow zones of micro-cracking often the precursor of macro-failure, we formulate a gradient model to capture appropriately localization phenomena. To describe the translational and rotational motion of material particles on the macro-level, a theory of oriented media is formulated including a length scale associated with the size of material particles. Taking into account that inhomogeneities on the mesolevel as micro-shearbands, -cracks, -voids, dislocations and disclinations and their time changes are caused by dissipative driving forces and couples in the material space, we postulate two additional balance laws of material forces and couples introducing in this way two additional length scales. Next, the first and second law of thermodynamics for physical and material space are formulated with the entropy production as a measure of non-equilibrium. Furthermore, general constitutive equations consistent with the second law of thermodynamics are formulated and some special cases are considered.

1 INTRODUCTION

It is well-known that numerical solutions based on local theories suffer from pathological mesh-dependency, if localization of the deformation occurs (e.g. Schieck et al. 1999). Also local theories are not able to capture well-known size effects: in testing of brittle solids a bigger but otherwise identical structure has a lower failure load than a small specimen (Bažant and Ožbolt 1990). To overcome these deficiencies various models were proposed in the literature:

- nonlocal and gradient enhancements of local theories (e.g. Aifantis 1984, Triantafyllidis and Aifantis 1986, de Borst and Mühlhaus 1992),
- micro-polar continua (e.g. Mühlhaus 1986, Mühlhaus and Vardoulakis 1987, de Borst 1991, Steinmann 1994),
- continua with micro-structure based on balance laws of material forces on micro-defects (e.g. Naghdi and Srinivasa 1994, Maugin 1995, Le and Stumpf 1996, Makowski and Stumpf 2001, Stumpf et al. 2001, Stumpf and Hackl 2003, Saczuk et al. 2003).

To derive a thermodynamically consistent nonlocal (gradient type) theory appropriate to analyse a wide range of materials such as concrete, rocks, sands and metals, we have to describe the translational and rotational motion of material particles in the physical space and the change of the internal structure due to dissipative driving forces and couples on the meso-level in the material space.

In Section 2 the kinematics of oriented media, the balance laws of physical macroscopic forces and couples and the principal of virtual velocities are formulated. In Section 3 a consistent linearization for small strains and rotations is performed. In Section 4 balance laws of micro-forces and micro-couples in the material space and the first and second law of thermodynamics for physical and material space are formulated. In Section 5 the constitutive modeling is investigated under the assumption of small strains and rotations.

2 LAWS OF MECHANICS IN PHYSICAL SPACE

2.1 *Kinematics of oriented media*

Throughout this work standard vector and tensors notation is consistently used strictly following the definitions and theorems which can be found, e.g. in Silhavy (1997). In particular, vectors and second order tensors are denoted by lower and upper case bold letters, respectively, while third order tensors are denoted by $\mathbb{T}$, $\mathbb{H}$, etc. Moreover, a material body under consideration (possibly with additional kinematical and dynamical structure) will be identified with a region B in the physical space which the body occupies in a fixed reference configuration. This configuration serves to identify material particles, and the complete theory will be formulated in the referential description. For various time dependent fields defined on B, the gradient, divergence and curl operators are denoted by $\nabla \equiv$ Grad, Div and Curl, respectively. These operators are defined for vector and tensor fields (Šilhavý 1997). The superimposed dot will be used for the material time derivative, i.e. the time derivative following the motion of material particles. In the referential description, the material time derivative coincides with the usual partial derivative with respect to time t keeping the material particle $\mathbf{X} \in B$ fixed.

Relative to the reference configuration B, the motion of a body in the physical space is described by a time dependent mapping which carries each material particle $\mathbf{X}$ into its spatial place $\mathbf{x}$ in the deformed configuration at time t:

$$\mathbf{x} = \mathbf{x}(\mathbf{X}, t) = \mathbf{X} + \mathbf{u}(\mathbf{X}, t), \tag{1}$$

where $\mathbf{u}(\mathbf{X}, t)$ denotes the displacement vector. Then the velocity $\mathbf{v}(\mathbf{X}, t)$ and the deformation gradient $\mathbf{F}(\mathbf{X}, t)$ are given by

$$\mathbf{v} = \dot{\mathbf{x}} = \dot{\mathbf{u}}, \quad \mathbf{F} = \nabla \mathbf{x} = \mathbf{1} + \nabla \mathbf{u}. \tag{2}$$

With the aim to model granular media and materials with texture at the meso-level, it is further assumed that material particles possess orientation specified by a triad of vectors $\mathbf{t}_a(\mathbf{X})$, $a = 1, 2, 3$, associated with each particle $\mathbf{X}$. During the motion of the body in the physical space, the orientation of a particle $\mathbf{X}$ will change so that the same material particle at time t is equipped with a triad $\mathbf{d}_a(\mathbf{x}, t)$ obtained from the reference triad by a rigid rotation (without stretching):

$$\mathbf{d}_a(\mathbf{x}, t) = \mathbf{Q}(\mathbf{X}, t)\mathbf{t}_a(\mathbf{X}), \quad a = 1, 2, 3. \tag{3}$$

Here $\mathbf{Q}$ denotes the rotation tensor, $\mathbf{Q}^{\mathrm{T}}\mathbf{Q} = \mathbf{1}$, and the angular velocity of the particle is determined by time derivative of (3) yielding

$$\dot{\mathbf{d}}_a = \dot{\mathbf{Q}}\mathbf{t}_a = \dot{\mathbf{Q}}\mathbf{Q}^{\mathrm{T}}\mathbf{d}_a = \boldsymbol{\Omega}\mathbf{d}_a = \boldsymbol{\omega} \times \mathbf{d}_a. \tag{4}$$

The skew-symmetric tensor $\boldsymbol{\Omega}$ is the spin of the rotating triad associated with moving particles, and $\boldsymbol{\omega}$ denotes the angular velocity vector defined by

$$\boldsymbol{\omega} = \mathrm{ax}\boldsymbol{\Omega}, \quad \boldsymbol{\Omega} = \dot{\mathbf{Q}}\mathbf{Q}^{\mathrm{T}}. \tag{5}$$

Here and throughout the paper, the axial vector $\boldsymbol{\omega} = \mathrm{ax}\boldsymbol{\Omega}$ of the skew-symmetric tensor $\boldsymbol{\Omega}$ is defined by $\boldsymbol{\Omega}\mathbf{c} = \boldsymbol{\omega} \times \mathbf{c}$ for every vector $\mathbf{c}$.

On the macro-scale, the rotation of material particles as described by (3) can be considered as mean rotation of the particles in a representative volume element (RVE) at the meso-scale.

2.2 *Macroscopic balance laws of physical forces and couples*

We assume that the referential mass density $\rho_0(\mathbf{X})$, measured per unit volume of the reference configuration, is independent of time so that the balance law of mass is satisfied identically. Then the balance laws of macro-momentum and angular macro-momentum in the referential description are

$$\begin{aligned}\int_P \mathbf{b}\,dv + \int_{\partial P} \mathbf{Tn}\,da &= \frac{d}{dt}\int_P \rho_0 \mathbf{v}\,dv,\\ \int_P (\mathbf{c} + \mathbf{x} \times \mathbf{b})dv + \int_{\partial P} (\mathbf{Mn} + \mathbf{x} \times \mathbf{Tn})da &= \frac{d}{dt}\int_P \rho_0 (\mathbf{J}\boldsymbol{\omega} + \mathbf{x} \times \mathbf{v})dv.\end{aligned} \tag{6}$$

The balance law $(6)_1$ is classical with $\mathbf{b}(\mathbf{X}, t)$ denoting the body macro-force, $\mathbf{T}(\mathbf{X}, t)$ the stress tensor (first Piola-Kirchhoff type) and $\mathbf{n}(\mathbf{X})$ the outward unit normal vector to the boundary ∂P of an arbitrary subdomain $P \subset B$ of the material body in the reference configuration. The balance of angular macro-momentum $(6)_2$ is assumed here in general form with $\mathbf{c}(\mathbf{X}, t)$ denoting the body couple, $\mathbf{M}(\mathbf{X}, t)$ the couple stress tensor and $\mathbf{J}(\mathbf{X}, t)$ the inertia tensor assumed to satisfy the conditions discussed in detail by Cowin and Leslie (1980) to ensure the existence of the rotational kinetic energy.

As in the classical theories of continuum mechanics, it is assumed here that the integral laws of mechanics (6) hold for every subbody $P \subset B$. Then under the usual regularity assumptions the following local balance laws can be derived:

$$\mathrm{Div}\,\mathbf{T} + \mathbf{b} = \rho_0 \dot{\mathbf{v}}, \quad \mathrm{Div}\,\mathbf{M} + \mathrm{ax}(\mathbf{TF}^{\mathrm{T}} - \mathbf{FT}^{\mathrm{T}}) + \mathbf{c} = \rho_0 (\mathbf{J}\dot{\boldsymbol{\omega}})^{\boldsymbol{\cdot}}. \tag{7}$$

It should be noted that the integral laws (6) and the resulting local equations (7) constitute the mechanical basis for the so-called unconstrained and constrained Cosserat theory. In the former case, the rotation tensor $\mathbf{Q}$ is regarded as kinematical variable independent of the macroscopic translational motion of the body. In the latter case, this tensor is enforced to coincide with the local rotation of the body determined by the rotation tensor $\mathbf{R}$ resulting from the polar decomposition, $\mathbf{F} = \mathbf{RU}$, of the deformation gradient defined by $(2)_2$, i.e. $\mathbf{Q} = \mathbf{R}$. Thus, the two theories are considered here within the same framework.

2.3 *Principle of virtual velocities*

In order to complete the general description of oriented media it remains to introduce the appropriate strain measures. This can be achieved through an analysis of the local deformation taking into account the rotational motion of the body. Such an analysis leads to many possibilities in defining local strain measures (see e.g. Steinmann 1994 and references cited therein). It is then interesting to see that among various possible strain measures for oriented media, there are strain measures naturally determined by the laws of mechanics.

As a direct implication of the equations of motion (7) we have

$$\int_P \Big(\mathrm{Div}(\mathbf{T} + \mathbf{b} - \rho_0 \dot{\mathbf{v}}) \bullet \mathbf{v} + (\mathrm{Div}\,\mathbf{M} + \mathrm{ax}(\mathbf{TF}^{\mathrm{T}} - \mathbf{FT}^{\mathrm{T}}) + \mathbf{c} - \rho_0 (\mathbf{J}\dot{\boldsymbol{\omega}})^{\boldsymbol{\cdot}}) \bullet \boldsymbol{\omega}\Big) dv = 0 \tag{8}$$

for every part P of the body. The use of the divergence theorem together with the aforementioned assumption on the inertia tensor $\mathbf{J}$ allows us to rewrite equation (8) in the form

$$\frac{d}{dt}\int_P k^{\mathrm{c}} dv + \int_P \sigma^{\mathrm{c}} dv = \int_P (\mathbf{b} \bullet \mathbf{v} + \mathbf{c} \bullet \boldsymbol{\omega})dv + \int_{\partial P} (\mathbf{Tn} \bullet \mathbf{v} + \mathbf{Mn} \bullet \boldsymbol{\omega})da, \tag{9}$$

which expresses the local laws of mechanics (7) in the weak form known as the principle of virtual velocities. In (9), k^c is the kinetic energy density given by

$$k^c \equiv \frac{1}{2}(\rho_0 \dot{\mathbf{x}} \bullet \dot{\mathbf{x}} + \rho_0 \mathbf{J}\boldsymbol{\omega} \bullet \boldsymbol{\omega}) \tag{10}$$

and σ^c is the stress power of physical forces and couples given by

$$\sigma^c \equiv \mathbf{T} \bullet \nabla \mathbf{v} + (\mathrm{ax}(\mathbf{T}\mathbf{F}^{\mathrm{T}} - \mathbf{F}\mathbf{T}^{\mathrm{T}})) \bullet \boldsymbol{\omega} + \mathbf{M} \bullet \nabla \boldsymbol{\omega}. \tag{11}$$

Taking further into account that $\dot{\mathbf{F}} = \nabla \mathbf{v}$ and making use of the tensor identity

$$(\mathrm{ax}(\mathbf{T}\mathbf{F}^{\mathrm{T}} - \mathbf{F}\mathbf{T}^{\mathrm{T}})) \bullet \boldsymbol{\omega} = \frac{1}{2}(\mathbf{T}\mathbf{F}^{\mathrm{T}} - \mathbf{F}\mathbf{T}^{\mathrm{T}}) \bullet \boldsymbol{\Omega} = -\mathbf{T} \bullet \boldsymbol{\Omega}\mathbf{F}, \tag{12}$$

the stress power (11) can be rewritten as

$$\sigma^c = \mathbf{T} \bullet (\dot{\mathbf{F}} - \boldsymbol{\Omega}\mathbf{F}) + \mathbf{M} \bullet \nabla \boldsymbol{\omega}. \tag{13}$$

It then follows that the natural strain measures for oriented media consist of two tensors $\mathbf{E}$ and $\mathbf{K}$, in general non-symmetric, whose appropriate rates are equal to $\dot{\mathbf{F}} - \boldsymbol{\Omega}\mathbf{F}$ and $\nabla\boldsymbol{\omega}$, respectively. It can be shown that such rates are defined by

$$\mathbf{E}^{\circ} \equiv \mathbf{Q}\left(\frac{d}{dt}(\mathbf{Q}^{\mathrm{T}}\mathbf{E})\right) = \dot{\mathbf{F}} - \boldsymbol{\Omega}\mathbf{F}, \quad \mathbf{K}^{\circ} \equiv \mathbf{Q}\left(\frac{d}{dt}(\mathbf{Q}^{\mathrm{T}}\mathbf{K})\right) = \nabla\boldsymbol{\omega}, \tag{14}$$

so that the stress power may finally be written as

$$\sigma^c = \mathbf{T} \bullet \mathbf{E}^{\circ} + \mathbf{M} \bullet \mathbf{K}^{\circ}. \tag{15}$$

Since our subsequent considerations will be restricted to a linearized theory, we do not present here a detailed analysis of the kinematics of oriented media in the fully nonlinear case.

Reversing the derivation outlined above it is easily seen that the local equations (7) can also be derived taking the principle of virtual velocities (9) as the basic postulate of mechanics instead of the integral balance laws. Moreover, it is obvious that this principle applies also to the whole body B. In this case, the inspection of the boundary integral in (9) naturally leads to a consistent formulation of the boundary conditions for oriented media. Such boundary conditions require that either the force $\mathbf{Tn} = \mathbf{t}$ and couple $\mathbf{Mn} = \mathbf{m}$ or the displacement $\mathbf{u}$ and rotation $\mathbf{Q}$ must be prescribed at the boundary of the whole body. Typically, it is assumed that the boundary ∂B consists of two mutually disjoint parts: ∂B_f, where the forces and couples are prescribed, and the complementary part ∂B_d, where the displacements and rotations are given. Then the boundary integral in the principle of virtual velocities written for the whole body takes the form

$$\int_{\partial B}(\mathbf{Tn} \bullet \mathbf{v} + \mathbf{Mn} \bullet \boldsymbol{\omega})da = \int_{\partial B_f}(\mathbf{t} \bullet \mathbf{v} + \mathbf{m} \bullet \boldsymbol{\omega})da + \int_{\partial B_d}(\mathbf{Tn} \bullet \mathbf{v} + \mathbf{Mn} \bullet \boldsymbol{\omega})da \tag{16}$$

with the second boundary integral vanishing for kinematically admissible displacements and rotations.

2.4 *Consistent linearization*

In the remaining part of this paper we shall be concerned with the linear theory of oriented media. Such a theory models physical phenomena under the assumption that the displacements and rotations are small in some sense. Stating this assumption in precise form, consistently linearized equations and relations can be derived as follows.

If the rotations are restricted in magnitude, then the rotation tensor $\mathbf{Q}$ can be defined uniquely in terms of the exponential function (this representation is not unique for arbitrarily large rotations)

$$\mathbf{Q}(\mathbf{X},t)=\exp \mathbf{\Phi}(\mathbf{X},t), \quad \mathbf{\Phi}(\mathbf{X},t)=\mathrm{ax}^{-1}\boldsymbol{\varphi}(\mathbf{X},t), \tag{17}$$

where $\mathbf{\Phi}$ is a skew-symmetric tensor and $\boldsymbol{\varphi}$ denotes its axial vector. Let

$$\varepsilon_1(\mathbf{X},t)=\|\nabla\mathbf{u}(\mathbf{X},t)\|, \quad \varepsilon_2(\mathbf{X},t)=\|\boldsymbol{\varphi}(\mathbf{X},t)\|, \tag{18}$$

and set $\varepsilon(\mathbf{X}, t) = \max\{\varepsilon_1(\mathbf{X}, t), \varepsilon_2(\mathbf{X}, t)\}$. Then the linear theory is based on the assumption $\varepsilon(\mathbf{X}, t) \ll 1$ at all points $\mathbf{X}$ and for all time instants t. Under this assumption, it follows from the definition of the exponential function that the rotation tensor $\mathbf{Q}$ is given by

$$\mathbf{Q}=\mathbf{1}+\mathbf{\Phi}=\mathbf{1}-\mathrm{ax}^{-1}\boldsymbol{\varphi} \tag{19}$$

to within an error of $\mathbf{O}(\varepsilon^2)$ as $\varepsilon \to 0$. Thus, within the indicated error, the motion of the body is completely determined by the displacement vector $\mathbf{u}$ and the rotation vector $\boldsymbol{\varphi}$ through the formulae:

$$\mathbf{x}=\mathbf{X}+\mathbf{u}, \quad \mathbf{d}_a=(\mathbf{1}+\mathbf{\Phi})\mathbf{t}_a=\mathbf{t}_a+\boldsymbol{\varphi}\times\mathbf{t}_a, \quad a=1,2,3. \tag{20}$$

The field equations of the linear theory of oriented media can next be derived from (7) taking into account the small strain-rotation assumption. With the deformation gradient given by $(2)_2$, the second term in the nonlinear equation $(7)_2$ reads

$$\mathbf{T}\mathbf{F}^{\mathrm{T}}-\mathbf{F}\mathbf{T}^{\mathrm{T}}=(\mathbf{1}+\nabla\mathbf{u})^{\mathrm{T}}\mathbf{T}-(\mathbf{1}+\nabla\mathbf{u})\mathbf{T}^{\mathrm{T}}=\mathbf{T}-\mathbf{T}^{\mathrm{T}}+(\nabla\mathbf{u})^{\mathrm{T}}\mathbf{T}-(\nabla\mathbf{u})\mathbf{T}^{\mathrm{T}}. \tag{21}$$

Thus to within a small relative error of order $\mathbf{O}(\varepsilon)$ we have

$$\mathbf{T}\mathbf{F}^{\mathrm{T}}-\mathbf{F}\mathbf{T}^{\mathrm{T}}=\mathbf{T}-\mathbf{T}^{\mathrm{T}} \tag{22}$$

and the nonlinear equations of motion (7) reduce to the linear forms

$$\operatorname{Div}\mathbf{T}+\mathbf{b}=\rho_0\ddot{\mathbf{u}}, \quad \operatorname{Div}\mathbf{M}+\mathrm{ax}(\mathbf{T}-\mathbf{T}^{\mathrm{T}})+\mathbf{c}=\rho_0\mathbf{J}\ddot{\boldsymbol{\varphi}}, \tag{23}$$

where the rotational inertia tensor $\mathbf{J}$ is assumed to be time independent.

Within a small relative error of order $\mathbf{O}(\varepsilon^2)$, the rotation tensor $\mathbf{Q}$ is given by (19) and the use of this formula in the definition (14) of the strain rates yields

$$\mathbf{E}^{\circ}=\mathbf{Q}\left(\frac{d}{dt}(\mathbf{Q}^{\mathrm{T}}\mathbf{E})\right)=(\mathbf{1}+\mathbf{\Phi})\left(\frac{d}{dt}((\mathbf{1}-\mathbf{\Phi})\mathbf{E})\right)=(\mathbf{1}+\mathbf{\Phi})\left(\dot{\mathbf{E}}+\mathbf{O}(\varepsilon^2)\right)=\dot{\mathbf{E}}+\mathbf{O}(\varepsilon^2), \tag{24}$$

with the corresponding result for the rate $\mathbf{K}^{\circ}$ of the curvature tensor. Moreover, within the same relative error, the right hand sides of (14) simplify to the forms

$$\begin{gathered}\dot{\mathbf{F}}-\mathbf{\Omega}\mathbf{F}=\nabla\dot{\mathbf{u}}-\mathbf{\Omega}(\mathbf{1}+\mathbf{O}(\varepsilon))=\nabla\dot{\mathbf{u}}-\dot{\mathbf{\Phi}}+\mathbf{O}(\varepsilon^2)=\frac{d}{dt}(\nabla\mathbf{u}-\mathbf{\Phi})+\mathbf{O}(\varepsilon^2),\\ \nabla\boldsymbol{\omega}=\nabla(\mathbf{1}+\dot{\boldsymbol{\varphi}}+\mathbf{O}(\varepsilon^2))=\nabla\dot{\boldsymbol{\varphi}}+\mathbf{O}(\varepsilon^2)=\frac{d}{dt}(\nabla\boldsymbol{\varphi})+\mathbf{O}(\varepsilon^2).\end{gathered} \tag{25}$$

Thus, the strain measures of the linear theory of oriented media are obtained as

$$\mathbf{E}=\nabla\mathbf{u}-\mathbf{\Phi}=\nabla\mathbf{u}-\mathrm{ax}^{-1}\boldsymbol{\varphi}, \quad \mathbf{K}=\nabla\boldsymbol{\varphi}, \tag{26}$$

and the stress power (15) reads

$$\sigma^c = \mathbf{T} \bullet \dot{\mathbf{E}} + \mathbf{M} \bullet \dot{\mathbf{K}}. \tag{27}$$

It may be shown that although $\nabla\mathbf{u}$ and $\boldsymbol{\Phi}$ separately are not infinitesimaly objective, the strain measures (26) are infinitesimally objective as it is required in an objective linear theory.

3 THERMOMECHANICS OF INELASTIC DEFORMATIONS

3.1 *Micro-structural variables, balance of micro-forces and micro-couples*

The nonlinear theory for oriented media considered so far and its consistently linearized version are based only on the general laws of Newtonian mechanics. As such these theories apply to all materials and all types of deformations, elastic or inelastic, which can be modeled properly within the basic assumptions underlying the adopted form of the laws of mechanics. When considering inelastic behavior of materials, it is typically required to introduce into the theory certain measures of irreversible deformation in the material space, e.g. a plastic deformation tensor in the theory of elasto-plasticity or a damage tensor in continuum damage mechanics. Such additional measures of irreversible deformations in the material space have their source in the evolution of the material substructure during the macroscopic deformation of the body. Their role in the macroscopic theory is to represent such micro-structural effects, which give rise to the macroscopically observed irreversible deformation of the body.

The type of appropriate measures of inelastic deformation and its physical interpretation depend on the kind of material under consideration and the kind of deformation to be modeled. For example, plastic deformations in metals are caused by creation and motion of dislocations. These effects are macroscopically represented by the plastic deformation "gradient" $\mathbf{F}^p$, which is a locally defined linear map, and its spatial gradient $\nabla\mathbf{F}^p$ (see e.g. Le and Stumpf, 1996). It is important to note that the type and number of such material strain measures depend also on the macroscopic theory. In particular, while for metals treated as Cauchy continua there is one physical strain measure on the macro-level, e.g. the Green strain tensor $\mathbf{E}$, and one material strain measure, e.g. a plastic strain tensor $\mathbf{E}^p$, there are for polar media two physical strain measures on the macro-level, i.e. $\mathbf{E}$ and $\mathbf{K}$, which in the linear theory are defined by (26) in terms of the displacement vector $\mathbf{u}$ and the small rotation vector $\boldsymbol{\varphi}$. Accordingly, under the assumption of small strains and rotations, the plastic strain measures in the material space must consist of two second order tensors $\mathbf{E}^p$ and $\mathbf{K}^p$, both non-symmetric in general. However, while the macroscopic strains $\mathbf{E}$ and $\mathbf{K}$ are derivable from the displacement and rotation vectors through the relations (26), the microscopic strains $\mathbf{E}^p$ and $\mathbf{K}^p$ are not defined in terms of other field variables but they constitute additional field variables of the theory.

Since Leibfried (1949) it is known that there are forces on defects causing their motion. In Le and Stumpf 1996 a balance law for dissipative driving forces on dislocations was postulated to formulate a micro-theory of finite elastoplasticity for Cauchy continua. Correspondingly, for the modeling of an elastic-plastic theory of granular media, we postulate two balance laws of material micro-forces and micro-couples:

$$\begin{aligned} &\int_P (-\mathbf{T}^p + \mathbf{B}^p)dv + \int_{\partial P} \mathbb{T}^p\mathbf{n}\,da = \frac{d}{dt}\int_P \rho_0 \mathbf{P}^p\,dv, \\ &\int_P (-\mathbf{M}^p + \mathbf{C}^p)dv + \int_{\partial P} \mathbb{M}^p\mathbf{n}\,da = \frac{d}{dt}\int_P \rho_0 \mathbf{S}^p\,dv. \end{aligned} \tag{28}$$

In these two balance equations, $\mathbf{T}^p(\mathbf{X}, t)$ and $\mathbf{M}^p(\mathbf{X}, t)$ are intrinsic micro-force and micro-couple, $\mathbf{B}^p(\mathbf{X}, t)$ and $\mathbf{C}^p(\mathbf{X}, t)$ are extrinsic micro-force and micro-couple, while $\mathbb{T}^p(\mathbf{X}, t)$ and $\mathbb{M}^p(\mathbf{X}, t)$ are micro-stress and micro-couple stress tensors, respectively. Moreover, $\mathbf{P}^p(\mathbf{X}, t)$ and $\mathbf{S}^p(\mathbf{X}, t)$ are micro-momentum and micro-spin, both defined per unit mass. Let us note that $\mathbb{T}^p$ and $\mathbb{M}^p$

are third order tensors while the other tensors in (28) are second order tensors. Their physical interpretation will become clear when we consider the stress power derived below.

Analogous to the classical theories of continuum mechanics, it is assumed that the integral laws of micro-forces and micro-couples (28) hold for every subbody $P \subset B$. Then under the usual regularity requirements the following local balance laws can be derived:

$$\mathrm{Div}\, \mathbb{T}^{\mathrm{p}} - \mathbf{T}^{\mathrm{p}} + \mathbf{B}^{\mathrm{p}} = \rho_0 \dot{\mathbf{P}}^{\mathrm{p}}, \quad \mathrm{Div}\, \mathbb{M}^{\mathrm{p}} - \mathbf{M}^{\mathrm{p}} + \mathbf{C}^{\mathrm{p}} = \rho_0 \dot{\mathbf{S}}^{\mathrm{p}}. \tag{29}$$

In order to obtain a further insight into the physical interpretation of the material microscopic field variables we may apply the same procedure which has been used in the formulation of the balance laws of physical forces and couples in the form (8). Thus, as a straightforward implication of the local laws (29) we have

$$\int_B \Big((\mathrm{Div}\, \mathbb{T}^{\mathrm{p}} - \mathbf{T}^{\mathrm{p}} + \mathbf{B}^{\mathrm{p}} - \rho_0 \dot{\mathbf{P}}^{\mathrm{p}}) \bullet \dot{\mathbf{E}}^{\mathrm{p}} + (\mathrm{Div}\, \mathbb{M}^{\mathrm{p}} - \mathbf{M}^{\mathrm{p}} + \mathbf{C}^{\mathrm{p}} - \rho_0 \dot{\mathbf{S}}^{\mathrm{p}}) \bullet \dot{\mathbf{K}}^{\mathrm{p}} \Big) dv = 0. \tag{30}$$

Now, it is necessary to make certain assumptions concerning the nature of the micro-momentum $\mathbf{P}^{\mathrm{p}}$ and micro-spin $\mathbf{S}^{\mathrm{p}}$. Here we simply assume that there exists a kinetic energy in the material space $k^{\mathrm{p}} = k^{\mathrm{p}}(\dot{\mathbf{E}}^{\mathrm{p}}, \dot{\mathbf{K}}^{\mathrm{p}})$ such that its time derivative is given by

$$\dot{k}^{\mathrm{p}} = \rho_0 \dot{\mathbf{P}}^{\mathrm{p}} \bullet \dot{\mathbf{E}}^{\mathrm{p}} + \rho_0 \dot{\mathbf{S}}^{\mathrm{p}} \bullet \dot{\mathbf{K}}^{\mathrm{p}}. \tag{31}$$

This assumption together with the use of the divergence theorem makes it possible to transform the integral identity (30) to

$$\frac{d}{dt}\int_B k^P dv + \int_B \sigma^P dv = \int_B \Big(\mathbf{B}^{\mathrm{p}} \bullet \dot{\mathbf{E}}^{\mathrm{p}} + \mathbf{C}^{\mathrm{p}} \bullet \dot{\mathbf{K}}^{\mathrm{p}} \Big) dv + \int_{\partial B} \Big(\mathbb{T}^{\mathrm{p}}\mathbf{n} \bullet \dot{\mathbf{E}}^{\mathrm{p}} + \mathbb{M}^{\mathrm{p}}\mathbf{n} \bullet \dot{\mathbf{K}}^{\mathrm{p}} \Big) dv, \tag{32}$$

where the micro-stress power σ^{p} is given by

$$\sigma^{\mathrm{p}} \equiv \mathbf{T}^{\mathrm{p}} \bullet \dot{\mathbf{E}}^{\mathrm{p}} + \mathbf{M}^{\mathrm{p}} \bullet \dot{\mathbf{K}}^{\mathrm{p}} + \mathbb{T}^{\mathrm{p}} \bullet \nabla \dot{\mathbf{E}}^{\mathrm{p}} + \mathbb{M}^{\mathrm{p}} \bullet \nabla \dot{\mathbf{K}}^{\mathrm{p}}. \tag{33}$$

Equation (32) represents the local balance equations of material micro-forces and micro-couples (30) in weak form and thus it may be called the principle of virtual micro-velocities. Moreover, it follows from (33) that micro-stress and micro-couple tensors $\mathbf{T}^{\mathrm{p}}$ and $\mathbf{M}^{\mathrm{p}}$ are power-conjugate to $\mathbf{E}^{\mathrm{p}}$ and $\mathbf{K}^{\mathrm{p}}$, respectively, while the third order micro-stress and micro-couple stress tensors $\mathbb{T}^{\mathrm{p}}$ and $\mathbb{M}^{\mathrm{p}}$ are power-conjugate to $\nabla\mathbf{E}^{\mathrm{p}}$ and $\nabla\mathbf{K}^{\mathrm{p}}$. Corresponding to the approach already exposed in the formulation of the theory of polar media, the inspection of the boundary integral in (32) makes it possible to deduce a consistent form of the boundary conditions for material micro-field variables.

3.2 *Balance of energy – first law of thermodynamics*

Any theory of inelastic deformation requires a firm thermodynamical basis in order to formulate the constitutive equations, which are consistent with the basic principles of physics. Such a thermodynamical basis is provided by an appropriate statement of the balance of energy and the principle of irreversibility. Although the theory considered in this paper is far from being standard, the detailed formulation of the balance laws of mechanics of physical and material forces and couples presented in the previous sections allows us to conclude that the balance of energy must take the form

$$\begin{aligned} \frac{d}{dt}\Big\{\int_P (\varepsilon + k) dv\Big\} &= \int_P \Big(\mathbf{b} \bullet \mathbf{v} + \mathbf{c} \bullet \boldsymbol{\omega} + \mathbf{B}^{\mathrm{p}} \bullet \dot{\mathbf{E}}^{\mathrm{p}} + \mathbf{C}^{\mathrm{p}} \bullet \dot{\mathbf{K}}^{\mathrm{p}} \Big) dv \\ &+ \int_{\partial P} \Big(\mathbf{Tn} \bullet \mathbf{v} + \mathbf{Mn} \bullet \boldsymbol{\omega} + \mathbb{T}^{\mathrm{p}}\mathbf{n} \bullet \dot{\mathbf{E}}^{\mathrm{p}} + \mathbb{M}^{\mathrm{p}}\mathbf{n} \bullet \dot{\mathbf{K}}^{\mathrm{p}} \Big) da \\ &+ \int_P r\, dv - \int_{\partial P} \mathbf{q} \bullet \mathbf{n}\, da. \end{aligned} \tag{34}$$

Here, $\varepsilon(\mathbf{X}, t)$ is the internal energy, $r(\mathbf{X}, t)$ the external body heating and $\mathbf{q}(\mathbf{X}, t)$ the referential heat flux vector. Moreover, $k = k^c + k^p$ denotes the total kinetic energy which is the sum of two terms, the kinetic energy k^c due to the macro-motion of the body in the physical space given by (10) and the kinetic energy k^p due to the plastic and irreversible, respectively, evolution of the material substructure according to (33).

It should be noted here that the macroscopic laws of mechanics (6) and the balance laws of micro-forces and micro-couples (28) are two sets of uncoupled postulates. The coupling between the macroscopic and microscopic forces and couples appears only in the balance of energy (34). Assuming now that the local balance laws of physical macro-forces and macro-couples (7) and the balance laws of material micro-forces and micro-couples (29) hold, the local equation of the energy balance implied by (34) is obtained in the form

$$\dot{\varepsilon} = \sigma^c + \sigma^p + r - \operatorname{Div} \mathbf{q}, \tag{35}$$

where the physical and material stress powers σ^c and σ^p are given by (27) and (33), respectively.

3.3 *Principle of irreversibility – second law of thermodynamics*

In order to complete the formulation of the laws of mechanics and thermodynamics for physical and material space, the above balance laws must be supplemented by an appropriate form of the second law of thermodynamics. Without loss of generality we assume that the Clausius-Duhem inequality

$$\frac{d}{dt}\int_P \eta\, dv \geq \int_P \theta^{-1} r\, dv - \int_{\partial P} \theta^{-1}\mathbf{q} \bullet \mathbf{n}\, da \tag{36}$$

is an appropriate statement of the second law of thermodynamics. In (36), $\eta(\mathbf{X}, t)$ is the entropy and $\theta(\mathbf{X}, t) > 0$ the absolute temperature.

Assuming now that the balance laws of forces and couples for physical and material space and the balance of energy hold, the following local dissipation inequality

$$\mathcal{D} \equiv \sigma^c + \sigma^p - \dot{\Psi} - \eta\dot{\theta} - \dot{\theta}^{-1}\mathbf{q} \bullet \nabla\dot{\theta} \geq 0 \tag{37}$$

can be derived from (36). In (37), $\Psi(\mathbf{X}, t)$ is the free energy measured per unit volume of the reference configuration given by

$$\Psi = \varepsilon - \theta\eta, \tag{38}$$

and $\mathcal{D}(\mathbf{X}, t)$ denotes the dissipation function.

4 CONSTITUTIVE MODELING

4.1 *General constitutive equations*

The independent thermo-kinematical variables of the theory presented in this paper consist of the displacement and rotation vectors, $\mathbf{u}$ and $\boldsymbol{\varphi}$, with the associated strain measures $(\mathbf{E}, \mathbf{K})$, the plastic strain measures $(\mathbf{E}^p, \mathbf{K}^p)$ and the temperature θ. These variables must be determined by the solution of the field equations (23) and (29), supplemented by appropriate constitutive equations and prescribed boundary and initial conditions. It follows from (23) and (29) that the unknown dependent field variables $(\mathbf{T}, \mathbf{M}, \mathbf{T}^p, \mathbf{M}^p, \mathbb{T}^p, \mathbb{M}^p, \Psi, \eta, \mathbf{q})$ must be given by constitutive equations as functions of the strain measures and their rates together with $(\theta, \nabla\theta)$.

We introduce the following notation for the physical and material strains

$$\boldsymbol{\varepsilon} \equiv (\mathbf{E}, \mathbf{K}), \quad \boldsymbol{\varepsilon}^p \equiv (\mathbf{E}^p, \mathbf{K}^p), \quad \nabla\boldsymbol{\varepsilon}^p \equiv (\nabla\mathbf{E}^p, \nabla\mathbf{K}^p), \tag{39}$$

and for the power conjugate physical and material stresses

$$\boldsymbol{\sigma} \equiv (\mathbf{T}, \mathbf{M}), \quad \boldsymbol{\sigma}^{\mathrm{p}} \equiv (\mathbf{T}^{\mathrm{p}}, \mathbf{M}^{\mathrm{p}}), \quad \boldsymbol{\mu}^{\mathrm{p}} \equiv (\mathbb{T}^{\mathrm{p}}, \mathbb{M}^{\mathrm{p}}), \tag{40}$$

leading with (27) and (33) to the stress power in the form

$$\begin{aligned} \sigma \equiv \sigma^{\mathrm{c}} + \sigma^{\mathrm{p}} &= \boldsymbol{\sigma} \bullet \dot{\boldsymbol{\varepsilon}} + \boldsymbol{\sigma}^{\mathrm{p}} \bullet \dot{\boldsymbol{\varepsilon}}^{\mathrm{p}} + \boldsymbol{\mu}^{\mathrm{p}} \bullet \nabla \dot{\boldsymbol{\varepsilon}}^{\mathrm{p}} \\ &= \mathbf{T} \bullet \dot{\mathbf{E}} + \mathbf{M} \bullet \dot{\mathbf{K}} + \mathbf{T}^{\mathrm{p}} \bullet \dot{\mathbf{E}}^{\mathrm{p}} + \mathbf{M}^{\mathrm{p}} \bullet \dot{\mathbf{K}}^{\mathrm{p}} + \mathbb{T}^{\mathrm{p}} \bullet \nabla \dot{\mathbf{E}}^{\mathrm{p}} + \mathbb{M}^{\mathrm{p}} \bullet \nabla \dot{\mathbf{K}}^{\mathrm{p}}. \end{aligned} \tag{41}$$

To formulate the constitutive equations in a compact form, we use the following abbreviations

$$\boldsymbol{\alpha} \equiv (\boldsymbol{\varepsilon}, \ \boldsymbol{\varepsilon}^{\mathrm{p}}, \ \nabla \boldsymbol{\varepsilon}^{\mathrm{p}}), \quad \dot{\boldsymbol{\alpha}} = (\dot{\boldsymbol{\varepsilon}}, \ \dot{\boldsymbol{\varepsilon}}^{\mathrm{p}}, \ \nabla \dot{\boldsymbol{\varepsilon}}^{\mathrm{p}}). \tag{42}$$

For the description of strain hardening an additional variable $\kappa(\mathbf{X}, t) \geqslant 0$ must be introduced into the theory. This variable characterizes e.g. the accumulation of defects during inelastic deformation and as such it does not represent an independent kinematical variable but it must be specified by an appropriate constitutive law.

The general constitutive equations are assumed here as

$$\begin{gathered} \Psi = \tilde{\Psi}(\boldsymbol{\alpha}, \kappa, \theta, \dot{\boldsymbol{\alpha}}, \nabla\theta), \\ \boldsymbol{\sigma} = \tilde{\boldsymbol{\sigma}}(\boldsymbol{\alpha}, \kappa, \theta, \dot{\boldsymbol{\alpha}}, \nabla\theta), \quad \boldsymbol{\sigma}^{\mathrm{p}} = \tilde{\boldsymbol{\sigma}}^{\mathrm{p}}(\boldsymbol{\alpha}, \kappa, \theta, \dot{\boldsymbol{\alpha}}, \nabla\theta), \quad \boldsymbol{\mu}^{\mathrm{p}} = \tilde{\boldsymbol{\mu}}^{\mathrm{p}}(\boldsymbol{\alpha}, \kappa, \theta, \dot{\boldsymbol{\alpha}}, \nabla\theta), \\ \eta = \tilde{\eta}(\boldsymbol{\alpha}, \kappa, \theta, \dot{\boldsymbol{\alpha}}, \nabla\theta), \quad \mathbf{q} = \tilde{\mathbf{q}}(\boldsymbol{\alpha}, \kappa, \theta, \dot{\boldsymbol{\alpha}}, \nabla\theta). \end{gathered} \tag{43}$$

They have to satisfy the dissipation inequality (37), which can be written with (41) in compact form as

$$\mathcal{D} \equiv \boldsymbol{\sigma} \bullet \dot{\boldsymbol{\varepsilon}} + \boldsymbol{\sigma}^{\mathrm{p}} \bullet \dot{\boldsymbol{\varepsilon}}^{\mathrm{p}} + \boldsymbol{\mu}^{\mathrm{p}} \bullet \nabla \dot{\boldsymbol{\varepsilon}}^{\mathrm{p}} - \dot{\Psi} - \eta\dot{\theta} - \theta^{-1}\mathbf{q} \bullet \nabla\dot{\theta} \geq 0. \tag{44}$$

The differentiation of the free energy Ψ with respect to time and substituting the result into (44) yields

$$\begin{aligned} \mathcal{D} \equiv {} & \left(\boldsymbol{\sigma} - \frac{\partial\Psi}{\partial\boldsymbol{\varepsilon}}\right) \bullet \dot{\boldsymbol{\varepsilon}} + \left(\boldsymbol{\sigma}^{\mathrm{p}} - \frac{\partial\Psi}{\partial\boldsymbol{\varepsilon}^{\mathrm{p}}}\right) \bullet \dot{\boldsymbol{\varepsilon}}^{\mathrm{p}} + \left(\boldsymbol{\mu}^{\mathrm{p}} - \frac{\partial\Psi}{\partial\nabla\boldsymbol{\varepsilon}^{\mathrm{p}}}\right) \bullet \nabla\dot{\boldsymbol{\varepsilon}}^{\mathrm{p}} - \frac{\partial\Psi}{\partial\kappa}\dot{\kappa} - \left(\eta + \frac{\partial\Psi}{\partial\theta}\right)\dot{\theta} - \theta^{-1}\mathbf{q} \bullet \nabla\theta \\ & - \frac{\partial\Psi}{\partial\dot{\boldsymbol{\varepsilon}}} \bullet \ddot{\boldsymbol{\varepsilon}} - \frac{\partial\Psi}{\partial\dot{\boldsymbol{\varepsilon}}^{\mathrm{p}}} \bullet \ddot{\boldsymbol{\varepsilon}}^{\mathrm{p}} - \frac{\partial\Psi}{\partial\nabla\dot{\boldsymbol{\varepsilon}}^{\mathrm{p}}} \bullet \nabla\ddot{\boldsymbol{\varepsilon}}^{\mathrm{p}} - \frac{\partial\Psi}{\partial\nabla\theta} \bullet \nabla\dot{\theta} \geq 0. \end{aligned} \tag{45}$$

According to the approach generally accepted in thermodynamics of continuous media the inequality (45) must be satisfied identically by all response functions. This implies that the most general form of the constitutive equation for the free energy is

$$\Psi = \tilde{\Psi}(\boldsymbol{\alpha}, \kappa, \theta) = \tilde{\Psi}(\boldsymbol{\varepsilon}, \ \boldsymbol{\varepsilon}^{\mathrm{p}}, \ \nabla\boldsymbol{\varepsilon}^{\mathrm{p}}, \kappa, \theta) \tag{46}$$

and the constitutive equation for the entropy is determined by the free energy (46) through the general relation

$$\eta = \eta(\boldsymbol{\alpha}, \kappa, \theta) = \frac{\partial\tilde{\Psi}(\boldsymbol{\alpha}, \kappa, \theta)}{\partial\theta} \tag{47}$$

An important result which also follows from (45) is that the constitutive equations for macro- and micro-stresses can be written as the sum of two terms

$$\begin{aligned}
\boldsymbol{\sigma} &= \frac{\partial\tilde{\Psi}(\boldsymbol{\alpha}, \kappa, \theta)}{\partial\boldsymbol{\varepsilon}} + \tilde{\boldsymbol{\sigma}}_*(\boldsymbol{\alpha}, \kappa, \theta, \dot{\boldsymbol{\alpha}}, \nabla\theta),\\
\boldsymbol{\sigma}^{\mathrm{p}} &= \frac{\partial\tilde{\Psi}(\boldsymbol{\alpha}, \kappa, \theta)}{\partial\boldsymbol{\varepsilon}^{\mathrm{p}}} + \tilde{\boldsymbol{\sigma}}^{\mathrm{p}}_*(\boldsymbol{\alpha}, \kappa, \theta, \dot{\boldsymbol{\alpha}}, \nabla\theta),\\
\boldsymbol{\mu}^{\mathrm{p}} &= \frac{\partial\tilde{\Psi}(\boldsymbol{\alpha}, \kappa, \theta)}{\partial\nabla\boldsymbol{\varepsilon}^{\mathrm{p}}} + \tilde{\boldsymbol{\mu}}^{\mathrm{p}}_*(\boldsymbol{\alpha}, \kappa, \theta, \dot{\boldsymbol{\alpha}}, \nabla\theta).
\end{aligned} \tag{48}$$

The first terms on the right side of the constitutive equation (48) define the thermodynamical forces derivable from the free energy, the second terms marked by an asterisk, define dissipative driving forces which have to satisfy the reduced dissipation inequality

$$\mathcal{D} \equiv \boldsymbol{\sigma}_* \bullet \dot{\boldsymbol{\varepsilon}} + \boldsymbol{\sigma}^{\mathrm{p}}_* \bullet \dot{\boldsymbol{\varepsilon}}^{\mathrm{p}} + \boldsymbol{\mu}^{\mathrm{p}}_* \bullet \nabla\dot{\boldsymbol{\varepsilon}}^{\mathrm{p}} - \frac{\partial\Psi}{\partial\kappa}\dot{\kappa} - \theta^{-1}\mathbf{q} \bullet \nabla\theta \geq 0. \tag{49}$$

The restrictions on the constitutive equations derived here are direct consequences of the laws of thermodynamics and as such they are entirely general in the sense that they must be satisfied by all materials within the class of the assumed constitutive equations (43). Their further specifications can be obtained only through the introduction of additional constitutive assumptions.

4.2 *Quasi-static isothermal theory*

When considering quasi-static deformation processes, the inertia terms in (23) and (29) can be neglected so that the governing field equations take the form

$$\operatorname{Div}\mathbf{T} + \mathbf{b} = \mathbf{0}, \quad \operatorname{Div}\mathbf{M} + \operatorname{ax}(\mathbf{T} - \mathbf{T}^{\mathrm{T}}) + \mathbf{c} = \mathbf{0} \tag{50}$$

and

$$\operatorname{Div}\mathbb{T}^{\mathrm{p}} - \mathbf{T}^{\mathrm{p}} + \mathbf{B}^{\mathrm{p}} = \mathbf{0}, \quad \operatorname{Div}\mathbb{M}^{\mathrm{p}} - \mathbf{M}^{\mathrm{p}} + \mathbf{C}^{\mathrm{p}} = \mathbf{0}. \tag{51}$$

Moreover, in the analysis of special problems it is often convenient to represent all field variables and the relevant equations in component form relative to fixed Cartesian coordinates $\{x_i\} \equiv \{x_1, x_2, x_3\}$ in the physical space. Denoting by $\{\mathbf{e}_i\}$ an orthonormal basis for the chosen Cartesian coordinates, the stress tensors have the following component representations

$$\mathbf{T} = T_{ij}\mathbf{e}_i \otimes \mathbf{e}_j, \quad \mathbf{T}^{\mathrm{p}} = T^{\mathrm{p}}_{ij}\mathbf{e}_i \otimes \mathbf{e}_j, \quad \mathbb{T}^{\mathrm{p}} = T^{\mathrm{p}}_{ijk}\mathbf{e}_i \otimes \mathbf{e}_j \otimes \mathbf{e}_k. \tag{52}$$

With corresponding component representations of the remaining vector and tensor variables, the equations (50) and (51) in component form are

$$T_{ij},_j + b_i = 0, \quad M_{ij,j} + \varepsilon_{ijk}T_{jk} + c_i = 0 \tag{53}$$

and

$$T^{\mathrm{p}}_{ijk,k} - T^{\mathrm{p}}_{ij} - B^{\mathrm{p}}_{ij} = 0, \quad M^{\mathrm{p}}_{ijk,k} - M^{\mathrm{p}}_{ij} - C^{\mathrm{p}}_{ij} = 0. \tag{54}$$

Here ε_{ijk} denotes the permutation symbols, the comma stands for partial derivatives with respect to the Cartesian coordinates and the standard summation convention applies. In the same manner, introducing the components of the displacement and small rotation vector, $\mathbf{u} = u_i\mathbf{e}_i$ and $\boldsymbol{\varphi} = \varphi_i,\mathbf{e}_i$, the Cartesian components of the strain tensors (26) are obtained as

$$E_{ij} = u_{i,j} - \varepsilon_{ijk}\varphi_k, \quad K_{ij} = \varphi_{i,j}. \tag{55}$$

The equations (53), (54) and the kinematical relations (55) can be rewritten also in a vector-matrix notation (see e.g. de Borst 1991).

Under the restriction to isothermal processes the temperature field is constant and its gradient vanishes. In effect, the temperature may be omitted in the list of constitutive variables so that the constitutive equation (46) for the free energy reduces to the form

$$\Psi = \tilde{\Psi}(\boldsymbol{\varepsilon}, \boldsymbol{\varepsilon}^{\mathrm{p}}, \nabla\boldsymbol{\varepsilon}^{\mathrm{p}}, \kappa), \tag{56}$$

and we can omit the constitutive equations for entropy and heat flux vector. Also, the constitutive equations (48) for macro- and micro-stresses can be reduced correspondingly.

4.3 *Elastic strain and decomposition of total strain*

In the constitutive equations (43) the dependence of the response functions on the macroscopic strain rates $\dot{\boldsymbol{\varepsilon}}$ is needed, if viscous effects have to be taken into account. However, in the analysis of rate independent elastic-plastic material behavior, such effects are typically small and consequently the total strain rate $\dot{\boldsymbol{\varepsilon}}$ may be suppressed from the list of constitutive variables. Moreover, in the classical theories of elastic-plastic materials it is common to assume that the linear physical strain tensor can be regarded as the sum of elastic and plastic parts. In the approach of this paper it is possible to deduce this decomposition formally through the definition of elastic strains

$$\boldsymbol{\varepsilon}^{\mathrm{e}} = \boldsymbol{\varepsilon} - \boldsymbol{\varepsilon}^{\mathrm{p}}, \quad (\mathbf{E}^{\mathrm{e}} = \mathbf{E} - \mathbf{E}^{\mathrm{p}}, \mathbf{K}^{\mathrm{e}} = \mathbf{K} - \mathbf{K}^{\mathrm{p}}). \tag{57}$$

Then we can reformulate the theory in such a way as to outline the role played by the elastic strains.

With the use of (57), the stress power (41) can be rewritten as

$$\begin{aligned} \sigma &= \boldsymbol{\sigma} \bullet \dot{\boldsymbol{\varepsilon}}^{\mathrm{e}} + (\boldsymbol{\sigma} + \boldsymbol{\sigma}^{\mathrm{p}}) \bullet \dot{\boldsymbol{\varepsilon}}^{\mathrm{p}} + \boldsymbol{\mu}^{\mathrm{p}} \bullet \nabla\dot{\boldsymbol{\varepsilon}}^{\mathrm{p}} \\ &= \mathbf{T} \bullet \dot{\mathbf{E}}^{\mathrm{e}} + \mathbf{M} \bullet \dot{\mathbf{K}}^{\mathrm{e}} + \bar{\mathbf{T}}^{\mathrm{p}} \bullet \dot{\mathbf{E}}^{\mathrm{p}} + \bar{\mathbf{M}}^{\mathrm{p}} \bullet \dot{\mathbf{K}}^{\mathrm{p}} + \mathbb{T}^{\mathrm{p}} \bullet \nabla\dot{\mathbf{E}}^{\mathrm{p}} + \mathbb{M}^{\mathrm{p}} \bullet \nabla\dot{\mathbf{K}}^{\mathrm{p}}, \end{aligned} \tag{58}$$

where the new stress and couple stress tensors $\bar{\mathbf{T}}^{\mathrm{p}}$ and $\bar{\mathbf{M}}^{\mathrm{p}}$ are defined by

$$\bar{\mathbf{T}}^{\mathrm{p}} = \mathbf{T} + \mathbf{T}^{\mathrm{p}}, \quad \bar{\mathbf{M}}^{\mathrm{p}} = \mathbf{M} + \mathbf{M}^{\mathrm{p}}. \tag{59}$$

Then the micro-structural material field equations (51) can be rewritten in the form

$$\mathrm{Div}\, \mathbb{T}^{\mathrm{p}} - \bar{\mathbf{T}}^{\mathrm{p}} + \mathbf{T} + \mathbf{B}^{\mathrm{p}} = \mathbf{0}, \quad \mathrm{Div}\, \mathbb{M}^{\mathrm{p}} - \bar{\mathbf{M}}^{\mathrm{p}} + \mathbf{M} + \mathbf{C}^{\mathrm{p}} = \mathbf{0}. \tag{60}$$

It is interesting to observe that, unlike in (51), the macroscopic stress and couple stress tensors $\boldsymbol{\sigma} \equiv (\mathbf{T}, \mathbf{M})$ appear explicitly in (60). It then follows that the stresses $\bar{\mathbf{T}}^{\mathrm{p}}$ and $\bar{\mathbf{M}}^{\mathrm{p}}$, instead of $\boldsymbol{\sigma}^{\mathrm{p}} \equiv (\mathbf{T}^{\mathrm{p}}, \mathbf{M}^{\mathrm{p}})$ should be prescribed constitutively besides the constitutive equations for stresses $\boldsymbol{\sigma}$ and $\boldsymbol{\mu}^{\mathrm{p}}$.

Moreover, with the use of (57), the total strain tensors $\boldsymbol{\varepsilon}$ may be eliminated from the list of constitutive variables in favor of the elastic strains $\boldsymbol{\varepsilon}^{\mathrm{e}}$. Then it is also possible to consider the constitutive equation for the free energy function in an uncoupled form

$$\Psi = \tilde{\Psi}(\boldsymbol{\varepsilon}^{\mathrm{e}}, \kappa, \boldsymbol{\varepsilon}^{\mathrm{p}}, \nabla\boldsymbol{\varepsilon}^{\mathrm{p}}) = \tilde{\Psi}^{\mathrm{e}}(\boldsymbol{\varepsilon}^{\mathrm{e}}, \kappa) + \tilde{\Psi}^{\mathrm{p}}(\boldsymbol{\varepsilon}^{\mathrm{p}}, \nabla\boldsymbol{\varepsilon}^{\mathrm{p}}). \tag{61}$$

Under these assumptions, the constitutive equations for the material stresses $\boldsymbol{\sigma}^{p} \equiv (\mathbf{T}^{p}, \mathbf{M}^{p})$ and the material microstresses $\boldsymbol{\mu}^{p} \equiv (\mathbb{T}^{p}, \mathbb{M}^{p})$ are given by

$$\boldsymbol{\sigma} = \tilde{\boldsymbol{\sigma}}(\boldsymbol{\varepsilon}^{e}, \kappa) = \frac{\partial \Psi^{e}(\boldsymbol{\varepsilon}^{e}, \kappa)}{\partial \boldsymbol{\varepsilon}^{e}}, \quad \boldsymbol{\mu}^{p} = \boldsymbol{\mu}^{p}(\boldsymbol{\varepsilon}^{p}, \nabla\boldsymbol{\varepsilon}^{p}) = \frac{\partial \Psi(\boldsymbol{\varepsilon}^{p}, \nabla\boldsymbol{\varepsilon}^{p})}{\partial \nabla\boldsymbol{\varepsilon}^{p}}, \tag{62}$$

while the constitutive equations for the modified micro-stresses $\bar{\mathbf{T}}^{p}$ and $\bar{\mathbf{M}}^{p}$ take the form

$$\bar{\boldsymbol{\sigma}}^{p} \equiv (\bar{\mathbf{T}}^{p}, \bar{\mathbf{M}}^{p}) = \tilde{\bar{\boldsymbol{\sigma}}}^{p}(\boldsymbol{\varepsilon}^{p}, \nabla\boldsymbol{\varepsilon}^{p}, \dot{\boldsymbol{\varepsilon}}^{p}) = \frac{\partial \Psi(\boldsymbol{\varepsilon}^{p}, \nabla\boldsymbol{\varepsilon}^{p})}{\partial \boldsymbol{\varepsilon}^{p}} + \tilde{\bar{\boldsymbol{\sigma}}}^{p}_{*}(\boldsymbol{\varepsilon}^{p}, \nabla\boldsymbol{\varepsilon}^{p}, \dot{\boldsymbol{\varepsilon}}^{p}) \tag{63}$$

with the dissipative driving stresses $\bar{\boldsymbol{\sigma}}^{p}_{*}$ restricted by the dissipation inequality.

4.4 *Simplified models, non-gradient theories*

For illustration of the general theory let us assume that the part $\tilde{\Psi}^{p}$ of the free energy depends only on the gradient $\nabla\boldsymbol{\varepsilon}^{p}$ of material micro-strains and takes the very simple form

$$\tilde{\Psi}^{p}(\nabla\boldsymbol{\varepsilon}^{p}) = \frac{1}{2} c_1 \nabla\mathbf{E}^{p} \bullet \nabla\mathbf{E}^{p} + \frac{1}{2} c_2 \nabla\mathbf{K}^{p} \bullet \nabla\mathbf{K}^{p}, \tag{64}$$

so that the corresponding constitutive equations are

$$\mathbb{T}^{p} = c_1 \nabla\mathbf{E}^{p}, \quad \mathbb{M}^{p} = c_2 \nabla\mathbf{K}^{p}, \tag{65}$$

where c_1 and c_2 are material parameters. Substituting these constitutive equations into the micro-structural field equations (60) and assuming that $\mathbf{G}^{p} = \mathbf{0}$ and $\mathbf{C}^{p} = \mathbf{0}$ we obtain

$$c_1 \operatorname{Div}(\nabla\mathbf{E}^{p}) - \bar{\mathbf{T}}^{p} + \mathbf{T} = \mathbf{0}, \quad c_2 \operatorname{Div}(\nabla\mathbf{K}^{p}) - \bar{\mathbf{M}}^{p} + \mathbf{M} = \mathbf{0}. \tag{66}$$

Taking into account that $\operatorname{Div}(\nabla\mathbf{D}) = \Delta\mathbf{D}$, where Δ denotes the Laplacian, equations (66) take the form

$$c_1 \Delta\mathbf{E}^{p} - \bar{\mathbf{T}}^{p} + \mathbf{T} = \mathbf{0}, \quad c_2 \Delta\mathbf{K}^{p} - \bar{\mathbf{M}}^{p} + \mathbf{M} = \mathbf{0}, \tag{67}$$

where the macro- and micro-stresses are given by the general constitutive equations consistent with the assumptions (61) and (64).

The general constitutive models can be simplified also by alternative assumptions. It can be assumed that the third order micro-stress and micro-couple stress vanish whenever the gradients of plastic strains are small and can be neglected, $\boldsymbol{\mu}^{p} \equiv (\mathbb{T}^{p}, \mathbb{M}^{p}) = \mathbf{0}$ whenever $\nabla\boldsymbol{\varepsilon}^{p} \equiv (\nabla\mathbf{E}^{p}, \nabla\mathbf{K}^{p}) = \mathbf{0}$. In this case the partial differential equations (60) reduce to the algebraic equations

$$\mathbf{T} - \bar{\mathbf{T}}^{p}(\boldsymbol{\varepsilon}, \boldsymbol{\varepsilon}^{p}, \dot{\boldsymbol{\varepsilon}}^{p}) = \mathbf{0}, \quad \mathbf{M} - \bar{\mathbf{M}}^{p}(\boldsymbol{\varepsilon}, \boldsymbol{\varepsilon}^{p}, \dot{\boldsymbol{\varepsilon}}^{p}, \kappa) = \mathbf{0}. \tag{68}$$

For rate independent materials, the response functions $\bar{\mathbf{T}}^{p}$ and $\bar{\mathbf{M}}^{p}$ must be homogenous functions of degree one. Taking their simplest possible forms $\tilde{\mathbf{T}}^{p} = \lambda\dot{\mathbf{E}}^{p}$ and $\tilde{\mathbf{M}}^{p} = \lambda\dot{\mathbf{K}}^{p}$, the equations (68) yield the classical flow rules with the parameter $\lambda \geq 0$ to be determined by the consistency condition.

4.5 *Yield conditions*

The rate-independent theory of elasto-plasticity requires the concept of a yield surface, either in stress or strain space. Classically, this concept has been developed for non-polar, pressure-insensitive and plastically incompressible materials. In general, the yield surface and yield criteria are determined by the yield function of the form

$$f(\boldsymbol{\sigma}, \kappa) \equiv f(\mathbf{T}, \mathbf{M}, \kappa) \leq \mathbf{0}. \tag{69}$$

Generalizations of the classical concept of yield surface to Cosserat continua were presented by Mühlhaus (1986), Mühlhaus and Vardoulakis (1987), de Borst (1991), Vardoulakis and Sulem (1995) and references cited therein. The following more general form of the yield function

$$f(\boldsymbol{\sigma}, \kappa) \equiv J_2(\mathbf{T}, \mathbf{M}) - \xi(\kappa) \tag{70}$$

with

$$J_2(\mathbf{T}, \mathbf{M}) \equiv \sqrt{\alpha_1 \mathbf{T} \bullet \mathbf{T} + \alpha_2 \mathbf{T} \bullet \mathbf{T}^{\mathrm{T}} + \beta_1 \mathbf{M} \bullet \mathbf{M} + \beta_2 \mathbf{M} \bullet \mathbf{M}^{\mathrm{T}}} \tag{71}$$

includes various possible models as special cases. Here $\alpha_1, \alpha_2, \beta_1, \beta_2$ are material parameters. For pressure insensitive materials, the stress $\mathbf{T}$ in (71) should be replaced by its deviatoric part

$$\operatorname{dev} \mathbf{T} \equiv \mathbf{T} - \frac{1}{3}(\operatorname{tr} \mathbf{T})\mathbf{1}. \tag{72}$$

Moreover, with $\beta_2 = 0$ we obtain

$$J_2(\mathbf{T}, \mathbf{M}) \equiv \sqrt{\alpha_1 \operatorname{dev}\mathbf{T} \bullet \operatorname{dev}\mathbf{T} + \alpha_2 \operatorname{dev}\mathbf{T} \bullet (\operatorname{dev}\mathbf{T}^{\mathrm{T}}) + \beta_1 \mathbf{M} \bullet \mathbf{M}}. \tag{73}$$

The yield function (73) takes the form first proposed by Mühlhaus and Vardoulakis (1987). However, in applications to granular media, polycrystalline metals, cellular solids and other porous materials, none of the classical assumptions is strictly valid and the general form of the yield surface need be considered furthermore.

5 CLOSING REMARKS

A thermodynamically consistent weakly-nonlocal (gradient type) theory for the analysis of granular media including possibly localization phenomena is presented. The theory is based on the balance laws of physical macro-forces and macro-couples, two balance laws of material micro-forces and micro-couples and the first and second law of thermodynamics for physical and material space. These laws are then linearized and general constitutive equations are given for small strains and rotations.

Various simplified models are considered, e.g. by assuming that the gradients of the two material micro-strain measures can be neglected, what enables to derive evolution laws for these micro-strains. Also, by introducing simplifying assumptions various types of the yield function for granular media are obtained and compared with results presented in the literature.

ACKNOWLEDGEMENT

The financial support, provided by Deutsche Forschungsgemeinschaft (DFG) under Grant SFB 398-A7/B8, is gratefully acknowledged.

REFERENCES

Aifantis, E.C. 1984. On the microstructural origin of certain inelastic models. *ASME J. Engng Mat Tech.* 106: 326–330.

Bažant, Z.P. and Ožbolt, J. 1990. Nonlocal microplane model for fracture, damage, and size effect in structures. *ASCE J. Engng Mech*, 116: 2485–2505.

Cowin, S.C. and Leslie, F.M. 1980. On kinetic energy and momenta in Cosserat continua. *ZAMP* 31: 247–260.

De Borst, R. 1991. Simulation of strain localization: a reappraisal of the Cosserat continuum. *Engng Comp.* 8: 317–332.

De Borst, R. and Mühlhaus, H.-B. 1992. Gradient-dependent plasticity: formulation and algorithmic aspects. *Int. J. Numer. Meths Engng* 35: 521–539.

Le, K.C. and Stumpf, H. 1996. A model of elastoplastic bodies with continuously distributed dislocations. *Int. J. Plasticity* 12: 611–627.

Leibfried, G. 1949. Über die auf eine Versetzung wirkenden Kräfte. *Zeitschrift für Physik* 126: 781–789.

Lippmann, H. 1995. Cosserat plasticity and plastic spin. *Applied Mechanics Review* 48: 753–762.

Makowski, J. and Stumpf, H. 2001. Thermodynamically based concept for the modelling of continua with microstructure and evolving defects. *Int. J. Solids Structures* 38: 1943–1961.

Maugin, G.A. 1995. Material forces: concepts and applications. *Applied Mechanics Review* 48: 213–245.

Mühlhaus, H.-B. 1986. Scherfugenanalyse bei granularem Material in Rahmen der Cosserat-Theorie. *Ingenieur-Archiv* 56: 389–399.

Mühlhaus, H.-B. and Vardoulakis, I. 1987. The thickness of shear bands in granular materials. *Geotechnique* 7: 271–283.

Naghdi, P.M. and Srinivasa, A.R. 1993. A dynamical theory of structured solids. *Phil. Trans. R. Soc. London* A 345: 425–458.

Saczuk, J., Hackl, K. and Stumpf, H. 2003. Rate theory of nonlocal gradient damage – gradient viscoinelasticity. *Int. J. Plasticity* 19: 675–706.

Schieck, B., Smolenski, W.M. and Stumpf, H. 1999. A shell finite element for large strain elastoplasticity with anisotropies – Part I: Shell theory and variational principle. – Part II: Constitutive equations and numerical applications. *Int. J. Solids Structures* 36: 5399–5424; 5425–5451.

Šilhavý, M. 1997. *The Mechanics and Thermodynamics of Continuous Media*. Springer-Verlag, Berlin.

Steinmann, P. 1994. A micropolar theory of finite deformation and finite rotation multiplicative elastoplasticity. *Int. J. Solids Structures* 31: 1063–1084.

Stumpf, H. and Hackl, K. 2003. Micromechanical concept for the analysis of damage evolution in thermo-viscoelastic and quasi-brittle materials. *Int. J. Solids Structures* 40: 1567–1584.

Stumpf, H., Saczuk, J. and Hackl, K. 2001. On a general concept for the analysis of crack growth and material damage. *Int. J. Plasticity* 17: 991–1028.

Triantafyllidis, T., Aifantis, E.C. 1986. A gradient approach to localization of deformation: 1. Hyperelastic materials. *J. Elasticity* 16: 225–237.

Vardoulakis, I. and Sulem, J. 1995. *Bifurcation Analysis in Geomechanics*. Blackie London.

Configurational Mechanics, Kalpakides & Maugin (eds)
© 2004, Taylor & Francis Group, London, ISBN 90 5809 667 X

Piezomagnetic tensors symmetries: an unifying tentative approach

Thibaut Weller & Giuseppe Geymonat
Laboratoire de Mécanique et Génie Civil, Université de Montpellier II, France

ABSTRACT: A classical topic is the determination of the form of a tensor expressing a physical property and invariant under the point group of a crystal. For example, one finds listings showing 16 distinct forms of the piezomagnetic tensor that can arise in a crystal. Alongside each of these 16 distinct forms is listed the point groups which give rise to that specific form (see Birss (1966) for example). This is a classification of the form of piezomagnetic tensors in crystals. We claim however that this classification is quite ambiguous in the sense that it hides the very difference between microscopic and macroscopic symmetries. The topic of the paper is a clarification of the difference between *material* and *behaviour* symmetries and a classification of piezomagnetic tensors in *behaviour equivalences classes*, found to be 15.

1 INTRODUCTION

All physical properties of crystals are related to their structure. This elementary observation is at the start of crystal physics, where symmetries play a fundamental role. However, as one moves away from the crystal to the deformable crystal in body, the crystallographic point of view is often forsaken especially as far as it concerns symmetry. For example, in elasticity, where symmetries play an eminent role in the mathematical formulation of its principles, only the macroscopic point of view is retained. Within this framework, one finds in Huo & Del Piero (1991) the first attempt to classify hyperelastic tensors according to their type of symmetry by building *equivalence classes*. This classification was then definitively established in Forte & Vianello (1996) and, with regard to photoelastic tensors, in Forte & Vianello (1997). See Forte & Vianello (1996) for a detailed history in elasticity.

The method initiated in these two references is particularly adapted to all types of mechanical behavior. In Geymonat & Weller (2002) it is extended to piezoelectricity. In the present paper the method is applied to piezomagnetism, where the analogies and the differences between the crystallographic and the macroscopic point of view will appear more clearly.

Piezomagnetic crystals are characterized by the appearance of a magnetic moment under the application of a stress. In addition to the usual symmetry operations which one meets in crystallography (namely: *orthogonal transformations*, *i.e.* elements of the *orthogonal group* denoted **O(3)**) we consider a generally denoted $\Re$ operation whose action is to change the sign of the magnetic moment. This is why it is often associated to time-reversal (*cf.* Landau & Lifchitz, 1969; Laughlin *et al.*, 2000; Sivardière, 1995). The symmetry operations which act on piezomagnetic crystals thus require the definition of the *extended orthogonal group* denoted $\mathbf{O}^{\mu}\mathbf{(3)}$ (*cf.* Sirotine & Chaskolskaïa (1984) as well as (9) formula).

The common background between microscopic and macroscopic approaches is the construction of *equivalence classes*. On the one hand, each crystalline structure is associated to a subgroup of the (eventually extended) orthogonal group: its *point group*. Of course, one can also build its *Bravais lattice* or its *space group*, which correspond to other kinds of equivalence classes. On the other hand, if we consider that tensors accurately translate the physical properties of matter, we are brought to build actions which, as for crystals, will enable us to associate a subgroup of **O(3)**, or eventually $\mathbf{O}^{\mu}\mathbf{(3)}$, to a given tensor. This subgroup is the tensor *symmetry class*. We will use the Cauchy-Born hypothesis to bind these two types of symmetry.

In order to fix the notations we recall in §2 some classical results of crystallography by using the mathematical tools developed in Pitteri & Zanzotto (1996, 1998, 2003). *Crystal lattices* are then rigorously classified in equivalence classes. Those can be *arithmetic* or *geometric*, then defining respectively the 14 *Bravais lattices* and the 7 *crystal systems*. Crystals are then ranged in 32 point groups and 230 space groups. These numbers increase to 90 point groups and 1421 space groups in the piezomagnetic crystals case, due to the additional time-reversal operation.

These results are very traditional. We nevertheless insist on the manner that is used to define and range these equivalence classes, because these methods are then transposed to piezomagnetic tensors by the use of the Cauchy-Born hypothesis, introduced in §3. Piezomagnetic tensors are *axial tensors*. We show in §3.4 the way we consider the action of the extended orthogonal group on such tensors. It is shown in particular that, via an isomorphism, the characterization of piezomagnetic tensors symmetries is equivalent to the characterization of piezoelectric tensors symmetries. The mathematical tools which make it possible to exhaust symmetry classes of such tensors are the *harmonic and Cartan decompositions*.

Harmonic decomposition maps a tensor of any order into a *n*-uplet of totally symmetric tensors (invariant by any permutations of its indexes) which are also traceless (the trace with respect to any pair of indexes is zero). These traceless totally symmetric tensors are then identified via an isomorphism to harmonic homogeneous polynomials (*i.e.* whose *Laplacian* is null) of corresponding degree. This is why a traceless totally symmetric tensor is also called harmonic (*cf.* Forte & Vianello, 1996, 1997; Spencer, 1970).

Cartan decomposition (*cf.* Golubitsky *et al.*, 1998) is an operation by which harmonic polynomials of a given degree are written as a linear combination of terms for which symmetry operations become elementary.

Each piezomagnetic tensor is then associated to a subgroup of $\mathbf{O}^{\mu}\mathbf{(3)}$, *i.e.* to a kind of behavior which is connected, in §4, to the magnetic point groups which give it birth. It is in particular shown that these point groups should not be classified in 16 but in 15 symmetry classes.

2 PRELIMINARIES: FORMULATION OF MICROSCOPIC EQUIVALENCE CLASSES

2.1 *Crystal lattices*

Mathematical modelings of crystal lattices and, in general, of crystallography, have been explored by a rich and various literature (see for instance Coxeter, 1989; Pitteri & Zanzotto, 1996, 1998, 2003; Weigel, 1972; Yale, 1968). We here take the way as it has been opened by Pitteri & Zanzotto (1996, 1998, 2003) within the framework of phase transition in solid mechanics.

We note **Lin** the space of second-order tensors, **I** the unit element in **Lin** and $\mathbf{L}^t$ the transpose of the element $\mathbf{L} \in$ **Lin**. We recall that $\mathbf{O(3)} = \{\mathbf{L} \in \mathbf{Lin} : \mathbf{L}\,\mathbf{L}^t = \mathbf{L}^t\,\mathbf{L} = \mathbf{I}\}$ is the space of *orthogonal transformations* and that the subgroup of all elements of **O(3)** with determinant equal to 1 (or *rotations* group) is denoted **SO(3)**.

Let $\mathbb{R}^3$ be the three-dimensional Euclidean vector space. Three linearly independent vectors $\mathbf{e}_a$ (a = 1, 2, 3) of $\mathbb{R}^3$ define a basis denoted $\{\mathbf{e}_a\} = (\mathbf{e}_1, \mathbf{e}_2, \mathbf{e}_3)$. The convention of summation over repeated indexes being understood, we define the (simple) *lattice* $\mathcal{R}(\{\mathbf{e}_a\})$ by:

$$\mathcal{R}(\{\mathbf{e}_a\}) := \{\mathbf{x} \in \mathbb{R}^3 : \mathbf{x} = M^a\mathbf{e}_a,\quad a = 1,2,3,\ M^a \in \mathbb{Z}\}. \tag{1}$$

We also say that the 3 linearly independent vectors $\mathbf{e}_a$ (a = 1, 2, 3) are the *basis vectors* of $\mathcal{R}(\{\mathbf{e}_a\})$. The parallelepiped $\mathcal{P}(\{\mathbf{e}_a\})$ spanned by $\mathbf{e}_a$ (a = 1, 2, 3) is called the *unit cell* of $\mathcal{R}(\{\mathbf{e}_a\})$.

It is well known that lattices can be ranged in *7 crystal systems* or *14 Bravais lattices*. These two types of classification correspond to two types of equivalence classes: 'geometrical' equivalence classes and "arithmetic" equivalence classes. These two kinds of equivalence classes arise on the one hand from geometrical actions (namely: elements $\mathbf{Q}$ of $\mathbf{O(3)}$) and on the other hand from arithmetic actions (elements m of $GL(3, \mathbb{Z})$, the space of 3×3 invertible unimodular matrices with integral entries). We thus define in the following the way these elements act on lattices and then build the equivalence classes which result from these actions.

An element $\mathbf{Q} \in \mathbf{O(3)}$ acts on $\{\mathbf{e}_a\}$ in the following way:

$$\mathbf{Q}\{\mathbf{e}_a\} := (\mathbf{Q}\,\mathbf{e}_1, \mathbf{Q}\,\mathbf{e}_2, \mathbf{Q}\,\mathbf{e}_3) = \{\mathbf{Q}\,\mathbf{e}_a\}. \tag{2}$$

As to arithmetic actions $m = (m_a^b) \in GL(3, \mathbb{Z})$, they are defined by:

$$m\{\mathbf{e}_a\} := (m_1^b\,\mathbf{e}_b, m_2^b\,\mathbf{e}_b, m_3^b\,\mathbf{e}_b) = \{m_a^b\,\mathbf{e}_b\}. \tag{3}$$

These actions are quite different: notice that elements $\mathbf{Q} \in \mathbf{O(3)}$ act on each basis vector independently, while it is not the case of elements $m \in GL(3, \mathbb{Z})$.

Of course, a lattice is not attached to a unique basis. A lattice basis is only determined up to arithmetic transformations. More precisely, let $\{\mathbf{e}_a\}$ and $\{\mathbf{f}_a\}$ be two basis of $\mathbb{R}^3$, we have:

$$\mathcal{R}(\{\mathbf{e}_a\}) = \mathcal{R}(\{\mathbf{f}_a\}) \iff \exists m \in GL(3, \mathbb{Z}), \{\mathbf{f}_a\} = m\{\mathbf{e}_a\}. \tag{4}$$

This invariance property is not the one which enables us to build interesting equivalence classes, but it is at the source of the various types of lattice symmetries.

We thus call *geometrical holohedry* $P(\{\mathbf{e}_a\})$ of $\mathcal{R}(\{\mathbf{e}_a\})$ the finite subgroup of all orthogonal transformations $\mathbf{Q} \in \mathbf{O(3)}$ leaving $\mathcal{R}(\{\mathbf{e}_a\})$ unchanged. It is written:

$$P(\{\mathbf{e}_a\}) := \{\mathbf{Q} \in \mathbf{O(3)} : \mathbf{Q}\{\mathbf{e}_a\} = m\{\mathbf{e}_a\}, \quad m \in GL(3, \mathbb{Z})\}. \tag{5}$$

As to the *lattice group* (or arithmetic holohedry), it is defined by:

$$L(\{\mathbf{e}_a\}) := \{m \in GL(3, \mathbb{Z}) : m\{\mathbf{e}_a\} = \mathbf{Q}\{\mathbf{e}_a\}, \quad \mathbf{Q} \in \mathbf{O(3)}\}. \tag{6}$$

Roughly speaking, the geometric holohedry represents the "macroscopic family", while the lattice group informs about the manner of laying out triply periodically points in space.

Taking into account that

$$P(m\{\mathbf{e}_a\}) = P(\{\mathbf{e}_a\}) \text{ if } m \in GL(3, \mathbb{Z}), \quad P(\mathbf{Q}\{\mathbf{e}_a\}) = \mathbf{Q}\, P(\{\mathbf{e}_a\})\, \mathbf{Q}^t \text{ if } \mathbf{Q} \in \mathbf{O(3)}, \tag{7}$$

we see that $P(\{\mathbf{e}_a\})$ only depends on the lattice $\mathcal{R}(\{\mathbf{e}_a\})$ itself, and not on its basis. Since $P(\{\mathbf{e}_a\})$ describes geometric properties of $\mathcal{R}(\{\mathbf{e}_a\})$ regarded as identical in any (geometrically) equivalent lattice, it is natural to put the initial $\mathbb{R}^3$ basis aside. Thus, rather than the holohedry $P(\{\mathbf{e}_a\})$, we consider its *equivalence class* in $\mathbf{O(3)}$. These equivalence classes are the *crystal systems*. Therefore, we say that two lattices are (geometrically) equivalent when their holohedries are *conjugate* in $\mathbf{O(3)}$, which means that they belong to the same crystal system. There are 7 such equivalence classes, widely known under the following denomination: *triclinic, monoclinic, orthorhombic, tetragonal, trigonal, hexagonal* and *cubic*.

We also have:

$$L(\mathbf{Q}\{\mathbf{e}_a\}) = L(\{\mathbf{e}_a\}) \text{ if } \mathbf{Q} \in \mathbf{O(3)}, \quad L(m\{\mathbf{e}_a\}) = m^{-1}\, L(\{\mathbf{e}_a\})\, m \text{ if } m \in GL(3, \mathbb{Z}). \tag{8}$$

This means that an orthogonal transformation does not modify the lattice group $L(\{\mathbf{e}_a\})$. Moreover, a change of lattice basis transforms $L(\{\mathbf{e}_a\})$ into a conjugate within $GL(3, \mathbb{Z})$. Thus, it will be said that two lattices $\mathcal{R}(\{\mathbf{e}_a\})$ and $\mathcal{R}(\{\mathbf{f}_a\})$ (or two lattice basis $\{\mathbf{e}_a\}$ and $\{\mathbf{f}_a\}$) are arithmetically equivalent if their lattice groups are conjugate within $GL(3, \mathbb{Z})$. They then belong to the same *Bravais class* and we write $\{\mathbf{e}_a\} \overset{\mathbb{Z}}{\sim} \{\mathbf{f}_a\}$.

The arithmetic criterion is more precise than the data of an equivalence class in **O(3)**. The reason is that conjugacy within $GL(3, \mathbb{Z})$ is more stringent than within **O(3)**. Thus, a geometric holohedry $P \subset \mathbf{O(3)}$ determines most of the time various lattice groups (the 14 Bravais lattices are obtained from the 7 crystal systems by introducing a *node point* at the center of the unit cell or at the center of the cell faces).

2.2 *Magnetic crystals as bicolor multilattices*

We here consider the three-dimensional real affine space $\mathbb{A}^3$. The subgroup of affine isometries of $\mathbb{A}^3$ is denoted $E(3)$. Once an origin 0 is chosen, any isometry $e \in E(3)$ can be represented by a couple $(\mathbf{t}, \mathbf{Q})$ where $\mathbf{t}$ is the vector from 0 to $e(0)$ and where $\mathbf{Q} \in \mathbf{O(3)}$.

Now, studying magnetic crystals, we need to introduce an additional operation whose action correspond to reversing the sign of the magnetic moment. This operation is generally denoted $\mathfrak{R}$ and can be regarded as reversing the *course* of time (not its sign) (*cf.* Landau & Lifchitz, 1969; Laughlin *et al.*, 2000; Sivardière, 1995; Sirotine & Chaskolskaïa, 1984). This internal degree of freedom is the reason why one often speaks of black and white symmetry (see Shubnikov, 1964). Each color then represents the sign of the magnetic moment.

We therefore consider in the following the extended orthogonal group (*cf.* Sirotine & Chaskolskaïa, 1984) of all orthogonal and anti-orthogonal (*i.e.* associated to time-reversal) operations:

$$\mathbf{O}^{\mu}\mathbf{(3)} := \mathbf{O(3)} \times \{\pm 1\}. \tag{9}$$

An element of $\mathbf{O}^{\mu}\mathbf{(3)}$ is thus a couple $(\mathbf{Q}, \delta)$ where $\mathbf{Q}$ is an orthogonal transformation associated (if $\delta = -1$) or not (if $\delta = 1$) to time-reversal. Let $(\mathbf{Q}, \delta)$ and $(\mathbf{R}, \theta)$ be two elements of $\mathbf{O}^{\mu}\mathbf{(3)}$. The composition law in $\mathbf{O}^{\mu}\mathbf{(3)}$ is given by

$$(\mathbf{Q}, \delta)(\mathbf{R}, \theta) = (\mathbf{QR}, \delta\theta). \tag{10}$$

Moreover, for later use (*cf.* relation (50) for instance), we denote ${}^{\mu}\mathbb{I}\mathrm{d}$ the subgroup $\{(\mathbf{I}, +1), (-\mathbf{I}, +1)\}$. We also denote $E^{\mu}(3)$ the group of all affine isometries and anti-isometries of $\mathbb{A}^3$, *i.e.* the set of all element of the form $(\mathbf{t}, (\mathbf{Q}, \delta)) \in \mathbb{R}^3 \times \mathbf{O}^{\mu}\mathbf{(3)}$ equipped with the composition law

$$(\mathbf{t}, (\mathbf{Q}, \delta))(\mathbf{s}, (\mathbf{R}, \theta)) = (\mathbf{t} + \mathbf{s}, (\mathbf{QR}, \delta\theta)). \tag{11}$$

Following Pitteri & Zanzotto (2003), it is a simple matter of notation to extend to the magnetic case a translate of a simple lattice:

$$\mathcal{L}((P, \theta), \mathbf{e}_a) := (P, \theta) + \mathcal{R}(\{\mathbf{e}_a\}), \tag{12}$$

where $\theta = \pm 1$ stands for the sign of the magnetic moment attached to the particle P in $\mathbb{A}^3$. It is then clear that $\mathfrak{R}\,(P, \theta) = (P, -\theta)$. This simply means: when reversing the course of time, the sign of the magnetic moment attached to the particle $P \in \mathbb{A}^3$ changes. Moreover, the action of an element $e = (\mathbf{t}, (\mathbf{Q}, \delta)) \in E^{\mu}(3)$ on $\mathcal{L}((P, \theta), \mathbf{e}_a)$ is defined by:

$$e(\mathcal{L}((P, \theta), \mathbf{e}_a)) := \mathcal{L}((P + \mathbf{t}, \delta\theta), \mathbf{Q}\mathbf{e}_a). \tag{13}$$

We can also define a magnetic crystal $\mathcal{M}^\mu$ as the subset of$\mathbb{A}^3 \times \{\pm 1\}$ such that

$$\mathcal{M}^\mu := \cup_{i=0}^{n-1} \mathcal{L}((P_i, \theta_i), \mathbf{e}_a). \tag{14}$$

A magnetic crystal is thus a *n-lattice* and we have

$$e(\mathcal{M}^\mu) = \cup_{i=0}^{n-1} \mathcal{L}((P_i + \mathbf{t}, \delta\theta_i), \mathbf{Q}\mathbf{e}_a)), \tag{15}$$

for all $e = (\mathrm{t}, (\mathbf{Q}, \delta)) \in E^\mu(3)$.

2.3 *Space and point groups for magnetic crystals*

To get informations about the crystal structure, we introduce two types of symmetries for magnetic crystals which are the natural generalization of point and space groups symmetries.

We call *magnetic space group* of a multilattice $\mathcal{M}^\mu((P_i, \theta_i), \mathbf{e}_a)$ the subgroup of $E^\mu(3)$ denoted by $\mathcal{S}(\mathcal{M}^\mu)$ and defined as follows:

$$\mathcal{S}(\mathcal{M}^\mu) := \{e \in E^\mu(3) : e(\mathcal{M}^\mu) = \mathcal{M}^\mu\}. \tag{16}$$

We then say that $e \in \mathcal{S}(\mathcal{M}^\mu)$ is an affine symmetry of the multilattice $\mathcal{M}^\mu$. As in $(7)_2$ and $(8)_2$ we have

$$\mathcal{S}(e\,\mathcal{M}^\mu) = e\,\mathcal{S}(\mathcal{M}^\mu)\,e^{-1}. \tag{17}$$

Therefore, an affine symmetry of $\mathcal{M}^\mu$ changes $\mathcal{S}(\mathcal{M}^\mu)$ into a conjugate in E^μ (3).

We now consider the group $\mathcal{P}(\mathcal{M}^\mu)$ collecting all (anti)orthogonal transformations $(\mathbf{Q}, \delta) \in \mathbf{O}^\mu(3)$ such that $(\mathbf{t}, (\mathbf{Q}, \delta))$ belongs to $\mathcal{S}(\mathcal{M}^\mu)$ for a given $\mathbf{t} \in \mathbb{R}^3$. We call $\mathcal{P}$ $(\mathcal{M}^\mu)$ the *magnetic point group* of $\mathcal{M}^\mu$.

The following result is known from a long time (see *e.g.* Bhagavantam (1966), Birss (1966) or Sivardière (1995)):

Theorem 1 *There are 1421 conjugacy classes of magnetic space groups in the set of all the orientation-preserving (anti)affine transformations of* E^μ (3).

There are 90 conjugacy classes of magnetic point groups in $\mathbf{O}^\mu(3)$.

3 A MACROSCOPIC POINT OF VIEW: TENSORS SYMMETRY CLASSES

3.1 *A bridge: the Cauchy-Born hypothesis*

The Cauchy-Born hypothesis is, roughly speaking, the statement that lattices vectors behave as *material vectors*. It simply means that when a crystal in the reference configuration with lattice $\mathcal{R}(\{\mathbf{e}_a^0\})$ experiences a homogeneous deformation whose gradient is $\mathbf{F}$, we have:

$$\mathbf{e}_a = \mathbf{F}\mathbf{e}_a^0, \tag{18}$$

where $\mathbf{e}_a$ are the lattice vectors of the crystal in the deformed configuration.

This quite simple hypothesis allows us to transfer at the macroscopic level the symmetry properties studied up to now at the crystal level.

To this aim we first take for granted that the structure of crystals can be described by means of a simple lattice in all their allowed configurations. Second, we assume that the free energy w per unit cell of a deformable crystalline body whose current configuration is a simple lattice $\mathcal{R}(\{\mathbf{e}_a\})$ can be written as (in the following, we neglect the temperature θ):

$$w = \overline{w}(\mathbf{e}_a). \tag{19}$$

We just have to write that

$$w = \overline{w}(\mathbf{e}_a) = \overline{w}(\mathbf{F}\mathbf{e}_a^0) = \widetilde{w}(\mathbf{F}) = \hat{w}(\mathbf{C})^1, \tag{20}$$

to obtain, by classical arguments

$$\hat{w}(\mathbf{C}) = \hat{w}(m^t\,\mathbf{C}\,m), \tag{21}$$

for all symmetric and positive definite tensor **C** and for all m in $GL(3, \mathbb{Z})$.

3.2 *Application to piezomagnetism*

For the sake of simplicity, we won't speak about the Pitteri-Ericksen neighborhoods in the following and we will just focus on the main step of the derivation of macroscopic equivalence classes.

In the piezomagnetic case, the energy is a function of the strain tensor **E** (whose Cauchy-Green tensor is **C**) and of the magnetization polar[2] vector **b**:

$$w = \tilde{w}(\mathbf{E}, \mathbf{b}) = \hat{w}(\mathbf{C}, \mathbf{b}). \tag{22}$$

Restricting the invariance relation in (21) to the orthogonal transformations, we have:

$$\hat{w}(\mathbf{C}, \mathbf{b}) = \hat{w}(\mathbf{Q}^t\mathbf{C}\mathbf{Q},\ \det(\mathbf{Q})\,\mathbf{Q}\mathbf{b}),\quad \forall\mathbf{Q} \in P(\{\mathbf{e}_a^0\}), \tag{23}$$

because **b** is an axial vector (see relation (30)). Moreover, as the sign of the magnetization vector **b** directly depends on the course of time, we have $\Re\mathbf{b} = -\mathbf{b}$ and therefore:

$$\hat{w}(\mathbf{C}, \mathbf{b}) = \hat{w}(\mathbf{C}, -\mathbf{I}\Re\,\mathbf{b}). \tag{24}$$

Now, the piezomagnetic tensor **p** is given by:

$$\mathbf{p} = \frac{\partial^2 \hat{w}}{\partial\mathbf{E}\partial\mathbf{b}}. \tag{25}$$

So, in the linear case, since $\mathbf{E} = ½(\mathbf{C} - \mathbf{I})$, relations (23) and (24) lead to:

$$\begin{aligned} \mathbf{Q}\,\mathbf{p}[\mathbf{b}]\,\mathbf{Q}^t &= \det(\mathbf{Q})\,\mathbf{p}[\mathbf{Q}\,\mathbf{b}], \\ \mathbf{Q}^t\,\mathbf{p}[\mathbf{E}] &= \mathbf{p}[\mathbf{Q}^t\,\mathbf{E}\,\mathbf{Q}], \end{aligned} \tag{26}$$

[1] Here, **C** stands for the symmetric, positive definite Cauchy-Green tensor $\mathbf{F}^t\mathbf{F}$ classically obtained in the polar decomposition of **F**.
[2] See §3.3 for further details on polar variables.

for all $\mathbf{Q}$ in $P(\{\mathbf{e}_a^0\})$, all $\mathbf{b}$ in $\mathbb{R}^3$ and all second-order symmetric tensor $\mathbf{E}$. Furthermore,

$$\mathfrak{R}\,\mathbf{p} = \text{-}\mathbf{I}\,\mathbf{p} = -\mathbf{p}, \tag{27}$$

which means that the piezomagnetic tensor is odd according to time-reversal (this is meaningless in $(26)_2$ because in this relation, the piezomagnetic tensor acts on $\mathbf{E}$ which do not depend on the course of time).

It is important here to compare $(26)_1$ and (27) to the action that will arise in (35). It is clearly show how, in this way, some of the symmetry elements present in a crystal can be brought across the microscopic level to the macroscopic one, where another equivalence classes can be built. In fact, $(26)_1$ says that each element of the geometrical holohedry of a studied crystal is an element of the symmetry group of the piezomagnetic tensor of this crystal. This has to be compared to the *Neumann's Principle*, which states that *"the symmetry group of any property of a crystal must include the point group of this last"*. We emphasize on the fact that all hypothesis that have been made since we introduced the Cauchy-Born hypothesis concern the crystal viewed as a *lattice*. The point groups therefore cannot appear in this approach.The important fact, anyway, is how the $\star$ action that will be introduced in the next section, and which is of fundamental use in the tensors symmetries study, naturally arises from the crystal structure via the Cauchy-Born hypothesis (see also Pitteri & Zanzotto §6.6.2).

3.3 *Polar and axial variables*

The distinction between *polar* and *axial* variables is related to the need of defining, when necessary, an orientation in the three dimensional Euclidean space. We shall speak of polar variable when it is defined without ambiguity by the nature. In particular, its sign does not depend on any anthropomorphic convention, nor on any reference to a basis. On the other hand, a variable is known as axial if its sign cannot be fixed without a human reference to space orientation.

In order to formalize this, we consider a tensor of any order noted $\mathbf{T}$. Its components in a given basis $R = (O;\, \mathbf{u}, \mathbf{v}, \mathbf{w})$ of $\mathbb{R}^3$ are denoted $\mathbf{T}_{\dots ijk\dots}$. Let $\mathbf{Q} \in \mathbf{O(3)}$ be an orthogonal transformation. In the new basis $R' = (O;\, \mathbf{Qu}, \mathbf{Qv}, \mathbf{Qw}) = \mathbf{Q}\,R$, the components of $\mathbf{T}$ are mapped into

$$(\mathbf{Q} \star \mathbf{T})_{\dots ijk\dots} := \cdots Q_{ip} Q_{jq} Q_{kr} \cdots T_{\dots pqr \dots}. \tag{28}$$

Let us now consider the alternating tensor ϵ_{ijk}. It is easy to see that for $\mathbf{Q} \in \mathbf{O(3)}$ we have

$$(\mathbf{Q} \star \epsilon)_{ijk} = Q_{ip} Q_{jq} Q_{kr}\, \epsilon_{pqr} = \det(\mathbf{Q})\, \epsilon_{ijk}. \tag{29}$$

The fact that the determinant of the orthogonal transformation $\mathbf{Q}$ appears in this relation means that the alternating tensor *components* are sensitive to space orientation. We shall say that they are *pseudo-scalars* or *axial scalars* and note them ${}^{\epsilon}$*scalars*. More precisely: as these scalars represent the alternating tensor, they are ${}^{\epsilon}$scalars. On the other hand, variables which are not sensitive to space orientation, in the sense that they transform as $\mathbf{T}$ in (28), are called polar variables.

Let us take another example: let $\mathbf{V}$ be a skew-symmetric second-order tensor and $\mathbf{v} \in \mathbb{R}^3$ its associated vector, *i.e.* the vector such that $V_{ij} = \epsilon_{ijk} v_k$, or, equivalently, $v_k = \frac{1}{2} \epsilon_{ijk} V_{ij}$. We write $\mathbf{v} = f(\mathbf{V})$. Consequently:

$$f(\mathbf{Q} \star \mathbf{V}) = \det(\mathbf{Q})\, f(\mathbf{V}). \tag{30}$$

This relation means that f, which maps $\mathbf{V}$ into $\mathbf{v}$, is sensitive to space orientation. We in fact say that $\mathbf{v}$ associated in this way to $\mathbf{V}$ is axial, and we note it ${}^{\epsilon}\mathbf{v}$. However, hence $\mathbf{v}$ defined above can also be viewed as a polar vector disconnected of any skew-symmetric second-order tensor, we will say that $\mathbf{v}$ and ${}^{\epsilon}\mathbf{v}$ are *twin vectors*. We shall write that

$$\mathbf{v} = {}^{\epsilon}\mathrm{Id}({}^{\epsilon}\mathbf{v}), \tag{31}$$

to express that their *nature* is different but their components are the same.

To generalize all these concepts, we denote ${}^\epsilon\mathbf{T}$ the tensors which do not transform as $\mathbf{T}$ in (28) but as the $\boldsymbol{\epsilon}$ components in (29), translating that they are pseudotensors, or axial tensors.

Finally, for two tensors $\mathbf{T}$ (polar) and ${}^\epsilon\mathbf{T}$ (axial) of same order and with the same components, we shall write $\mathbf{T} = {}^\epsilon\mathrm{Id}({}^\epsilon\mathbf{T})$. Such tensors are called *twin tensors* in the following.

3.4 *The piezomagnetic tensors symmetries*

From now on $\mathbb{R}^3$ is associated to a fixed orthonormal basis denoted $(\mathbf{i}, \mathbf{j}, \mathbf{k})$.

Let ${}^\epsilon\mathbf{Magn}$ be the space of third-order pseudotensors which are symmetric according to their first two indexes and odd according to time-reversal:

$$
{}^\epsilon\mathbf{Magn} := \{{}^\epsilon\mathbf{M} : {}^\epsilon M_{ijk} = {}^\epsilon M_{jik} \,, \mathfrak{R}\, {}^\epsilon\mathbf{M} = \text{-}{}^\epsilon\mathbf{M}\}. \tag{32}
$$

We also consider the set denoted **Piez** of all third-order polar tensors, symmetric according to their first two indexes:

$$
\mathbf{Piez} := \{\mathbf{P} : P_{ijk} = P_{jik}\}. \tag{33}
$$

A tensor $\mathbf{P} \in \mathbf{Piez}$ is called *piezoelectric* tensor.

In a piezomagnetic material, the magnetic polarization pseudo-vector ${}^\epsilon b$ is connected to the stress tensor σ via the relation

$$
{}^\epsilon b_k = {}^\epsilon M_{ijk}\, \sigma_{ij}, \tag{34}
$$

where ${}^\epsilon\mathbf{M} \in {}^\epsilon\mathbf{Magn}$.

We define the action of $\mathbf{O}^\mu\mathbf{(3)}$ on ${}^\epsilon\mathbf{Magn}$ by:

$$
(\mathbf{Q}, \delta) \circledast {}^\epsilon\mathbf{M} := \delta \det(\mathbf{Q})\, \mathbf{Q} \star {}^\epsilon\mathbf{M}, \quad \text{for } (\mathbf{Q}, \delta) \in \mathbf{O}^\mu\mathbf{(3)}, \tag{35}
$$

where $\star$ is defined in (28).

The *symmetry group* $\mathcal{G}({}^\epsilon\mathbf{M})$ of ${}^\epsilon\mathbf{M} \in {}^\epsilon\mathbf{Magn}$ is:

$$
\mathcal{G}({}^\epsilon\mathbf{M}) := \{(\mathbf{Q}, \delta) \in \mathbf{O}^\mu\mathbf{(3)} : (\mathbf{Q}, \delta) \circledast {}^\epsilon\mathbf{M} = {}^\epsilon\mathbf{M}\}, \tag{36}
$$

For all ${}^\epsilon\mathbf{M} \in {}^\epsilon\mathbf{Magn}$ we observe that the symmetry group $\mathcal{G}({}^\epsilon\mathbf{M})$ contains the element $(-\mathbf{I}, +1)$ of $\mathbf{O}^\mu\mathbf{(3)}$. Therefore, recalling (*cf.* §2.2.) that ${}^\mu\mathbb{I}\mathrm{d}$ is the subgroup $\{(\mathbf{I}, +1), (-\mathbf{I}, +1)\}$ and that $\oplus$ is the direct sum of subgroups in $\mathbf{O}^\mu\mathbf{(3)}$, we necessarily have

$$
\mathcal{G}({}^\epsilon\mathbf{M}) = \widetilde{g}({}^\epsilon\mathbf{M}) \oplus {}^\mu\mathbb{I}\mathrm{d}, \tag{37}
$$

where $\widetilde{g}\,({}^\epsilon\mathbf{M})$ is defined by:

$$
\widetilde{g}({}^\epsilon\mathbf{M}) := \{(\mathbf{Q}, \delta) \in \mathbf{SO}^\mu\mathbf{(3)} : (\mathbf{Q}, \delta) \circledast {}^\epsilon\mathbf{M} = \delta\, \mathbf{Q} \star {}^\epsilon\mathbf{M} = {}^\epsilon\mathbf{M}\}, \tag{38}
$$

and where $\mathbf{SO}^\mu\mathbf{(3)} : = \mathbf{SO(3)} \times \{\pm 1\}$.

We now define the symmetry group $g(\mathbf{P})$ of an element $\mathbf{P} \in \mathbf{Piez}$ by:

$$
g(\mathbf{P}) := \{\mathbf{Q} \in \mathbf{O(3)} : \mathbf{Q} \star \mathbf{P} = \mathbf{P}\}. \tag{39}
$$

Proposition 1 *1. Every ${}^{\epsilon}\boldsymbol{M} \in {}^{\epsilon}Magn$ has a twin tensor $\boldsymbol{P} \in$ Piez and vice-versa.*
2. If ${}^{\epsilon}\boldsymbol{M} \in {}^{\epsilon}Magn$ and $\boldsymbol{P} \in$ Piez are twins, then $\tilde{g}({}^{\epsilon}\boldsymbol{M})$ and $g(\boldsymbol{P})$ are isomorphic.

Proof – Part 1 is obvious: each piezomagnetic tensor can be considered as a piezoelectric one if we neglect the magnetic moment, *i.e.* if we don't take time-reversal into account.

To prove part 2, we introduce the isomorphism Iso defined by:

$$\mathbf{SO}^{\mu}\mathbf{(3)} \ni (\mathbf{Q}, \delta) \quad \overset{\mathsf{Iso}}{\longmapsto} \quad \delta\,\mathbf{Q} \in \mathbf{O(3)}, \tag{40}$$

and:

$$\mathbf{O(3)} \ni \mathbf{Q} \quad \overset{\mathsf{Iso}^{-1}}{\longmapsto} \quad (\det(\mathbf{Q})\,\mathbf{Q}, \det(\mathbf{Q})) \in \mathbf{SO}^{\mu}\mathbf{(3)}. \tag{41}$$

We then have:

$$\begin{aligned} \mathsf{Iso}(\tilde{g}({}^{\epsilon}\mathbf{M})) &= \{\mathbf{R} \in \mathbf{O(3)}, \mathbf{R} = \delta\,\mathbf{Q} : \ \delta\,\mathbf{Q} \star {}^{\epsilon}\mathbf{M} = {}^{\epsilon}\mathbf{M}\} \\ &= \{\mathbf{R} \in \mathbf{O(3)} : \mathbf{R} \star \mathbf{P} = \mathbf{P}\}, \end{aligned} \tag{42}$$

where ${}^{\epsilon}\mathbf{M}$ and $\mathbf{P}$ are twins. Then $\mathrm{Iso}(\tilde{g}({}^{\epsilon}\mathbf{M})) \subset g(\mathbf{P})$. The reverse inclusion follows from (41) and the two symmetry groups are thus isomorphic. ■

This Proposition means that the characterization of piezomagnetic tensors symmetries is strictly equivalent to the characterization of piezoelectric tensors symmetries.

In this direction, we first of all observe that, by continuity, the group $g(\mathbf{P})$ defined in (39) is a closed subgroup of $\mathbf{O(3)}$. Following Forte & Vianello (1996, 1997), we observe that $g(\mathbf{P})$ is *conjugate* through $\mathbf{Q} \in \mathbf{O(3)}$ to $g(\mathbf{Q} \star \mathbf{P})$:

$$g(\mathbf{Q} \star \mathbf{P}) = \mathbf{Q}\,g(\mathbf{P})\,\mathbf{Q}^t, \tag{43}$$

and say that two tensors $\mathbf{P}_1$ and $\mathbf{P}_2$ in **Piez** are *equivalent* if there is an element $\mathbf{Q}$ of $\mathbf{O(3)}$ such that $g(\mathbf{P}_1) = g(\mathbf{Q} \star \mathbf{P}_2)$. In other words, $\mathbf{P}_1$ and $\mathbf{P}_2$ are equivalent when their symmetry groups are conjugate:

$$\mathbf{P}_1 \overset{\star}{\sim} \mathbf{P}_2 \iff \{\exists \mathbf{Q} \in \mathbf{O(3)} : g(\mathbf{P}_1) = \mathbf{Q}\,g(\mathbf{P}_2)\,\mathbf{Q}^t\}. \tag{44}$$

Since $g(\mathbf{P}) = g(-\mathbf{P})$, it is easily checked that:

$$\mathbf{P}_1 \overset{\star}{\sim} \mathbf{P}_2 \iff \{\exists \mathbf{Q} \in \mathbf{SO(3)} : g(\mathbf{P}_1) = \mathbf{Q}\,g(\mathbf{P}_2)\,\mathbf{Q}^t\}. \tag{45}$$

This relation translates the physical intuition that whenever two material bodies can be rigidly rotated so that their symmetry groups become identical, they share the same "symmetry".

Therefore, the equivalence classes of **Piez** which result from $\overset{\star}{\sim}$ are called *symmetry classes* and we note $\{\mathbf{P}\}$ the symmetry class of $\mathbf{P} \in$ **Piez**. In the same direction, for $\mathbb{G}$ a subgroup of $\mathbf{O(3)}$, we note $\{\mathbb{G}\}$ the collection of all its conjugate among the subgroups of $\mathbf{O(3)}$. The question we ask now is: how many symmetry classes are there in **Piez** and which are there?

In order to give an accurate answer, we first of all point out some traditional results concerning the closed subgroups of $\mathbf{O(3)}$. It is well known (see for instance Golubitsky *et al.*, 1988) that they fall in three categories (*cf.* Appendix **A**):

1. The closed subgroups of $\mathbf{SO(3)}$ or *class I* closed subgroups: $\mathbb{I}d$, $\mathbb{Z}_m$ $(m \geqslant 2)$, $\mathbb{D}_m$ $(m \geqslant 2)$, $\mathcal{I}$, $\mathcal{O}$, $\mathcal{T}$, $\mathbf{SO(2)}$, $\mathbf{O(2)}$ and $\mathbf{SO(3)}$.

2. The *class II* closed subgroups: $\mathbb{G} \oplus \{\pm\mathbf{I}\}$, where $\mathbb{G}$ is a class I closed subgroup.
3. The *class III* closed subgroups. They are not contained in **SO(3)** and do not contain $-\mathbf{I}$: $\mathbb{Z}_{2m}^{-}$ $(m \geqslant 1)$, $\mathbb{D}_{2m}^{d}$ $(m \geqslant 2)$, $\mathbb{D}_{m}^{z}$ $(m \geqslant 2)$, $\mathcal{O}^{-}$ and $\mathbf{O(2)}^{-}$.

Since the symmetry groups of the elements $\mathbf{P} \in \mathbf{Piez}$ are closed subgroups of **O(3)**, each $\mathbf{P} \in \mathbf{Piez}$ is therefore associated to a unique subgroup $\mathbb{G}$ defined above such that $\mathbb{G}$ is conjugate to $g(\mathbf{P})$.We write $\mathbf{Piez}(\mathbb{G})$ for the collection of all tensors whose symmetry group is conjugate to $\mathbb{G}$:

$$\mathbf{Piez}(\mathbb{G}) := \{\mathbf{P} \in \mathbf{Piez} : g(\mathbf{P}) \in \{\mathbb{G}\}\}. \tag{46}$$

Obviously, we can apply the above results to tensors ${}^{\epsilon}\mathbf{M} \in {}^{\epsilon}\mathbf{Magn}$ and to subgroups ${}^{\mu}\mathbb{G} \subset \mathbf{O}^{\mu}\mathbf{(3)}$. We then have:

$${}^{\epsilon}\mathbf{Magn}({}^{\mu}\mathbb{G}) := \{{}^{\epsilon}\mathbf{M} \in {}^{\epsilon}\mathbf{Magn} : \mathcal{G}({}^{\epsilon}\mathbf{M}) \in \{{}^{\mu}\mathbb{G}\}\}. \tag{47}$$

In other words, the symmetry classes of **Piez** and ${}^{\epsilon}$**Magn** are related by the relations:

$$\forall\, \mathbb{G} \subset \mathbf{O(3)}, \quad \mathbf{Piez}(\mathbb{G}) = {}^{\epsilon}\mathrm{Id}({}^{\epsilon}\mathbf{Magn}(\mathsf{Iso}^{-1}(\mathbb{G}) \oplus {}^{\mu}\mathbb{I}\mathrm{d})), \tag{48}$$

and:

$$\forall\, {}^{\mu}\mathbb{G} \subset \mathbf{O}^{\mu}\mathbf{(3)}, \quad {}^{\epsilon}\mathbf{Magn}({}^{\mu}\mathbb{G}) = {}^{\epsilon}\mathrm{Id}(\mathbf{Piez}(\mathsf{Iso}({}^{\mu}\mathbb{G} \cap \mathbf{SO}^{\mu}\mathbf{(3)}))). \tag{49}$$

In the following, for any subgroup $\mathbb{G}$ of **O(3)**, we will note ${}^{\mu}\mathbb{G}$ the subgroup of $\mathbf{O}^{\mu}\mathbf{(3)}$ defined by

$${}^{\mu}\mathbb{G} := \mathsf{Iso}^{-1}(\mathbb{G}) \oplus {}^{\mu}\mathbb{I}\mathrm{d}. \tag{50}$$

3.4.1 *Harmonic decomposition*

Any second-order tensor $\mathbf{E}$ can be seen as the sum of a symmetric tensor $\mathbf{S} = ½(\mathbf{E} + \mathbf{E}^{t})$ and of an antisymmetric tensor $\mathbf{W} = ½(\mathbf{E} - \mathbf{E}^{t})$ associated to an axial vector ${}^{\epsilon}\mathbf{w}$ such that $W_{ij} = \epsilon_{ijk}\,{}^{\epsilon}wk$, where ϵ_{ijk} are the components of the alternating tensor. We note tr for the *trace* operator. The relation $\mathbf{S} = \mathbf{S}^{D} + 1/3(\mathrm{tr}\,\mathbf{S})\mathbf{I} = \mathbf{S}^{D} + 1/3(\mathrm{tr}\,\mathbf{E})\mathbf{I}$, where $\mathbf{S}^{D}$ is the deviatoric part of $\mathbf{S}$, points out an isomorphism ϕ between **Lin** and $\mathbf{Dev} \times \mathbb{R}^3 \times \mathbb{R}$ where **Dev** is the space of second-order traceless and symmetric tensors. We therefore write:

$$\mathbf{Lin} \ni \mathbf{E} \approx (\mathbf{S}^{D}, {}^{\epsilon}\mathbf{w}, \frac{1}{3}(\mathrm{tr}\,\mathbf{E})) \in \mathbf{Dev} \times \mathbb{R}^3 \times \mathbb{R}. \tag{51}$$

Moreover, according to (28) and (29) we have:

$$\mathbf{Q} \star \mathbf{E} \approx (\mathbf{Q} \star \mathbf{S}^{D}, \det(\mathbf{Q})\,(\mathbf{Q} \star {}^{\epsilon}\mathbf{w}), \frac{1}{3}(\mathrm{tr}\,\mathbf{E})). \tag{52}$$

Thus the symmetry class of $\mathbf{E}$ verifies:

$$g(\mathbf{E}) = g(\mathbf{S}^{D}) \cap {}^{\epsilon}g({}^{\epsilon}\mathbf{w}), \tag{53}$$

where $g(\mathbf{E})$ is defined as in (42) and where

$$^{\epsilon}g(^{\epsilon}\mathbf{w}) = \{\mathbf{Q} \in \mathbf{O(3)} \,:\, \mathbf{Q} \star {}^{\epsilon}\mathbf{w} = \det(\mathbf{Q})\,{}^{\epsilon}\mathbf{w}\}. \tag{54}$$

The generalization of the correspondence ϕ as noted in (51) to tensors of any order (*cf.* Spencer (1970) and Appendix **B**) is called *harmonic decomposition*. A tensor is harmonic when it is *totally symmetric* (its components are unchanged under any permutations of indexes) and *traceless* (the trace with respect to any pair of indexes is null). Applied to **Piez**, this decomposition makes it possible to write:

$$\mathbf{Piez} \ni \mathbf{P} \approx (\mathbf{H}, {}^{\epsilon}\mathbf{C}, \mathbf{w}, \mathbf{v}) \in \mathbf{Hrm} \times {}^{\epsilon}\mathbf{Dev} \times \mathbb{R}^3 \times \mathbb{R}^3, \tag{55}$$

where **Hrm** is the space of third-order harmonic tensors and $^{\epsilon}\mathbf{Dev}$ the space of harmonic and axial second-order tensors. We then have:

$$\forall \mathbf{Q} \in \mathbf{O(3)}, \quad \mathbf{Q} \star \mathbf{P} \approx (\mathbf{Q} \star \mathbf{H}, \det(\mathbf{Q})\,(\mathbf{Q} \star {}^{\epsilon}\mathbf{C}), \mathbf{Q} \star \mathbf{w}, \mathbf{Q} \star \mathbf{v}), \tag{56}$$

and thus:

$$g(\mathbf{P}) = g(\mathbf{H}) \cap {}^{\epsilon}g({}^{\epsilon}\mathbf{C}) \cap g(\mathbf{w}) \cap g(\mathbf{v}), \quad \forall\, \mathbf{P} \in \mathbf{Piez}, \tag{57}$$

where $^{\epsilon}g(^{\epsilon}\mathbf{C})$ follows from (54).

3.4.2 *Cartan decomposition*

Let $\mathbf{r} = x\mathbf{i} + y\mathbf{j} + z\mathbf{k}$ be a vector of $\mathbb{R}^3$ and let $\mathcal{P}_n$ be the space of homogeneous polynomials of degree n in the three variables x, y and z. There is a classical isomorphism ψ between $\mathcal{P}_n$ and $\mathbf{Sym}^n$, the space of totally symmetric tensors of order n (*cf.* Forte & Vianello, 1997):

$$\mathbf{Sym}^n \ni \mathbf{T} \mapsto \psi(\mathbf{T}) := \mathbf{T}[\mathbf{r}, \mathbf{r}, \cdots, \mathbf{r}] = T_{i_1 i_2 \cdots i_n} r_{i_1} r_{i_2} \cdots r_{i_n} \in \mathcal{P}_n, \tag{58}$$

where the convention of summation over repeated indexes is understood. Thus, the second-order symmetric tensor $\mathbf{S} = (S_{ij})$ defined in the preceding paragraph is mapped into

$$\psi(\mathbf{S}) = S_{11}\,x^2 + 2\,S_{12}\,xy + 2\,S_{13}\,xz + S_{22}\,y^2 + 2\,S_{23}\,yz + S_{33}\,z^2. \tag{59}$$

A polynomial $h \in \mathcal{P}_n$ is *harmonic* when $\Delta h = 0$, where Δ is the *Laplacian* operator. We write $h \in \mathcal{H}_n$. Here, we have $\Delta\psi(\mathbf{S}) = 2S_{11} + 2S_{22} + 2S_{33}$ which implies that

$$\Delta\,\psi(\mathbf{S}) = 0 \iff \mathrm{tr}\,(\mathbf{S}) = 0. \tag{60}$$

More generally, it is easy to check that the space of harmonic tensors of order n is isomorphic via ψ to $\mathcal{H}_n$.

The isomorphism (58) enables us to extend the action $\star$ defined in (28) to $\mathcal{P}_n$: for $p = \psi(\mathbf{T}) \in \mathcal{P}_n$ and $\mathbf{Q} \in \mathbf{O(3)}$, we define

$$\begin{aligned}(\mathbf{Q} \star p)(x, y, z) &:= \mathbf{T}[\mathbf{Q}^t\mathbf{r}, \mathbf{Q}^t\mathbf{r}, \cdots, \mathbf{Q}^t\mathbf{r}] \\ &= T_{i_1 i_2 \cdots i_n} r_{j_1} r_{j_2} \cdots r_{j_n} Q_{j_1 i_1} Q_{j_2 i_2} \cdots Q_{j_n i_n}.\end{aligned} \tag{61}$$

Moreover, the linear mapping ψ is **O(3)**-invariant in the following sense

$$\forall \mathbf{Q} \in \mathbf{O(3)},\ \mathbf{Q} \star \psi(\mathbf{T}) = \psi(\mathbf{Q} \star \mathbf{T}). \tag{62}$$

An essential property of $\mathcal{P}_n$ is that it is the direct sum of $\mathcal{H}_n$ and $\mathcal{Q}_n$, the space of polynomials which are multiples (by polynomials of $\mathcal{P}_{n-2}$) of $\rho := x^2 + y^2 + z^2$ (it is a traditional result, *cf.* Golubitsky *et al.*, 1988 p.109). Therefore, for each $p \in \mathcal{P}_n$, there exists a unique $q \in \mathcal{P}_{n-2}$ such that $h := (p - \rho q) \in \mathcal{H}_n$. We call h the *harmonic part* of p. The link with the harmonic decomposition is obvious: for $n = 2$ and $p = \psi(\mathbf{S})$ we have $h = \psi(\mathbf{S}^D)$ and $q = \psi(1/3\,\mathrm{tr}\,(\mathbf{S})\mathbf{I})$.

The goal is now to map harmonic polynomials (and consequently harmonic tensors) into spaces where the action $\star$ will become elementary. This purpose is achieved by the *Cartan decomposition*, which is an **SO(2)**-invariant decomposition of $\mathcal{H}_n$. To give a preliminary idea of it, let us consider a second-order harmonic polynomial f in the three variables x, y and z. We then have $f \in \mathcal{H}_2$. It is clear that f can be expressed on the following "basis":

$$\begin{aligned} u &:= z^2 - \frac{1}{3}\rho, \\ (s_1, t_1) &:= (xz, yz), \\ (s_2, t_2) &:= (x^2 - y^2, 2xy). \end{aligned} \tag{63}$$

This decomposition can also be viewed as the decomposition of second-order deviatoric tensors on the following basis:

$$\mathbf{U} = \begin{pmatrix} -1/3 & 0 & 0 \\ 0 & -1/3 & 0 \\ 0 & 0 & 2/3 \end{pmatrix},$$

$$\mathbf{S}_1 = \begin{pmatrix} 0 & 0 & 1/2 \\ 0 & 0 & 0 \\ 1/2 & 0 & 0 \end{pmatrix}, \mathbf{T}_1 = \begin{pmatrix} 0 & 0 & 0 \\ 0 & 0 & 1/2 \\ 0 & 1/2 & 0 \end{pmatrix},$$

$$\mathbf{S}_2 = \begin{pmatrix} 1 & 0 & 0 \\ 0 & -1 & 0 \\ 0 & 0 & 0 \end{pmatrix}, \mathbf{T}_2 = \begin{pmatrix} 0 & 1 & 0 \\ 1 & 0 & 0 \\ 0 & 0 & 0 \end{pmatrix}.$$

All these tensors are traceless, symmetric and translate very simply the action of orthogonal transformations which operate on them.

In the general case, let $w := x + iy$ be a complex number. For $0 \leqslant l \leqslant n$ we note s_l the harmonic part of $z^{n-l}\Re(w^l)$ and t_l the harmonic part of $z^{n-l}\Im(w^l)$. In the following we write u instead of s_0. We also notice that $t_0 = 0$. We at last define

$$\begin{aligned} K_0 &:= span(u), \\ K_l &:= span(s_l, t_l), \quad 1 \leqslant l \leqslant n. \end{aligned} \tag{64}$$

The Cartan decomposition states that the space $\mathcal{H}_n$ of harmonic polynomials of degree n has an **SO(2)**-invariant decomposition denoted

$$K_0 \oplus K_1 \oplus \cdots \oplus K_n. \tag{65}$$

Moreover, each rotation $\mathbf{Q}(\mathbf{k}, \theta) \in \mathbf{SO(2)}$ about $\mathbf{k}$ of angle θ acts as the unit element on K_0 and like a rotation of angle θ on K_l for $1 \leq l \leq n$:

$$\begin{aligned} \mathbf{Q}(\mathbf{k},\theta) \star s_l &= s_l \cos(l\theta) + t_l \sin(l\theta), \\ \mathbf{Q}(\mathbf{k},\theta) \star t_l &= -s_l \sin(l\theta) + t_l \cos(l\theta). \end{aligned} \tag{66}$$

Finally, $\mathbf{Q}(\mathbf{i}, \pi)$ acts as a reflection on K_l $(0 \leqslant l \leqslant n)$:

$$\begin{aligned} \mathbf{Q}(\mathbf{i},\pi) \star u &= (-1)^n u, \\ \mathbf{Q}(\mathbf{i},\pi) \star s_l &= (-1)^{n-l} s_l, \\ \mathbf{Q}(\mathbf{i},\pi) \star t_l &= (-1)^{n-l+1} t_l. \end{aligned} \tag{67}$$

Arguments which amount to the above results are described in Golubitsky *et al.* (1988 pp.109–111).

One of the immediate consequences of the Cartan decomposition is that to each $h \in \mathcal{H}_n$ correspond $2n + 1$ real numbers λ, α_l, β_l $(1 \leqslant l \leqslant n)$ such that

$$h = \lambda u + \sum_{l=1}^{n} (\alpha_l s_l + \beta_l t_l). \tag{68}$$

The isomorphism ψ defined in (58) enables us to transfer the Cartan decomposition from $\mathcal{H}_n$ into the space of harmonic tensors of degree n. The latter is then the direct sum of $n + 1$ subspaces ψ^{-1} (K_l) $(0 \leqslant l \leqslant n)$ for which we use again the same symbol K_l to simplify the notations.

If $n = 3$, we consider an element $\mathbf{H} \in \mathbf{Hrm}$ and write $\psi(\mathbf{H}) = h = \lambda u + \sum_{l=1}^{3}(\alpha_l s_l + \beta_l t_l)$. Simple computations (see Golubitsky *et al.*, 1988) show that a basis for K_l $(0 \leqslant l \leqslant 3)$ is given by:

$$\begin{aligned} u &= z^3 - 3\rho z/5, \\ (s_1, t_1) &= (z^2 x - \rho x/5,\ z^2 y - \rho y/5), \\ (s_2, t_2) &= (z(x^2 - y^2), xyz), \\ (s_3, t_3) &= (x^3 - 3xy^2, 3x^2 y - y^3). \end{aligned} \tag{69}$$

Let $\mathbf{U} := \psi^{-1}(u)$, $\mathbf{S}_l := \psi^{-1}(s_l)$ and $\mathbf{T}_l := \psi^{-1}(t_l)$. For $\mathbf{H} \in \mathbf{Hrm}$, we can write $\mathbf{H} = \mathbf{H}_0 + \mathbf{H}_1 + \mathbf{H}_2 + \mathbf{H}_3$ where $\mathbf{H}_0 := \lambda\mathbf{U}$ and, for $1 \leqslant l \leqslant 3$, $\mathbf{H}_l := \alpha_l \mathbf{S}_l + \beta_l \mathbf{T}_l$. The straightforward consequence of results (65)–(67) is the description and characterization of the symmetry groups of $\mathbf{H} \in \mathbf{Hrm}$ through the properties of the "components" $\mathbf{H}_l$ (it will be said in the following that $\mathbf{H}_l$ is "horizontal" if it is a multiple of $\mathbf{S}_l$ and "vertical" if it is a multiple of $\mathbf{T}_l$):

Proposition 2 *For a third order harmonic tensor* $\boldsymbol{H} = \boldsymbol{H}_0 + \boldsymbol{H}_1 + \boldsymbol{H}_2 + \boldsymbol{H}_3$, *we have:*

(i) $g(\boldsymbol{H}) \supset \boldsymbol{O(2)}^- \iff \boldsymbol{H} = \boldsymbol{H}_0;$

(ii) $g(\boldsymbol{H}) \supset \boldsymbol{SO(2)} \iff \boldsymbol{H} = \boldsymbol{H}_0;$

(iii) $g(\boldsymbol{H}) \supset \mathbb{D}_k^z \iff \boldsymbol{H} = \boldsymbol{H}_0, (k \geq 4);$

(iv) $g(\boldsymbol{H}) \supset \mathbb{Z}_k \iff \boldsymbol{H} = \boldsymbol{H}_0, (k \geq 4)$;

(v) $g(\boldsymbol{H}) \supset$ ***O(2)*** $\iff \boldsymbol{H} = 0$;

(vi) $g(\boldsymbol{H}) \supset \mathbb{D}_{2k}^d \iff \boldsymbol{H} = 0, (k \geq 4)$;

(vii) $g(\boldsymbol{H}) \supset \mathbb{Z}_{2k}^- \iff \boldsymbol{H} = 0, (k \geq 4)$;

(viii) $g(\boldsymbol{H}) \supset \mathbb{D}_k \iff \boldsymbol{H} = 0, (k \geq 4)$;

(ix) $g(\boldsymbol{H}) \supset \mathbb{Z}_2 \iff \boldsymbol{H}_1 = \boldsymbol{H}_3 = 0$;

(x) $g(\boldsymbol{H}) \supset \mathbb{D}_2 \iff$ ***H*** *is vertical in* K_2;

(xi) $g(\boldsymbol{H}) \supset \mathcal{T} \iff$ ***H*** *is vertical in* K_2;

(xii) $g(\boldsymbol{H}) \supset \mathbb{D}_4^d \iff$ ***H*** *is vertical in* K_2;

(xiii) $g(\boldsymbol{H}) \supset \mathcal{O}^- \iff$ ***H*** *is vertical in* K_2;

(xiv) $g(\boldsymbol{H}) \supset \mathbb{Z}_3 \iff \boldsymbol{H}_1 = \boldsymbol{H}_2 = 0$;

(xv) $g(\boldsymbol{H}) \supset \mathbb{D}_3 \iff$ ***H*** *is horizontal in* K_3;

(xvi) $g(\boldsymbol{H}) \supset \mathbb{Z}_2^- \iff \boldsymbol{H}_0 = \boldsymbol{H}_2 = 0$;

(xvii) $g(\boldsymbol{H}) \supset \mathbb{D}_2^z \iff \boldsymbol{H} = \boldsymbol{H}_0 + \boldsymbol{H}_2$ *with* $\boldsymbol{H}_2$ *horizontal in* K_2;

(xviii) $g(\boldsymbol{H}) \supset \mathbb{Z}_4^- \iff \boldsymbol{H} = \boldsymbol{H}_2$;

(xix) $g(\boldsymbol{H}) \supset \mathbb{D}_6^d \iff \boldsymbol{H} = \boldsymbol{H}_3$ *with* $\boldsymbol{H}_3$ *horizontal in* K_3;

(xx) $g(\boldsymbol{H}) \supset \mathbb{D}_3^z \iff \boldsymbol{H} = \boldsymbol{H}_0 + \boldsymbol{H}_3$ *with* $\boldsymbol{H}_3$ *vertical in* K_3;

(xxi) $g(\boldsymbol{H}) \supset \mathbb{Z}_6^- \iff \boldsymbol{H} = \boldsymbol{H}_3$;

Remark 1 A third-order harmonic tensor **H** has 7 components (three couples $(\alpha_l, \beta_l)_{1 \leqslant l \leqslant 3}$ and λ). The Cartan decomposition is particularly adapted to the determination of the number of constants which rise from a given symmetry. To take an example, part (iii) above indicates that $\mathbb{Z}_2$ symmetry requires three constants (one for $\mathbf{H}_0$ and two for $\mathbf{H}_2$). By applying this method to the other elements of the harmonic decomposition, we obtain the minimum number of constants of a tensor **P** ∈ **Piez** having a given symmetry. The determination of the optimal basis in which a given tensor is indeed written with this minimal number of constants requires a specific study however (*cf.* Berthaud, François & Geymonat (1998) in the elasticity case).

Proof. – Relations concerning closed subgroups of **SO(3)** have been shown in Forte & Vianello (1997). They are an immediate consequence of the Cartan decomposition. The other results easily come from Golubitsky *et al.* (1988 pp. 122 and 123).■

3.4.3 *Symmetry classes of Piez*

We first show the

Lemma 1 *Let* ${}^\epsilon\boldsymbol{C} \in {}^\epsilon\boldsymbol{Dev}$ *and* $\boldsymbol{v}, \boldsymbol{w} \in \mathbb{R}^3$. *Then:*

(a) ${}^\epsilon g({}^\epsilon\boldsymbol{C})$ *is conjugate to* $\mathbb{D}_2$, $\mathbb{D}_4^d$, ***O(2)*** *or* ***O(3)***,

(b) *g(**v**) is conjugate to* ***O(2)***$^-$ *or* ***O(3)***,

(c) $g(\boldsymbol{w}) \cap g(\boldsymbol{v})$ *is conjugate to* $\mathbb{Z}_2^-$, $\mathbb{D}_2^z$, ***O(2)*** $\oplus\, \mathbb{Z}_2^c$, ***O(2)***$^-$ *or* ***O(3)***.

Proof – (a) It is an extension of Forte & Vianello (1996, Theorem 3): if the three eigenvalues of ${}^{\epsilon}\mathbf{C}$ are distinct and if none of them is zero then it does not exist an orthogonal transformation which maps ${}^{\epsilon}\mathbf{C}$ into $-{}^{\epsilon}\mathbf{C}$. We then are in the case of Forte & Vianello (1996, Theorem 3). If only one of the eigenvalues is zero, then the two others are opposite and different from zero. The rotations of angle π about the characteristic axes leave ${}^{\epsilon}\mathbf{C}$ unchanged while an inversion of the characteristic axes associated to non zero eigenvalues maps ${}^{\epsilon}\mathbf{C}$ into $-{}^{\epsilon}\mathbf{C}$. This subgroup is conjugate to $\mathbb{D}_4^d$.
(b) If $\mathbf{v}$ is different from zero then there is a rotation $\mathbf{Q}$ such that $\mathbf{Q} \star \mathbf{v}$ is parallel to $\mathbf{k}$. Thus $g(\mathbf{v}) = \mathbf{O(2)}^-$.■
(c) If one of the two vectors is null or if the two vectors are collinear but not opposite, then the result is as in (b) above. In other cases, there are three possibility:

(c_1) If the length of v and $\mathbf{w}$ are different and if they are not collinear, then there exists a plane P which contains them. The symmetry according to this plane P let unchanged v and $\mathbf{w}$. This symmetry group is conjugate to $\mathbb{Z}_2^-$.

(c_2) If v and $\mathbf{w}$ have the same length and are not opposite then the symmetry according to P and the symmetry according to the bisectrices of v and $\mathbf{w}$ let the figure unchanged. This group is conjugate to $\mathbb{D}_2^z$.

(c_3) If v and $\mathbf{w}$ are opposite then the operation above plus the inversion and any rotation of axis v (or $\mathbf{w}$, it is the same) let the figure unchanged. This group is conjugate to $\mathbf{O(2)} \oplus \mathbb{Z}_2^c$.■

From proposition 2 and from Lemma 1 rises the

Corollary 2 *Let $\boldsymbol{P} \in$ Piez, we have:*

(a) $g(\boldsymbol{P}) \supset \mathbb{Z}_k \iff g(\boldsymbol{P}) \supset \boldsymbol{SO(2)}$, $(k \geq 4)$,

(b) $g(\boldsymbol{P}) \supset \mathbb{D}_k \iff g(\boldsymbol{P}) \supset \boldsymbol{O(2)}$, $(k \geq 4)$,

(c) $g(\boldsymbol{P}) \supset \mathbb{D}_k^z \iff g(\boldsymbol{P}) \supset \boldsymbol{O(2)}^-$, $(k \geq 4)$,

(d) $g(\boldsymbol{P}) \supset \mathbb{Z}_{2k}^- \iff g(\boldsymbol{P}) \supset \mathbb{D}_{2k}^d \iff g(\boldsymbol{P}) \supset \boldsymbol{O(3)}$, $(k \geq 4)$,

(e) $g(\boldsymbol{P}) \supset \mathbb{Z}_6^- \Rightarrow \{\exists \boldsymbol{Q} \in \mathbb{O}_+^3; g(\boldsymbol{Q} \star \boldsymbol{P}) \supset \mathbb{D}_6^d\}$,

(f) $g(\boldsymbol{P}) \supset \mathcal{T} \iff g(\boldsymbol{P}) \supset \mathcal{O}^-$.

Proof – We just show part (e) with $\mathbf{P} = (\mathbf{H}, {}^{\epsilon}\mathbf{C}, \mathbf{w}, \mathbf{v})$: if $g(\mathbf{P}) \supset \mathbb{Z}_6^-$ then, according to Lemma 1, we have ${}^{\epsilon}g({}^{\epsilon}\mathbf{C}) \cap g(\mathbf{w}) \cap g(\mathbf{v}) \supset \mathbf{O(3)}$. The part *(xxi)* of the Proposition 2 imply that $\mathbf{H} = \mathbf{H}_3$. We know that $\mathbf{Q}(\mathbf{k}, \theta) \in \mathbf{SO(2)}$ act on K_3 as rotations of angle 3θ. Therefore there exists a $\bar{\theta}$ such that $\bar{\mathbf{Q}} := \mathbf{Q}(\mathbf{k}, \bar{\theta})$ makes $\mathbf{H}_3$ horizontal in K_3. Then, from part *(xvix)* of Proposition 2, we have that $g(\bar{\mathbf{Q}} \star \mathbf{H}) \supset \mathbb{D}_6^d$. Morerover, since $\bar{\mathbf{Q}} \in \mathbf{SO(2)}$, we have:

$$g(\bar{\mathbf{Q}} \star \mathbf{w}) \cap g(\bar{\mathbf{Q}} \star \mathbf{v}) = \{\bar{\mathbf{Q}} g(\mathbf{w}) \bar{\mathbf{Q}}^t\} \cap \{\bar{\mathbf{Q}} g(\mathbf{v}) \bar{\mathbf{Q}}^t\} = \bar{\mathbf{Q}}\{g(\mathbf{w}) \cap g(\mathbf{v})\}\bar{\mathbf{Q}}^t \supset \mathbf{O(3)}.$$

Therefore, $g(\bar{\mathbf{Q}} \star \mathbf{P}) = g(\bar{\mathbf{Q}} \star \mathbf{H}) \cap g(\bar{\mathbf{Q}} \star {}^{\epsilon}\mathbf{C}) \cap g(\bar{\mathbf{Q}} \star \mathbf{w}) \cap g(\bar{\mathbf{Q}} \star \mathbf{v}) \supset \mathbb{D}_6^d$.■

We consequently obtain trivially the

Corollary 3 *Let $k \geqslant 4$ and $m \geqslant 3$, then* ***Piez***$(\mathbb{Z}_k)$, ***Piez***$(\mathbb{D}_k)$, ***Piez***$(\mathbb{D}_{2k}^d)$, ***Piez***$(\mathbb{D}_k^z)$, ***Piez***$(\mathbb{Z}_{2m}^-)$, ***Piez***$(\mathcal{T})$, ***Piez***$(\mathcal{O})$, ***Piez***$(\mathcal{I})$ *and* ***Piez***$(\boldsymbol{SO(3)})$ *are empty. Moreover,* ***Piez***$(\boldsymbol{O(3)}) = \{\boldsymbol{0}\}$.

These results do not mean that there are no tensors in **Piez** that have, for example, a $\mathbb{Z}_k$ symmetry when $k \geqslant 4$, but such tensors then have necessarily a "higher" symmetry.

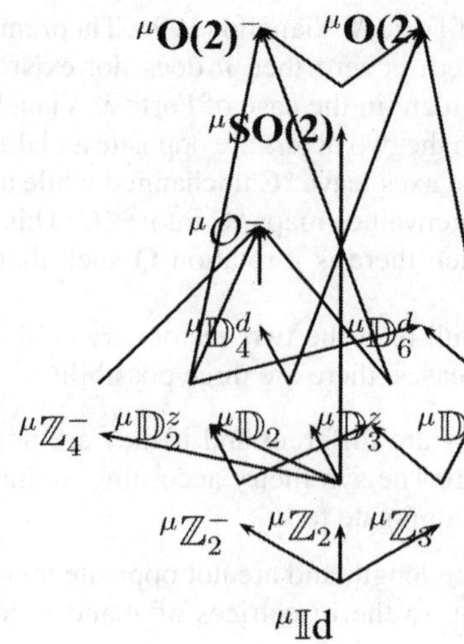

Figure 1. Characteristic tree of $^{\epsilon}$**M**agn symmetry classes.

We finally have

Theorem 4 *The following sets are not empty:* ***Piez***$(\mathbb{I}d)$, ***Piez***$(\mathbb{Z}_2)$, ***Piez***$(\mathbb{Z}_3)$, ***Piez***$(\mathbb{Z}_2^-)$, ***Piez***$(\mathbb{Z}_4^-)$, ***Piez***$(\mathbb{D}_2^z)$, ***Piez***$(\mathbb{D}_3^z)$, ***Piez***$(\mathbb{D}_2)$, ***Piez***$(\mathbb{D}_3)$, ***Piez***$(\mathbb{D}_4^d)$, ***Piez***$(\mathbb{D}_6^d)$, ***Piez***$(\mathcal{O}^-)$, ***Piez(SO(2))***, ***Piez(O(2))*** *and* ***Piez(O(2)***$^-$***)***.

Proof– The method consists in suitably choosing each element of the harmonic decomposition of $\mathbf{P} = (\mathbf{H}, {}^{\epsilon}\mathbf{C}, \mathbf{w}, \mathbf{v})$ in a similar way to that used in Forte &Vianello (1996, 1997). Thus, by for example choosing a vertical tensor **H** in K_2 as the only non zero element in the harmonic decomposition of $\mathbf{P} \in$ Piez, we have $g(\mathbf{P}) = \mathcal{O}^-$.■

Remark 1 *There is a misprint in the classification established in Geymonat & Weller (2002) where* $\mathbb{Z}_4^-$ *does not appear as a possible piezoelectric tensor symmetry class. We correct this misprint here (see also Fig. 1).*

4 CONCLUSIONS

By considering the relations (48)–(50), the results obtained above are extended to $^{\epsilon}$**M**agn. The containment relations between symmetry classes of $^{\epsilon}$**M**agn are presented on Figure 1. This characteristic tree shows the natural diagram of evolution of constitutive symmetries (relating to the behavior, *i.e.* to tensors (*cf.* Zheng & Boelher, 1994)).

We recapitulate in Table 1 the bond between constitutive symmetries and microstructure symmetries (concerning notations *cf.* Bhagavantam (1996) and Birss (1996) for example). Each of the magnetic point groups $\underline{6}$, $\underline{\bar{6}}$ and $\underline{6/m}$ are subgroups of $^{\mu}\mathbb{Z}_6^-$. However, $^{\mu}\mathbb{Z}_6^-$ is not a constitutive symmetry group of the piezomagnetic phenomenon. This is a consequence of parts (c) of corollary 2 which states that piezomagnetic tensors cannot have $^{\mu}\mathbb{Z}_6^-$ symmetry without having a larger one. Therefore, with an appropriate rotation, any piezomagnetic tensor $^{\epsilon}\mathbf{M}$ with symmetry group containing $^{\mu}\mathbb{Z}_6^-$ can be transformed into a piezomagnetic tensor with symmetry group containing $^{\mu}\mathbb{D}_6^d$. This is why $^{\mu}\mathbb{Z}_6^-$ do not appear as piezomagnetic constitutive symmetry.

We then reduce to 15 the number of equivalence classes of piezomagnetic tensors, up to now considered to be 16.

Table 1. Constitutive symmetries of the piezomagnetic phenomenon related to the magnetic point groups which give them birth.

Constitutive symmetries	Magnetic point groups
$^{\mu}\mathbb{I}\mathrm{d}$	1, 1
$^{\mu}\mathbb{Z}_2^-$	$\underline{m}, \underline{2}, \underline{2}/\underline{m}$
$^{\mu}\mathbb{Z}_4^-$	$\underline{4}, \underline{4}, \underline{4}/m$
$^{\mu}\mathbb{Z}_2$	$m, 2, 2/m$
$^{\mu}\mathbb{Z}_3$	3, 3
$^{\mu}\mathbb{D}_2^z$	$\underline{2}m\underline{m}, 2\underline{mm}, \underline{22}2, m\underline{mm}$
$^{\mu}\mathbb{D}_2$	$2mm, 222, mmm$
$^{\mu}\mathbb{D}_3^z$	$3\underline{2}, 3\underline{m}, 3\underline{m}$
$^{\mu}\mathbb{D}_3$	$32, 3m, 3m$
$^{\mu}\mathbb{D}_4^d$	$\underline{4}2\underline{2}, \underline{4}m\underline{m}, \underline{\bar{4}}2\underline{m}, \underline{\bar{4}}\underline{2}m, \underline{4}/mm\underline{m}$
$^{\mu}\mathbb{D}_6^d$	$\underline{6}, \underline{\bar{6}}, \underline{6}/\underline{m}, \underline{6}2\underline{2}, \underline{6m}m, \underline{\bar{6}m}2, \underline{\bar{6}}m\underline{2}, \underline{6}/\underline{m}m\underline{m}$
$^{\mu}\mathcal{O}^-$	$23, m3, \underline{4}3\underline{2}, \underline{4}3\underline{m}, m3\underline{m}$
$^{\mu}\mathbf{SO(2)}$	$4, 4, 4/m, 6, 6, 6/m$
$^{\mu}\mathbf{O(2)}$	$422, 4mm, 42m, 4/mmm, 622, 6mm, \bar{6}m2, 6/mmm$
$^{\mu}\mathbf{O(2)}^-$	$4\underline{22}, 4\underline{mm}, \underline{\bar{4}2m}, 4/m\underline{mm}, 6\underline{22}, 6\underline{mm}, 6\underline{m2}, 6/m\underline{mm}$

APPENDIXES

A – Closed subgroups of $\mathbf{O(3)}$

As it was announced in §3.4, the closed subgroups of **O(3)** fall in three categories.

The *class I* subgroups are closed subgroups of **SO(3)** and are conjugate to one of the following groups:

- $\mathbb{I}_d$, the trivial subgroup, formed by the unit element **I**.
- $\mathbb{Z}_m$ (for $m \geqslant 2$), the set of rotations about **k** of angle multiple of $2\pi/m$. It is a cyclic group with m elements generated by $\mathbf{Q}(\mathbf{k}, 2\pi/m)$.
- $\mathbb{D}_m$ (for $m \geqslant 2$), the *dihedral* group, generated by $\mathbb{Z}_m$ and $\mathbf{Q}(\mathbf{i}, \pi)$. It has $2m$ elements.
- $\mathcal{T}$, the *tetrahedral* group containing the 12 symmetry elements of the tetrahedron.
- $\mathcal{O}$, the *octahedral* group containing the 24 symmetry elements of the octahedron or of the cube.
- $\mathcal{I}$, the symmetry group of icosahedron or dodecahedron. It has 60 elements.
- **SO(2)**, the subgroup of rotations that leave **k** unchanged.
- **O(2)**, the subgroup of rotations which map **k** into $\pm\mathbf{k}$.
- **SO(3)**.

The *class II* subgroups simply derive from class I subgroups (*cf.* §3.3).

As to *class III* subgroups, they have the characteristic not to be included in **SO(3)** and not to contain **–I**. They are defined in Golubitsky *et al.* (1988) for example. We detail here the structure of the $\mathbb{Z}_{2m}^-$ ($m \geqslant 1$), $\mathbb{D}_{2m}^d$ ($m \geqslant 2$) and $\mathbb{D}_m^z$ ($m \geqslant 2$) subgroups.

The $\mathbb{Z}_2^-$ subgroup is generated by

$$\begin{pmatrix} 1 & 0 & 0 \\ 0 & 1 & 0 \\ 0 & 0 & -1 \end{pmatrix}.$$

Moreover, there are three nonconjugate subgroups of **O(3)** isomorphic to $\mathbb{D}_{2m}$, namely: $\mathbb{D}_{2m}$, $\mathbb{D}_{2m}^z$ and $\mathbb{D}_{2m}^d$. The cyclic subgroup $\mathbb{Z}_{2m}$ is present inside both $\mathbb{D}_{2m}$ and $\mathbb{D}_{2m}^z$. The $\mathbb{D}_{2m}^z$ group is generated by $\mathbb{Z}_{2m}$ and the reflection across the x-axis in the (x, y)–plane affected by

$$\begin{pmatrix} 1 & 0 & 0 \\ 0 & -1 & 0 \\ 0 & 0 & 1 \end{pmatrix}.$$

The cyclic subgroup of order $2m$ in $\mathbb{D}^d_{2m}$ is not the standard cyclic subgroup $\mathbb{Z}_{2m} \subset \mathbf{SO(3)}$ but the class III subgroup $\mathbb{Z}_{\bar{2}m}$ generated by the rotation

$$\begin{pmatrix} \cos\frac{\pi}{m} & \sin\frac{\pi}{m} & 0 \\ -\sin\frac{\pi}{m} & \cos\frac{\pi}{m} & 0 \\ 0 & 0 & 1 \end{pmatrix}.$$

The reflection across the x-axis in $\mathbb{D}^d_{2m}$ is the same as in $\mathbb{D}_{2m}$.

B – Harmonic Decomposition

As announced in §3.4.1, the harmonic decomposition is based on a method introduced by Spencer (1970). We apply it to the third-order tensors which are symmetric according to their first two indexes (in this direction, see Forte & Vianello, 1997). We first point out the general step.

Let $A_{i_1 i_2 i_3 \cdots i_n}$ be the components of a tensor $\mathbf{A}$ of order n. We note $A_{(i_1 i_2 \cdots i_r) i_{r+1} \cdots i_n}$ the tensor obtained by symmetrization of $\mathbf{A}$ according to its first r indexes. It is in fact the sum (divided by $r!$) of the $r!$ components obtained from $A_{i_1 i_2 i_3 \cdots i_n}$ by applying all the permutations to its first $r!$ indexes. The fundamental relation due to Spencer (1970) is:

$$\begin{aligned} r\,A_{(i_1 i_2 \cdots i_{r-1}) i_r i_{r+1} \cdots i_n} &= r\,A_{(i_1 i_2 \cdots i_r) i_{r+1} \cdots i_n} \\ &+ \epsilon_{i_1 i_r i_s}\,\epsilon_{i_p i_q i_s}\,A_{(i_p i_2 i_3 \cdots i_{r-1}) i_q i_{r+1} \cdots i_n} \\ &+ \epsilon_{i_2 i_r i_s}\,\epsilon_{i_p i_q i_s}\,A_{(i_p i_1 i_3 \cdots i_{r-1}) i_q i_{r+1} \cdots i_n} \\ &+ \cdots + \epsilon_{i_{r-1} i_r i_s}\,\epsilon_{i_p i_q i_s}\,A_{(i_p i_1 i_3 \cdots i_{r-2}) i_q i_{r+1} \cdots i_n}, \end{aligned} \tag{70}$$

where $\epsilon_{i_1 i_2 i_3}$ are the components of the alternating tensor.

We define

$$B_{i_1 i_2 i_3 \cdots i_{r-2} i_{r+1} \cdots i_n i_s} := \epsilon_{i_p i_q i_s}\,A_{(i_p i_1 i_2 \cdots i_{r-2}) i_q i_{r+1} \cdots i_n}. \tag{71}$$

Thus $A_{(i_1 i_2 \cdots i_{r-1}) i_r \cdots i_n}$ is the sum of the symmetrization of the tensor $\mathbf{A} = A_{i_1 i_2 i_3 \cdots i_n}$ with respect to its first $r!$ indexes and some terms obtained by contraction of $B_{i_1 i_2 i_3 \cdots i_{r-2} i_{r+1} \cdots i_n i_s}$ with the alternating tensor. These terms are therefore pseudo-tensors. Applying by recurrence the equality above, we are finally able to express $A_{i_1 i_2 i_3 \cdots i_n}$ as a sum of (pseudo-)tensors totally symmetric of order equal or less than n, suitably contracted with isotropic tensors.

Let $\mathbf{P} \in \mathbf{Piez}$. By definition, we have $\mathbf{P} = P_{(ij)k}$. We note it $P_{ijk} = P_{(ij)k}$. Applying the relation above we find:

$$3\,P_{(ij)k} = 3\,P_{(ijk)} + \epsilon_{ikp}\,\epsilon_{abp}\,P_{ajb} + \epsilon_{jkp}\,\epsilon_{abp}\,P_{aib}. \tag{72}$$

Thus $P_{(ijk)}$ is a third-order totally symmetric tensor. It again breaks up into a third-order harmonic tensor $\mathbf{H}$ and a vector $\mathbf{w} \in \mathbb{R}^3$. In addition, by letting ${}^\epsilon C_{jp} = \epsilon_{abp}\,P_{ajb}$, we obtain:

$$2\,{}^\epsilon C_{ij} = 2\,{}^\epsilon C_{(ij)} + \epsilon_{ijp}\,v_p, \tag{73}$$

where v_p are the components of a vector $\mathbf{v} \in \mathbb{R}^3$.

It is thus seen that ${}^\epsilon C_{(ij)}$ is a second-order symmetric pseudo-tensor. Simple calculations show that it is traceless: ${}^\epsilon\mathbf{C}$ is harmonic.

We then show that any tensor $\mathbf{P} \in \mathbf{Piez}$ corresponds to a quadruplet including a third-order harmonic tensor $\mathbf{H}$ (with 7 components), a second-order harmonic pseudo-tensor ${}^\epsilon\mathbf{C}$ (5 components) and two vectors **v** and **w** (3 components each). We therefore verify that $\mathbf{P} = P_{(ij)k} \in$ **Piez** has 18 components. This is the reason why such tensors are most of the time presented in the form of a 3×6 matrix. This correspondence, which can also be found in Schouten (1954), is linear like every operations that intervene in the harmonic decomposition. Its kernel is $\{0\}$: it is consequently an isomorphism. The property of **SO(3)**-invariance according to the $\star$ action rises simply from the following equality, due to (28) and (29):

$$\forall \mathbf{Q} \in \mathbf{SO(3)},\ \mathbf{Q} \star \mathbf{P} = (\mathbf{Q} \star \mathbf{H}, \det(\mathbf{Q})\,\mathbf{Q} \star {}^{\epsilon}\mathbf{C}, \mathbf{Q} \star \mathbf{v}, \mathbf{Q} \star \mathbf{w}). \tag{74}$$

REFERENCES

Berthaud Y. & François M. & Geymonat G. (1998). *Int. J. Solids Structures* **35**, 4091–4106.
Bhagavantam S. (1966). *Crystal Symmetry and Physical Properties*, Academic Press.
Birss R. (1966). *Symmetry and Magnetism*, North Holland.
Coxeter H.S.M. (1989). *Introduction to Geometry*, John Wiley and Sons.
Forte S. & Vianello M. (1996). *J. Elasticity* **43**, 81–108.
Forte S. & Vianello M. (1997). *Int. J. Engng Sci.* **35**, 1317–1326.
Geymonat G. & Weller T. (2002). *C. R. Acad. Sci. Série I* **335**, 847–852.
Golubitsky M. & Stewart I. & Schaeffer D.G. (1988). *Singularities and Groups in Bifurcation Theory, Vol. 2*, Springer Verlag.
Huo Y.-Z. & Del Piero G. (1991). *J. Elasticity* **25**, 203–246.
Landau L. & Lifchitz E. (1969). *Électrodynamique des Milieux Continus*, Éditions Mir.
Laughlin D.E. & Willard M.A. & McHenry M.E. (2000). *The Mineral, Metal and Material Society*, 121–137.
Nye J.F. (1957). *Physical Properties of Crystals*, Clarendon Press.
Pitteri M. & Zanzotto G. (1996). *Acta Cryst.* **A52**, 830–838.
Pitteri M. & Zanzotto G. (1998). *Acta Cryst.* **A54**, 359–373.
Pitteri M. & Zanzotto G. (2003). *Continuum Models for Phase Transitions and Twinning in Crystals*, Chapman & Hall/CRC.
Schouten J.A. (1954). *Tensor Analysis for Physicists*, Clarendon Press.
Shubnikov A.V. (1964). *Colored Symmetry*, Pergamon Press.
Sirotine Y. & Chaskolskaïa M. (1984). *Fondements de la Physique des Cristaux*, Éditions Mir.
Sivardière J. (1995). *La Symétrie en Mathématiques, Physique et Chimie*, Presses Universitaires de Grenoble.
Spencer A.J.M. (1970). *Int. J. Engng Sci.* **8**, 475–481.
Weigel D. (1972). *Cristallographie et Structure des Solides, Tome 1: algèbre et gèomètrie cristalline et moléculaire*, Masson.
Weyl H. (1952). *Symmetry*, Princeton University Press.
Yale P.B. (1968). *Geometry and Symmetry*, Holden-Day.
Zheng Q.-S. & Boelher J.P. (1994). *Acta Mechanica* **102**, 73–89.

Configurational Mechanics, Kalpakides & Maugin (eds)
© 2004, Taylor & Francis Group, London, ISBN 90 5809 667 X

Approaches to finite elasto-plasticity with dislocations

S. Cleja-Ţigoiu
Faculty of Mathematics and Informatics, University of Bucharest, Bucharest, Romania

ABSTRACT: In this paper we propose a macroscopic model for elasto-plastic materials with continuously distributed dislocations. The material response is elastic with respect to the configuration with torsion, in terms of elastic deformation measures, i.e. elastic Cauchy-Green strain tensor and metric non-Riemannian elastic connection. We emphasize some remarkable forms of the constitutive equations for the Cauchy stress, the stress momentum and the free energy density, based on the decomposition theorem for metric affine connections. Evolution laws have to be prescribed for both the plastic distorsion and the second order inhomogeneities.

1 INTRODUCTION

In this paper we describe the behaviour of the crystalline elasto-plastic materials with continuously distributed dislocations. The models are developed within the constitutive framework [6], [7] and are based on the multiplicative decomposition, see [17], of the deformation gradient $\mathbf{F}$ into its elastic $\mathbf{F}^e$ and plastic $\mathbf{F}^p$ components, called elastic and plastic distorsions.

These distorsions cannot be derived from a smooth mapping since the first integrability condition (1), written for $\mathbf{A} = \mathbf{F}^e$ and $\mathbf{A} = \mathbf{F}^p$, with $\mathbf{A} : \mathcal{U} \subset \mathcal{E} \longrightarrow Lin$, does not hold,

$$\mathbf{A}(\mathbf{x}) = \nabla\lambda(\mathbf{x}) \quad \Longleftrightarrow \quad \mathrm{curl}\mathbf{A}(\mathbf{x}) = 0, \quad \forall\, \mathbf{x} \in \mathcal{U}, \tag{1}$$

with $\mathcal{U}$ an open simple connected set in $\mathcal{E}$ an Euclidian space, does not hold. Here *Lin* denotes the set of all second order tensor.

Moreover (1) is equivalent with the symmetry condition (2), for the gradient of the tensor field $\mathbf{A}$

$$((\nabla\ \mathbf{A})\mathbf{u})\mathbf{v} - ((\nabla\ \mathbf{A})\mathbf{v})\mathbf{u} = 0, \quad \forall \mathbf{u}, \mathbf{v} \in \mathcal{V}, \tag{2}$$

if we take into account that

$$(\mathrm{curl}\ \mathbf{A}^T)(\mathbf{u} \times \mathbf{v}) = ((\nabla\ \mathbf{A})\mathbf{u})\mathbf{v} - ((\nabla\ \mathbf{A})\mathbf{v})\mathbf{u}, \quad \forall \mathbf{u}, \mathbf{v} \in \mathcal{V}. \tag{3}$$

$\mathcal{V}$ is a vector space of the translations of $\mathcal{E}$, and $\mathbf{A}^T$ is the transpose of the tensor $\mathbf{A}$.

If the symmetry condition written in (2) is not satisfied for the elastic and plastic distorsions, then the torsions of the appropriate elastic and plastic connections do not vanish.

That means the Noll's dislocations density, see [19], denoted here in (4) by $\bar{\boldsymbol{\alpha}}_{\mathcal{K}}$, is non-zero. The disolocations in continuum theory are modeled by the non-zero Burgers vectors. We recall the definition of the true Burgers vector $\mathbf{b}_{\mathcal{K}}$ as a measure of the material deformation failure, see [24].

Let C_0 be a closed curve in the reference configuration and $\mathcal{A}_0$ the surface with normal $\mathbf{N}$ bounded by C_0. In terms of the plastic distorsion we get

$$\begin{aligned}\mathbf{b}_{\mathcal{K}} &\equiv \int_{C_{\mathcal{K}}} d\mathbf{x}_{\mathcal{K}} = \int_{C_0} \mathbf{F}^p \, d\mathbf{X} = \int_{\mathcal{A}_0} \mathrm{curl}(\mathbf{F}^p)^T \mathbf{N} dA \\ &= \int_{\mathcal{A}_{\mathcal{K}}} \boldsymbol{\alpha}_{\mathcal{K}} \mathbf{n}_{\mathcal{K}} dA_{\mathcal{K}}, \quad \boldsymbol{\alpha}_{\mathcal{K}} \equiv \frac{1}{\det \mathbf{F}^p} \mathrm{curl}(\mathbf{F}^p)^T (\mathbf{F}^p)^T. \end{aligned} \tag{4}$$

Similarly, if C_t is a closed curve in the actual configuration and $\mathcal{A}_t$ the surface with normal $\mathbf{n}$ bounded by C_t, we derive the expression of the true Burgers vector in terms of the elastic distorsion, via the dislocation density $\bar{\boldsymbol{\alpha}}_{\mathcal{K}} \equiv \det\mathbf{F}^e \mathrm{curl}(\mathbf{F}^e)^{-T} (\mathbf{F}^e)^{-T}$. The equality $\bar{\boldsymbol{\alpha}}_{\mathcal{K}} = \boldsymbol{\alpha}_{\mathcal{K}}$ can be proved.

A continuum theory, using the differential geometry concepts, was constructed in Kondo and Yuki [14], Bilby [2], where a plastically deformed body was thought as a manifold, always embedded in the euclidian space. The fundamental geometrical problems related to metric, connection, torsion, curvature concepts were studied in Schouten [20], Kondo and Yuki [14]. The mathematical foundations for the theory of continuous distributions of dislocations was developed by Noll [19] and Wang [25].

Different views on the continuum theory of dislocations, which are motivated by the non-Riemannian geometry, can be found in [22], [16], [11], and so on.

In [16] a principle of the virtual work has been formulated when the free energy density depends on $\mathbf{F}$, $\mathbf{F}^p$, $\nabla\mathbf{F}^p$. The influence of the kinematically necessary dislocation densities on the hardening behaviour of crystalline is derived in [22] by exploiting the Clausius-Duhem inequality.

Within the framework of generalized Cosserat continua in [16], [18] the localized form of the balance equations in dynamics and the entropy production inequality are derived. The attention is focused on the expression of the free energy density. Explicit constitutive equations and hardening rules are proposed to close the theory in the case of Cosserat elastoviscoplastic medium in [11].

We should also mention that Epstein and Maugin [8] have developed a second order-theory which allows for growth diffusion.

On the other hand the strain gradient plasticity theory and its experimental motivations are presented in [9] within a non-linear generalized couple stress theory, but with small elastic and plastic deformations. In [10] a class of phenomenological strain gradient plasticity is formulated in order to accomodate more than one material length parametres, while in [1] only one parameter has been introduced.

Our models presented here extend the main ideas from [3], [5] to elasto-plastic materials with continuously distributed dislocations. The material response is elastic with respect to the configuration with torsion, in terms of the elastic deformation measures, i.e. elastic Cauchy-Green strain tensor and metric non-Riemannian elastic connection. Dynamic balance equations involve non-symmetric Cauchy stress tensor and stress momentum (a third order tensor field), but without Cosserat kinematic formalism. Based on work-conjugated variables, [6] and [4], we choose the coupling between the sets of the variables and we put into evidence the driving forces in our constitutive models. The Eshelby's or Mandel's stress tensors (see [4]) and the stress momentum are driving forces behind the evolution of both order inhomogeneities. Evolution laws have to be prescribed for both the plastic distorsion and the second order inhomogeneities. Discussion related to the possible evolution equations can be found in [6], [12], [11], [8], [23], [21], and so on.

I should mention that evolution equations are still to be developed in order to generate consistent models.

Finally we note that the form of the free energy density, used within the different approaches to the continuum theory of dislocations, is a comparition basis between our results and the results obtained for instance in [15], [10], [6], and so on.

2 CONSTITUTIVE FRAMEWORK OF ELASTO-PLASTIC MATERIAL WITH NON-RIEMANNIAN METRIC CONNECTION

The behaviour of the elasto-plastic material is described with respect to a configuration with torsion, or non-holonomic configuration, time dependent, that will be denoted by $\mathcal{K}$. The existence of such kind of configuration follows as a consequence of the constitutive assumptions pointed out below, within the constitutive framework developed in [6], [7].

Let $\mathcal{N}_X$ be a neighbourhood of the material point X of the body $\mathcal{B}$ and let $\chi : \mathcal{N}_X \times \mathcal{R} \rightarrow \mathcal{E}$ be a motion of the body restricted to $\mathcal{N}_X$. We denote by k a global configuration of the body and we refer to $\chi(\cdot, t)$, at any time t, as a global configuration on $\mathcal{N}_X$.

A1. For any local motion χ, at time t, there exists a pair $(\mathbf{F}^p, \Gamma_k^{(p)})$, with $\mathbf{F}^p$ an invertible second order tensor field, called plastic distorsion, and $\Gamma_k^{(p)}$ an affine plastic connection, with non-zero torsion, i.e. $\mathbf{S}_k^{(p)} \neq 0$,

$$(\overset{(p)}{\mathbf{S}}_k \mathbf{u})\mathbf{v} \equiv (\overset{(p)}{\mathbf{\Gamma}}_k \mathbf{u})\mathbf{v} - (\overset{(p)}{\mathbf{\Gamma}}_k \mathbf{v})\mathbf{u} \neq 0, \tag{5}$$

for all $\mathbf{u}, \mathbf{v} \in \mathcal{V}_k$, i.e. vectors in the reference configuration.

A2. The plastic distortion $\mathbf{F}^p$ and the plastic connection $\Gamma_k^{(p)}$ are compatible, that is

$$\overset{(p)}{\mathbf{\Gamma}}_k = (\mathbf{F}^p)^{-1} \nabla \mathbf{F}^p, \tag{6}$$

where ∇_k denotes a differential operator in $k(\mathcal{N}_X)$. The fields are defined for all $\mathbf{Z} \in k(\mathcal{N}_X)$.

At any time t the non-holonomic configuration $\mathcal{K}(\cdot, t)$, called also the configuration with torsion, is attached to the pair $(\mathbf{F}^p, \Gamma_k^{(p)})$, and it is defined by

$$\nabla(\mathcal{K}(\cdot,t)\circ k^{-1})\,|_{\mathbf{X}} = \mathbf{F}^p(\mathbf{X},t), \quad \nabla^2(\mathcal{K}(\cdot,t)\circ k^{-1})\,|_{\mathbf{X}} \neq \nabla \mathbf{F}^p(\mathbf{X},t). \tag{7}$$

Definition 1. For any motion χ, the pair $(\mathbf{F}^p, \Gamma_k^{(p)})$ of the plastic distorsion and plastic connection defines the elastic distorsion $\mathbf{F}^e$ and the elastic connection with respect to the actual configuration $\Gamma_\chi^{(e)}$ by

$$\begin{aligned} &\mathbf{F}^e = \mathbf{F}(\mathbf{F}^p)^{-1} \iff \mathbf{F} = \mathbf{F}^e\mathbf{F}^p, \quad \text{where} \\ &\mathbf{F}(\mathbf{X},t) = \nabla(\chi(\cdot,t)\circ k^{-1})\,|_{\mathbf{X}}, \end{aligned} \tag{8}$$

and

$$\begin{aligned} &(\overset{(e)}{\mathbf{\Gamma}}_\chi \mathbf{u})\mathbf{v} = -\mathbf{F}(\mathbf{\Gamma}_k - \overset{(p)}{\mathbf{\Gamma}}_k)\mathbf{F}^{-1}\mathbf{u})\mathbf{F}^{-1}\mathbf{v} \quad \text{with} \\ &\mathbf{\Gamma}_k = \mathbf{F}^{-1}\nabla\,\mathbf{F}, \end{aligned} \tag{9}$$

Γ_k is the motion connection defined with respect to the reference configuration k.

For any differential tensor field $\bar{\mathbf{F}}$ when we pass from one global configuration, say k, to an other one, say $\chi(\cdot, t)$, the following differential rule

$$(\nabla_k\bar{\mathbf{F}})\mathbf{u} = (\nabla_\chi\bar{\mathbf{F}})\nabla(\chi\circ k^{-1})\mathbf{u} \equiv (\nabla_\chi\bar{\mathbf{F}})\mathbf{F}\mathbf{u}. \tag{10}$$

For a non-holonomic configuration $\mathcal{K}$, we define, following Noll [19], the differential with respect to $\mathcal{K}$

$$(\nabla_{\mathcal{K}}\bar{\mathbf{F}})\tilde{\mathbf{u}} = (\nabla_k\bar{\mathbf{F}})((\mathbf{F}^p)^{-1}\tilde{\mathbf{u}}), \quad \text{or} \quad (\nabla_{\mathcal{K}}\bar{\mathbf{F}})\tilde{\mathbf{u}} = (\nabla_\chi\bar{\mathbf{F}})(\mathbf{F}^e\tilde{\mathbf{u}}), \tag{11}$$

for all $\tilde{\mathbf{u}}$ in the appropriate tangent space, denoted formally by $\mathcal{V}_{\mathcal{K}}$.

¿From (9)–(11) we can prove that

$$\overset{(e)}{\boldsymbol{\Gamma}}_{\chi}\,\mathbf{u} = \mathbf{F}^e(\nabla_{\chi}(\mathbf{F}^e)^{-1})\mathbf{u}. \tag{12}$$

Thus the plastic connection and elastic connection with respect to $\mathcal{K}$ the non-holonomic configuration can be defined by

$$\overset{(p)}{\boldsymbol{\Gamma}}_{\mathcal{K}}= \mathbf{F}^p\nabla(\mathbf{F}^p)^{-1}, \quad \overset{(e)}{\boldsymbol{\Gamma}}_{\mathcal{K}}= (\mathbf{F}^e)^{-1}\nabla_{\mathcal{K}}\mathbf{F}^e. \tag{13}$$

Proposition 1. 1. The following relationships between the elastic and plastic connections with respect to the non-holonomic configuration $\mathcal{K}$

$$\overset{(e)}{\boldsymbol{\Gamma}}_{\mathcal{K}}= \mathbf{F}^p\boldsymbol{\Gamma}_k[(\mathbf{F}^p)^{-1}, (\mathbf{F}^p)^{-1}]+ \overset{(p)}{\boldsymbol{\Gamma}}_{\mathcal{K}} \tag{14}$$

holds. Here we introduced the notation written in (15) for the third order object defined via the appropriate connection, say $\boldsymbol{\Gamma}$, and the second order tensor, say $\mathbf{F}$, by

$$(\boldsymbol{\Gamma}_k[\mathbf{F}, \mathbf{F}]\mathbf{u})\mathbf{v} \equiv (\boldsymbol{\Gamma}_k(\mathbf{F}\mathbf{u}))\mathbf{F}\mathbf{v} \quad \forall\, \mathbf{u}, \mathbf{v} \in \mathcal{V}_k. \tag{15}$$

2. Unlike Γ_k which is symmetric connection, $\Gamma^{(e)}_{\mathcal{K}}$ and $\Gamma^{(p)}_{\mathcal{K}}$ have non-zero torsion characterized by

$$\begin{aligned}
\mathbf{S}_{\mathcal{K}} &\equiv (\overset{(e)}{\boldsymbol{\Gamma}}_{\mathcal{K}}\,\tilde{\mathbf{u}})\tilde{\mathbf{v}} - (\overset{(e)}{\boldsymbol{\Gamma}}_{\mathcal{K}}\,\tilde{\mathbf{v}})\tilde{\mathbf{u}} \\
&= (\mathbf{F}^e)^{-1}[((\nabla_{\mathcal{K}}\mathbf{F}^e)\tilde{\mathbf{u}})\tilde{\mathbf{v}} - ((\nabla_{\mathcal{K}}\mathbf{F}^e)\tilde{\mathbf{v}})\tilde{\mathbf{u}}] \\
&= (\overset{(p)}{\boldsymbol{\Gamma}}_{\mathcal{K}}\,\tilde{\mathbf{u}})\tilde{\mathbf{v}} - (\overset{(p)}{\boldsymbol{\Gamma}}_{\mathcal{K}}\,\tilde{\mathbf{v}})\tilde{\mathbf{u}} = \mathbf{F}^p[(\nabla_{\mathcal{K}}(\mathbf{F}^p)^{-1}\tilde{\mathbf{u}})\tilde{\mathbf{v}} - (\nabla_{\mathcal{K}}(\mathbf{F}^p)^{-1}\tilde{\mathbf{v}})\tilde{\mathbf{u}}],
\end{aligned} \tag{16}$$

for all $\tilde{\mathbf{u}}, \tilde{\mathbf{v}} \in \mathcal{V}_{\mathcal{K}}$.

Proposition 2. The plastic torsion with respect to the reference configuration (5) and $\mathbf{S}_{\mathcal{K}}$ the torsion with respect to the non-holonomic configuration (16) are related by

$$\mathbf{S}_{\mathcal{K}} = -\mathbf{F}^p \overset{(e)}{\mathbf{S}}_k [(\mathbf{F}^p)^{-1}, (\mathbf{F}^p)^{-1}], \quad \text{or} \quad \overset{(p)}{\mathbf{S}}_k= -(\mathbf{F}^p)^{-1}\mathbf{S}_{\mathcal{K}}[\mathbf{F}^p, \mathbf{F}^p]. \tag{17}$$

A3. The linear momentum and angular momentum balance equations traditionally accepted in continuum mechanics are postulated but a surface element in the actual configuration, with normal $\mathbf{n}$, may transmit both force vector $\mathbf{T}^T\mathbf{n}$ and couple vector $\mathbf{M}^T\mathbf{n}$.

Here $\mathbf{T}$ is the Cauchy stress tensor and $\mathbf{M}$ is the couple stress tensor, both of them are generally non symmetric second order tensors.

Theorem 1. The virtual power in the deformed configuration at time t expanded by any $\mathcal{P}$ part of the body $\mathcal{B}$ is expressed by

$$\begin{aligned}
&\int_{\mathcal{P}_t} \rho\frac{d}{dt}\mathbf{v}\cdot\mathbf{w}d\mathbf{x} + \int_{\mathcal{P}_t}\{\mathbf{T}\}^s\cdot\boldsymbol{\epsilon}(\mathbf{w})d\mathbf{x} + \int_{\mathcal{P}_t}\mathbf{M}^T\cdot\nabla\boldsymbol{\theta}d\mathbf{x} = \\
&\int_{\partial\mathcal{P}_t}\mathbf{T}^T\mathbf{n}\cdot\mathbf{w}d\sigma + \int_{\partial\mathcal{P}_t}\mathbf{M}^T\mathbf{n}\cdot\boldsymbol{\theta}d\sigma + \int_{\mathcal{P}_t}\rho\mathbf{b}_f\cdot\mathbf{w}d\mathbf{x} + \int_{\mathcal{P}_t}\rho\mathbf{b}_m\cdot\boldsymbol{\theta}d\mathbf{x}.
\end{aligned} \tag{18}$$

The equation (18) holds for all generalized virtual velocities and for all the virtual rate of strain, defined by

$$(\mathbf{w}, \boldsymbol{\theta}),\ \boldsymbol{\theta} = \frac{1}{2}\text{curl}\,\mathbf{w}, \quad \boldsymbol{\epsilon}(\mathbf{w}) = \frac{1}{2}(\nabla\mathbf{w} + \nabla\mathbf{w}^T). \tag{19}$$

Here $\mathbf{b}_m$, $\mathbf{b}_f$ are densities of the body couples and body forces, $\mathbf{r}$ is the position vector of an arbitrary material point $\mathbf{x} \in \mathcal{P}_t$, with respect to a fixed point O, and $\mathcal{P}_t \equiv \chi(\mathcal{P}, t)$. We denote the mass densities in the actual and initial configurations by ρ and $\widetilde{\rho}_0$.

3 INTERNAL MECHANICAL POWER

Let us introduce the stress momentum $\boldsymbol{\mu}$, a third order tensor attached to the couple stresses $\mathbf{M}$, by

$$\boldsymbol{\mu} = -\frac{1}{2}\epsilon\mathbf{M}^T, \quad \mu_{ijk} = -\frac{1}{2}\epsilon_{ijm}\,(\mathbf{M}^T)_{mk}. \tag{20}$$

Theorem 2. 1. Two expressions for the density of mechanical power produced by the couple stresses can be introduced with respect to the actual configuration.

I. The stress momentum in the actual configuration is work conjugated to the gradient of the velocity gradient, i.e. the second order gradient of the velocity:

$$\frac{1}{2\rho}\mathbf{M}^T \cdot \nabla\text{curl}\,\mathbf{v} \equiv \frac{1}{\rho}\boldsymbol{\mu} \cdot \nabla_\chi\mathbf{L} \equiv \frac{1}{\rho}\boldsymbol{\mu} \cdot \nabla_\chi^2\mathbf{v}. \tag{21}$$

The velocity gradient can be calculated by the rates of elastic and plastic distortions

$$\mathbf{L} \equiv \nabla_\chi\mathbf{v} \equiv \text{grad}\,\mathbf{v} = \dot{\mathbf{F}}^e(\mathbf{F}^e)^{-1} + \mathbf{F}^e\mathbf{L}^p(\mathbf{F}^e)^{-1}, \quad \mathbf{L}^p = \dot{\mathbf{F}}^p(\mathbf{F}^p)^{-1}. \tag{22}$$

II. There exists a (1,2)-tensor field $\boldsymbol{\tau}$ attaced to the stress momentum, with the symmetry

$$(\boldsymbol{\tau}(\mathbf{u}))\mathbf{v} = (\boldsymbol{\tau}(\mathbf{v}))\mathbf{u}, \quad \text{with}, \tag{23}$$

which is work conjugated (in accordance with [9], [13])

$$\mathbf{M}^T \cdot \nabla\boldsymbol{\theta} = \boldsymbol{\tau} \cdot \boldsymbol{\eta} = \tau_{i(jk)}\eta_{i(jk)}. \tag{24}$$

to the rate of strain gradient tensor $\boldsymbol{\eta}$ defined for any $\mathbf{u}$, $\mathbf{w}$, $\mathbf{w} \in \mathcal{V}_\chi$ by

$$\begin{aligned}(\boldsymbol{\eta}\mathbf{u})\mathbf{w}\cdot\mathbf{z} &= (\nabla_\chi\mathbf{D}(\mathbf{v}))\mathbf{u})\mathbf{w}\cdot\mathbf{z} + \\ &+((\nabla_\chi\mathbf{D}(\mathbf{v}))\mathbf{w})\mathbf{u}\cdot\mathbf{z} - ((\nabla_\chi\mathbf{D}(\mathbf{v}))\mathbf{z})\mathbf{w}\cdot\mathbf{u}, \\ &\text{with} \quad \mathbf{D}(\mathbf{v}) = \frac{1}{2}(\nabla\mathbf{v} + \nabla\mathbf{v}^T),\end{aligned} \tag{25}$$

$\mathbf{D}(\mathbf{v})$ the rate of deformation tensor.

2. The pull back to the reference configuration of the stress momentum, defined by

$$\boldsymbol{\mu}_0 = (\det\mathbf{F})\mathbf{F}^T\boldsymbol{\mu}[\mathbf{F}^{-T}, \mathbf{F}^{-T}],\ (\boldsymbol{\mu}[\mathbf{F}^{-T}, \mathbf{F}^{-T}]\mathbf{u})\mathbf{v} = (\boldsymbol{\mu}(\mathbf{F}^{-T}\mathbf{u}))\mathbf{F}^{-T}\mathbf{v} \tag{26}$$

for all **u, v** vector fields in reference configuration is work conjugated to the rate of the motion connection $\mathbf{\Gamma}$

$$\frac{1}{\rho}\boldsymbol{\mu} \cdot \nabla_\chi \mathbf{L} = \frac{1}{\rho_0}\boldsymbol{\mu}_0 \cdot \dot{\mathbf{\Gamma}}_k. \tag{27}$$

The material derivative of the motion connection, relative to the reference configuration (see formula (9)), is related to the second order gradient of the velocity by

$$\dot{\mathbf{\Gamma}}_k = \mathbf{F}^{-1}(\nabla_\chi \mathbf{L})[\mathbf{F}, \mathbf{F}]. \tag{28}$$

We introduce now Piola-Kirchhoff stress tensor and the stress momentum as pulled back fields to the configuration $\mathcal{K}$, and Mandel's non-symmetric stress measure by

$$\mathbf{\Pi}_\mathcal{K} \equiv \mathbf{\Pi} = \det(\mathbf{F}^e)(\mathbf{F}^e)^{-1}\mathbf{T}^s(\mathbf{F}^e)^{-T}, \quad \det \mathbf{F}^e = \frac{\tilde{\rho}}{\rho}$$

$$\boldsymbol{\mu}_\mathcal{K} = (\det \mathbf{F}^e)(\mathbf{F}^e)^T\boldsymbol{\mu}[(\mathbf{F}^e)^{-T}, (\mathbf{F}^e)^{-T}], \tag{29}$$

$$\mathbf{\Sigma}_\mathcal{K} \equiv \mathbf{\Sigma} = \mathbf{C}^e\frac{\mathbf{\Pi}_\mathcal{K}}{\tilde{\rho}}, \qquad \mathbf{C}^e = (\mathbf{F}^e)^T\mathbf{F}^e.$$

$\widetilde{\rho}$ is the mass density in the non-holonomic configuration.

4 ELASTIC TYPE CONSTITUTIVE BEHAVIOUR IN NON-HOLONOMIC CONFIGURATION

We propose here the elastic constitutive functions with respect to the non-holonomic configurations, in terms of the elastic deformation measures

$$\mathbf{C}^e = (\mathbf{F}^e)^T\mathbf{F}^e, \quad \overset{(e)}{\mathbf{\Gamma}}_\mathcal{K} = (\mathbf{F}^e)^{-1}\nabla_\mathcal{K}\mathbf{F}^e. \tag{30}$$

$\mathbf{C}^e$ represents the Cauchy-Green elastic strain tensor and $\Gamma_\mathcal{K}^{(e)}$ measures the influence of the gradient of the elastic distorion, see formula (13). These elastic measures, as well as the stress measures defined in (29), are invariant with respect to a change of frame in the actual configurations, due to the objectivity assumptions introduced in [6].

Remarks. Certain work conjugated quantities have been put into evidence, see [6], in the expressions of the elastic power expanded by the stress and stress momentum

$$\frac{1}{2}\frac{\mathbf{\Pi}_\mathcal{K}}{\tilde{\rho}} \cdot \dot{\mathbf{C}}^e + \frac{1}{\rho}\boldsymbol{\mu} \cdot \nabla_\chi \mathbf{L}^e, \tag{31}$$

and by the plastic power

$$\mathbf{\Sigma}_\mathcal{K} \cdot \mathbf{L}^p + \frac{1}{\tilde{\rho}}\boldsymbol{\mu}_\mathcal{K} \cdot \nabla_\mathcal{K}\mathbf{L}^p. \tag{32}$$

We can prove that:

Theorem 3. The elastic type constitutive functions which give rise to Cauchy stress and to stress momentum, obey the objectivity assumptions if and only if the following reduced form of the constitutive relations.

$$\mathbf{T}^s(\mathbf{X},t) = \mathbf{R}^e \hat{\mathbf{f}}_{\mathcal{K}}(\mathbf{C}^e, \overset{(e)}{\mathbf{\Gamma}}_{\mathcal{K}})(\mathbf{R}^e)^T,$$
$$\boldsymbol{\mu}(\mathbf{X},t) = \mathbf{R}^e \hat{\mathbf{g}}_{\mathcal{K}}(\mathbf{C}^e, \overset{(e)}{\mathbf{\Gamma}}_{\mathcal{K}})[(\mathbf{R}^e)^T, (\mathbf{R}^e)^T] \tag{33}$$

hold. Here $\mathbf{R}^e \in Ort$ results from the polar decomposition applied to the elastic distortion $\mathbf{F}^e$.

2. An equivalent form of the elastic constitutive equations can be expressed in terms of Piola-Kirchhoff or Mandel's tensors, and stress momentum with respect to the configuration with torsion

$$\mathbf{\Pi}_{\mathcal{K}} = \hat{\mathbf{h}}_{\mathcal{K}}(\mathbf{C}^e, \overset{(e)}{\mathbf{\Gamma}}_{\mathcal{K}}), \quad \text{or} \quad \mathbf{\Sigma}_{\mathcal{K}} = \hat{\mathbf{s}}_{\mathcal{K}}(\mathbf{C}^e, \overset{(e)}{\mathbf{\Gamma}}_{\mathcal{K}}),$$
$$\boldsymbol{\mu}_{\mathcal{K}}(\mathbf{X},t) = \hat{\mathbf{g}}_{\mathcal{K}}(\mathbf{C}^e, \overset{(e)}{\mathbf{\Gamma}}_{\mathcal{K}}). \tag{34}$$

For an elasto-plastic body with continuously distributed dislocations, under the constant temperature, we postulate the existence of the free energy per unit crystal volume, i.e. with repect to the configuration with torsion, under the form

$$\psi_{\mathcal{K}} = \hat{\phi}_{\mathcal{K}}(\mathbf{C}^e, \overset{(e)}{\mathbf{\Gamma}}_{\mathcal{K}}). \tag{35}$$

Remark. If we start from the supposition that the elastic behaviour is described in term of the elastic state, $\mathbf{F}^e$, $\boldsymbol{\nabla}_{\mathcal{K}}\mathbf{F}^e$, see also [16], then the objectivity assumptions leads to the form (33), (34) and (35), respectively.

In our constitutive description, the plastic connection has the metricity property as a consequence of the assumption **A2.**:

$$(\overset{(p)}{\mathbf{\Gamma}}_k \mathbf{u})^T \mathbf{C}^p + \mathbf{C}^p(\overset{(p)}{\mathbf{\Gamma}}_k \mathbf{u}) - (\nabla \mathbf{C}^p)\mathbf{u} = 0, \quad \mathbf{C}^p = (\mathbf{F}^p)^T \mathbf{F}^p \tag{36}$$

is the right Cauchy-Green elastic tensor, i.e. a metric tensor in the reference configuration *k*.

Moreover, based on the definition (12) we can prove that the elastic connection $\overset{(e)}{\Gamma}_{\chi}$ has the metricity property (36). Thus

$$(\overset{(e)}{\mathbf{\Gamma}}_{\mathcal{K}} \tilde{\mathbf{u}})^T \mathbf{C}^e + \mathbf{C}^e(\overset{(e)}{\mathbf{\Gamma}}_{\mathcal{K}} \tilde{\mathbf{u}}) - (\nabla_{\mathcal{K}} \mathbf{C}^p)\tilde{\mathbf{u}} = 0, \tag{37}$$

for all $\tilde{\mathbf{u}} \in \mathcal{V}_{\mathcal{K}}$ the tangent space at X in $\mathcal{K}$.

In what follows we put into evidence the importance of **the decomposition theorem of the affine connection**, see [20], [22], applied here for the elastic metric connection with respect to the configuration $\mathcal{K}$.

Theorem 4. 1. The elastic connection $\overset{(e)}{\Gamma}_{\mathcal{K}}$ is expressed in terms of the metric tensor C^e and the contorsion $\mathbf{W}\mathcal{K}$ by

$$\overset{(e)}{\mathbf{\Gamma}}_{\mathcal{K}} = \gamma^e_{\mathcal{K}} + \mathbf{W}_{\mathcal{K}}. \tag{38}$$

The contorsion $\mathbf{W}_{\mathcal{K}}$ and the torsion $\mathbf{S}_{\mathcal{K}}$ third order tensor fields, determine each other by

$$(\mathbf{W}_{\mathcal{K}}\mathbf{u})\mathbf{v} = \frac{1}{2}((\mathbf{S}_{\mathcal{K}}\mathbf{u})\mathbf{v} - (\mathbf{S}_{\mathcal{K}}\mathbf{u})^T\mathbf{v} - (\mathbf{S}_{\mathcal{K}}\mathbf{v})^T\mathbf{u}),$$
$$(\mathbf{S}_{\mathcal{K}}\mathbf{u})\mathbf{v} = (\mathbf{W}_{\mathcal{K}}\mathbf{u})\mathbf{v} - (\mathbf{W}_{\mathcal{K}}\mathbf{v})\mathbf{u}. \tag{39}$$

2. The following skew-symmetry holds

$$(\mathbf{S}_{\mathcal{K}}\mathbf{u})\mathbf{v} = -(\mathbf{S}_{\mathcal{K}}\mathbf{v})\mathbf{u}, \quad (\mathbf{W}_{\mathcal{K}}\mathbf{u})^T = -\mathbf{W}_{\mathcal{K}}\mathbf{u}, \quad \forall\, \mathbf{u}, \mathbf{v} \in \mathcal{V}_{\mathcal{K}}, \tag{40}$$

the torsion $\mathbf{S}_{\mathcal{K}}$ is defined as skew-symmetric part in (16).

3. γ^e is the Riemannian connection, with components given by the second order Christoffel symbols corresponding to $\mathbf{C}^e$,

$$\begin{aligned}(\gamma^e_{\mathcal{K}}\mathbf{u})\mathbf{v}\cdot\mathbf{w} \equiv &\frac{1}{2}(\mathbf{C}^e)^{-1}\mathbf{w}\cdot[((\nabla_{\mathcal{K}}\mathbf{C}^e)\mathbf{v})\mathbf{u} + (\nabla_{\mathcal{K}}\mathbf{C}^e\mathbf{u})\mathbf{v}] \\ &-\frac{1}{2}(\nabla_{\mathcal{K}}\mathbf{C}^e(\mathbf{C}^e)^{-1}\mathbf{w})\mathbf{u}\cdot\mathbf{v},\end{aligned} \tag{41}$$

$(\mathbf{C}^e)^{-1}$ denotes the inverse tensor of $\mathbf{C}^e$.

As a consequence of the previous theorem we derive a remarkable representation for the constitutive functions.

Theorem 5. In our constitutive framework Cauchy stress, stress momentum and free energy are represented by

$$\begin{aligned}\mathbf{T}^s(\mathbf{X},t) &= \mathbf{R}^e\overline{\mathbf{f}}_{\mathcal{K}}(\mathbf{C}^e,\gamma^e;\mathbf{W}_{\mathcal{K}})(\mathbf{R}^e)^T, \\ \boldsymbol{\mu}(\mathbf{X},t) &= \mathbf{R}^e\overline{\mathbf{g}}_{\mathcal{K}}(\mathbf{C}^e,\gamma^e;\mathbf{W}_{\mathcal{K}})[(\mathbf{R}^e)^T,(\mathbf{R}^e)^T], \\ \psi_{\mathcal{K}}(\mathbf{X},t) &= \overline{\phi}_{\mathcal{K}}(\mathbf{C}^e,\gamma^e;\mathbf{W}_{\mathcal{K}}), \quad \text{where} \quad \gamma^e \equiv \gamma^e_{\mathcal{K}}.\end{aligned} \tag{42}$$

We mention that via the decomposition theorem (38) applied to $\overset{(e)}{\Gamma}_{\mathcal{K}}$ and (p) $\overset{(p)}{\Gamma}_{\mathcal{K}}$ the elastic and plastic contorsions coincide, due to the fact that the plastic and elastic torsion, see (16), are equal.

5 APPROACHES TO ELASTO-PLASTIC MATERIALS IN THE INITIAL CONFIGURATION

Following the mean ideas from [5], [3] we introduce the time-independent set of configurations with torsion, restricted to the assumption:

A5. The configurations with torsion and the constitutive functions satisfy the temporal invariance condition (written for instance for elastic constitutive function)

$$\overline{\mathbf{f}}_{\mathcal{K}(t)}(\mathbf{C}^e,\gamma^e;\boldsymbol{\alpha}) = \overline{\mathbf{f}}_{\mathcal{K}(t')}(\mathbf{C}^e,\gamma^e;\boldsymbol{\alpha}), \quad \forall t,t' \in R, \tag{43}$$

and for all ($\mathbf{C}^e$, γ^e; $\boldsymbol{\alpha}$) in the intersection of the definition domains of the constitutive functions.

Remark. The temporal invariance property describes the physical experimental evidence that the elastic constants remain unchanged when the material response is characterized with respect to the crystal reference (lattice structure).

When we pull back the constitutive description written in (34) to the initial (reference) configuration we compose the constitutive functions with the expressions of the elastic measures derived under the form

$$\begin{aligned}&\mathbf{C}^e \equiv (\mathbf{F}^p)^{-T}\mathbf{C}(\mathbf{F}^p)^{-1}, \quad \text{where} \quad \mathbf{C} = \mathbf{F}^T\mathbf{F}, \\ &\overset{(e)}{\Gamma}_{\mathcal{K}} = \mathbf{F}^p(\Gamma_k - \overset{(p)}{\Gamma}_k)[(\mathbf{F}^p)^{-1},(\mathbf{F}^p)^{-1}],\end{aligned} \tag{44}$$

as a consequence of (30) together with (8) and (9).

On the other hand the inverse of the plastic distorsion enters explicitely in the expression of Piola-Kirchoff stress tensor $\mathbf{\Pi}_0$ and stress momentum $\boldsymbol{\mu}_0$ written with respect to the initial configuration *k*, in term of the appropriate stress measure relative to the configuration with torsion.

$$\mathbf{\Pi}_0 = \det(\mathbf{F}^p)(\mathbf{F}^p)^{-1}\mathbf{\Pi}_{\mathcal{K}}(\mathbf{F}^p)^{-T},$$

$$\boldsymbol{\mu}_0 = (\det \mathbf{F}^p)(\mathbf{F}^p)^T\boldsymbol{\mu}_{\mathcal{K}}[(\mathbf{F}^p)^T, (\mathbf{F}^p)^{-T}]. \tag{45}$$

Theorem 6. The elastic behaviour of the elasto-plastic materials with elastic non-Riemannian metric connection is characterized relative to reference configuration by:

$$\mathbf{\Pi}_0 = \hat{\mathbf{h}}_0(\mathbf{C}, \mathbf{\Gamma}_k, \mathbf{Y}), \quad \boldsymbol{\mu}_0 = \hat{\mathbf{g}}_0(\mathbf{C}, \mathbf{\Gamma}_k, \mathbf{Y}),$$

$$\psi_0 = \hat{\phi}_0(\mathbf{C}, \mathbf{\Gamma}_k, \mathbf{Y}), \quad \text{where} \tag{46}$$
$$\mathbf{C} = \mathbf{F}^T\mathbf{F}, \quad \mathbf{\Gamma}_k = \mathbf{F}^{-1}\nabla_k\mathbf{F}, \quad \mathbf{Y} \equiv ((\mathbf{F}^p)^{-1}, \overset{(\mathrm{p})}{\mathbf{\Gamma}}_k).$$

Here $\psi_0 = (\det\mathbf{F}^p)\psi_{\mathcal{K}}$ is the free energy density in the initial configuration, $\mathbf{Y}$ is a measure of the irreversible behaviour of the material.

Remarks. We apply again the decomposition theorem of the metric connection, first for the motion connection and after that for the plastic connection also with respect to the initial configuration. Note that

$$\mathbf{\Gamma}_k = \boldsymbol{\gamma}, \quad \text{since} \quad \mathbf{F} = \nabla_k\chi, \quad \text{and} \quad \overset{(\mathrm{p})}{\mathbf{\Gamma}}_k = \boldsymbol{\gamma}_k + \mathbf{W}_k. \tag{47}$$

We emphasize that the plastic distorsion itself, $\mathbf{F}^p$; is involved in the constitutive functions, not only its symmetric and positive part $\mathbf{C}^p = (\mathbf{F}^p)^T\mathbf{F}^p$, as a consequence of the formulae (44).

Theorem 7. The elastic behaviour of the elasto-plastic materials with metric non-Riemannian connection is characterized by

$$\mathbf{\Pi}_0 = \overline{\mathbf{h}}_0(\mathbf{C}, \boldsymbol{\gamma}, (\mathbf{F}^p)^{-1}, \boldsymbol{\gamma}^p, \mathbf{W}_{\mathcal{K}}), \quad \boldsymbol{\mu}_0 = \overline{\mathbf{g}}_0(\mathbf{C}, \boldsymbol{\gamma}, (\mathbf{F}^p)^{-1}, \boldsymbol{\gamma}^p, \mathbf{W}_k)$$

$$\psi_0 = \overline{\phi}_0(\mathbf{C}, \boldsymbol{\gamma}, (\mathbf{F}^p)^{-1}, \boldsymbol{\gamma}^p, \mathbf{W}_k). \tag{48}$$

Here

$\mathbf{C} = \mathbf{F}^T\mathbf{F}$ is the strain measure, or the Cauchy-Green right tensor,

$\mathbf{F}^p$ is plastic distorsion, $\mathcal{W}_k$ the plastic contorsion with respect to the reference configuration k, calculated through (39) via the (plastic) torsion written in (5),

γ and γ^p are the Riemann-Cristoffel symbol of second order attached to the metric tensors $\mathbf{C}$, and $\mathbf{C}^p$, respectively, see formula (41).

Comparisons between different models via the expresssion of the free energy densities.

In [6] the elastic constitutive functions have been considered to be dependent on the elastic connection $\mathbf{C}^e$ and on the elastic connection only by its torsion, $\mathbf{S}_{\mathcal{K}}$, written in $(16)_1$. In this case the free energy function introduced in **Theorem 7** becomes

$$\psi_0 = \overline{\phi}_0(\mathbf{C}, (\mathbf{F}^p)^{-1}, \mathbf{W}_k), \quad \text{or} \quad \psi_0 = \overline{\phi}_0(\mathbf{C}, (\mathbf{F}^p)^{-1}, \mathbf{S}^p_k). \tag{49}$$

A similar result is derived in [16], [15], where the stored energy has been proved to be dependent on ($\mathbf{F}^p$, $\mathbf{C}$, $\mathbf{T}$), see formula (3.17) from [16], with $\mathbf{T} = -\mathbf{S}_k$ the torsion with respect to the initial configuration.

When there is no plastic deformation: $\mathbf{F}^p = \mathbf{I}$, $\mathbf{F}^e = \mathbf{F}$. Hence $\mathbf{C}^e = \mathbf{C}$ and the torsion with repect to the initial configuration $\mathbf{S}_k = 0$. From our constitutive representation given in **Theorem 7** we get

$$\mathbf{\Pi}_0 = \overline{\mathbf{h}}_0(\mathbf{C}, \boldsymbol{\gamma}), \quad \boldsymbol{\mu}_0 = \overline{\mathbf{g}}_0(\mathbf{C}, \boldsymbol{\gamma}), \quad \psi_0 = \overline{\phi}_0(\mathbf{C}, \boldsymbol{\gamma}). \tag{50}$$

Here $\mathbf{C} = \mathbf{F}^T\mathbf{F}$ is the Cauchy-Green tensor and γ is a measure of strain gradient tensor. We get a finite deformation constitutive relations of the type proposed in Fleck and Hutchinson [10], within the deformation theory.

6 SMALL ELASTIC STRAIN IN MODEL WITH NON-RIEMANNIAN METRIC CONNECTION

In [7] we derived the behaviour of the elasto-plastic materials with dislocations based on the assumptions of small elastic strains, but large plastic distorsions. We adopted the point of view developed by Mandel [17], based on the polar decomposition of the *elastic* distorsion under the form

$$\begin{aligned}&\mathbf{F}^e = \mathbf{R}^e\mathbf{U}^e, \quad \mathbf{R}^e \in Ort, \quad \mathbf{U}^e \simeq \mathbf{I} + \boldsymbol{\epsilon}^e, \quad \text{with} \quad |\, \boldsymbol{\epsilon}^e \,| \ll 1,\\ &\mathbf{C}^e \simeq \mathbf{I} + 2\boldsymbol{\epsilon}^e.\end{aligned} \tag{51}$$

Here $\boldsymbol{\epsilon}^e$ is symmetric tensor. The invariant elastic connection $\Gamma_{\mathcal{K}}^{(e)}$ with respect to the configuration with torsion, introduced in (13), can be expressed by

$$\overset{(e)}{\boldsymbol{\Gamma}}_{\mathcal{K}} \simeq \nabla_{\mathcal{K}}\boldsymbol{\epsilon}^e + (\mathbf{R}^e)^T\nabla_{\mathcal{K}}\mathbf{R}^e. \tag{52}$$

when $\boldsymbol{\epsilon}^e$ is neglected in the presence of the identity tensor, and so on.

The elastic metric connection $\Gamma_{\mathcal{K}}^{(p)}$ via the decomposition theorem (51), can be reduced to $\Gamma_{\mathcal{K}}^{(e)} = \gamma_{\mathcal{K}}^e + \mathbf{W}_{\mathcal{K}}$ with $\gamma_{\mathcal{K}}^e$, Riemann-Cristoffel symbol, given by

$$(\boldsymbol{\gamma}^e\mathbf{u})\mathbf{v}\cdot\mathbf{z} \simeq ((\nabla_{\mathcal{K}}\boldsymbol{\epsilon}^e\mathbf{v})\mathbf{u} + (\nabla_{\mathcal{K}}\boldsymbol{\epsilon}^e)\mathbf{u})\mathbf{v})\cdot\mathbf{z} - ((\nabla_{\mathcal{K}}\boldsymbol{\epsilon}^e)\mathbf{z})\mathbf{u})\cdot\mathbf{v}, \tag{53}$$

with $(\gamma^e\mathbf{u})\mathbf{v} = (\gamma^e\mathbf{v})\mathbf{u}$, while the (elastic) torsion $\mathbf{S}_{\mathcal{K}}$ is defined for all $\tilde{\mathbf{u}}, \tilde{\mathbf{v}}, \in \mathcal{V}_{\mathcal{K}}$ by

$$\begin{aligned}(\mathbf{S}_{\mathcal{K}}\tilde{\mathbf{u}})\tilde{\mathbf{v}} \;&= ((\nabla_{\mathcal{K}}\boldsymbol{\epsilon}^e)\tilde{\mathbf{u}})\tilde{\mathbf{v}} - ((\nabla_{\mathcal{K}}\boldsymbol{\epsilon}^e)\tilde{\mathbf{v}})\tilde{\mathbf{u}} +\\ &+(\mathbf{R}^e)^T[((\nabla_{\mathcal{K}}\mathbf{R}^e)\tilde{\mathbf{u}})\tilde{\mathbf{v}} - ((\nabla_{\mathcal{K}}\mathbf{R}^e)\tilde{\mathbf{v}})\tilde{\mathbf{u}}].\end{aligned} \tag{54}$$

The first part of the torsion $\mathbf{S}_{\mathcal{K}}$ is characterized by the skew-symmetric part of the gradient of $\boldsymbol{\epsilon}^e$ – the small elastic strain tensor, while the second part is the torsion attached to the affine connection generated by the elastic rotation only.

Now we pass to the contorsion $\mathbf{W}_{\mathcal{K}}$ calculated by

$$\begin{aligned}&(\mathbf{W}_{\mathcal{K}}\mathbf{u})\mathbf{v}\cdot\mathbf{z} \simeq ((\nabla_{\mathcal{K}}\boldsymbol{\epsilon}^e\mathbf{z})\mathbf{v}\cdot\mathbf{u} - (\nabla_{\mathcal{K}}\boldsymbol{\epsilon}^e)\mathbf{v})\mathbf{u})\cdot\mathbf{z} + (\mathbf{W}_{\mathcal{K}}^R\mathbf{u})\mathbf{v}\cdot\mathbf{z},\\ &\text{with} \quad \mathbf{W}_{\mathcal{K}}^R = (\mathbf{R}^e)^T\nabla_{\mathcal{K}}\mathbf{R}^e,\end{aligned} \tag{55}$$

due to the symmetry $(\boldsymbol{\epsilon}^e\mathbf{u})\cdot\mathbf{v} = (\boldsymbol{\epsilon}^e\mathbf{v})\cdot\mathbf{u}$, and to the skew-symmetry property $(\mathbf{R}^e)^T((\nabla_{\mathcal{K}}\mathbf{R}^e\mathbf{u})\mathbf{v}\cdot\mathbf{z} = -(\mathbf{R}^e)^T((\nabla_{\mathcal{K}}\mathbf{R}^e\mathbf{u})\mathbf{z}\cdot\mathbf{v}$, generated by the orthogonal tensor field $\mathbf{R}^e$.

In the case of small elastic distorsion, i.e. when the elastic strains as well as the elastic rotations are small, we introduce

$$\begin{aligned}&\mathbf{F}^e = \mathbf{I} + \boldsymbol{\beta}^e, \quad |\, \boldsymbol{\beta}^e \,| \ll 1,\\ &\mathbf{C}^e = \mathbf{I} + 2\boldsymbol{\epsilon}^e \text{ with } \boldsymbol{\epsilon}^e = \frac{1}{2}(\boldsymbol{\beta}^e + (\boldsymbol{\beta}^e)^T),\; \boldsymbol{\epsilon}^e = \frac{1}{2}(\boldsymbol{\beta}^e + (\boldsymbol{\beta}^e)^T).\end{aligned} \tag{56}$$

We get

$$\overset{(e)}{\boldsymbol{\Gamma}}_{\mathcal{K}} = \nabla_{\mathcal{K}}\boldsymbol{\beta}^e, \text{ and } (\mathbf{S}_{\mathcal{K}}\mathbf{u})\mathbf{v} = (\nabla_{\mathcal{K}}\boldsymbol{\beta}^e\mathbf{u})\mathbf{v} - (\nabla_{\mathcal{K}}\boldsymbol{\beta}^e\mathbf{v})\mathbf{u}. \tag{57}$$

It follows that

$$(\gamma^e \mathbf{u})\mathbf{v}\cdot\mathbf{z} \simeq ((\nabla_{\mathcal{K}}\boldsymbol{\epsilon}^e\mathbf{v})\mathbf{u} + (\nabla_{\mathcal{K}}\boldsymbol{\epsilon}^e)\mathbf{u})\mathbf{v})\cdot\mathbf{z} - ((\nabla_{\mathcal{K}}\boldsymbol{\epsilon}^e)\mathbf{z})\mathbf{u})\cdot\mathbf{v}, \tag{58}$$

while the (elastic) contorsion $\mathbf{W}_{\mathcal{K}}$

$$(\mathcal{W}_{\mathcal{K}}\tilde{\mathbf{u}})\tilde{\mathbf{v}}\cdot\mathbf{w} = ((\nabla_{\mathcal{K}}\boldsymbol{\omega}^e)\tilde{\mathbf{u}})\tilde{\mathbf{v}}\cdot\tilde{\mathbf{w}} - ((\nabla_{\mathcal{K}}\boldsymbol{\epsilon}^e)\tilde{\mathbf{v}})\tilde{\mathbf{u}}\cdot\tilde{\mathbf{w}} + ((\nabla_{\mathcal{K}}\boldsymbol{\epsilon}^e)\tilde{\mathbf{w}})\tilde{\mathbf{u}}\cdot\tilde{\mathbf{v}}. \tag{59}$$

Remark. The presence of dislocations is involved in the model if and only if the contorsion $\mathcal{W}_{\mathcal{K}}$, as a measure of the non-homogeneity, is a non-vanishing third order tensor.

Based on the decomposition theorem for the metric elastic connection we can propose another possible starting point than those utilized by Fleck and Hutchinson in [10]. We introduce the gradient of the small elastic distorsion and the gradient of the small elastic strain

$$\boldsymbol{\eta} \equiv \nabla_{\mathcal{K}}\boldsymbol{\beta}^e, \quad \boldsymbol{\rho} \equiv \nabla_{\mathcal{K}}\boldsymbol{\epsilon}^e, \tag{60}$$

and so on.

7 EVOLUTION EQUATIONS FOR IRREVERSIBLE VARIABLES

To define completely the mathematical model, evolution equations for time-dependent local configurations with torsion have to be introduced. This means the evolution equations for $\mathbf{F}^p$ and $\nabla\mathbf{F}^p$ will be prescribed.

A7. Evolution equations for $\mathbf{F}^p$, with respect to $\mathcal{K}$ will be described by

$$\mathbf{L}^p \equiv \dot{\mathbf{F}}^p(\mathbf{F}^p)^{-1} = \mathcal{B}_{\mathcal{K}}(\Sigma_{\mathcal{K}}, \boldsymbol{\mu}_{\mathcal{K}}, \kappa_1, \kappa_2, \dot{\mathbf{F}}, \frac{d}{dt}(\nabla_k\mathbf{F})), \tag{61}$$

related as usual to the yield surface defined in the appropriate stress space, say the space of stress and stress momentum, with respect to the set of variables ($\boldsymbol{\Sigma}_{\mathcal{K}}$, $\boldsymbol{\mu}_{\mathcal{K}}$, κ_1, κ_2).

A8. Evolution equations for $\nabla_k\mathbf{F}^p$, with respect to $\mathcal{K}$ will be prescribed by

$$\frac{d}{dt}(\nabla_k\mathbf{F})(\mathbf{F}^p)^{-1} = \mathcal{A}_{\mathcal{K}}(\Sigma_{\mathcal{K}}, \boldsymbol{\mu}_{\mathcal{K}}, \kappa_1, \kappa_2, \dot{\mathbf{F}}, \frac{d}{dt}(\nabla_k\mathbf{F})). \tag{62}$$

Remarks. 1. Here κ_1, κ_2, are hardening parameters, which have to be described also by the evolution equations, similar to the previous ones.

2. The relationship between the gradients of the plastic rate with respect to the initial and plastically deformed configurations $\mathcal{K}$, as well as between their rates, can be derived using **rules of calculus** given in formulae (11).

3. When the rate of the second pair of plastic distortion are given, then the rate of different measure of the inhomogeneities, say the Noll's densities of dislocations can be derived starting from their definitions.

4. In the case of metric affine connection, the plastic distorsion and the gradient of the plastic distorsion are related by the metricity conditions, via the plastic connection by (refE6). Hence the evolution equations for the plastic distorsion and the appropriate second order inhomogeneities are restricted to the compatibility relations.

REFERENCES

Aifantis E.C. (1984). On the microstructural origin of certain inelastic models. Trans. ASME, *J. Eng.Mat.Tech.*, 105:326–330.

Bilby B.A. (1960). Continuous distribution of dislocations. *In Progress in Solid Mechanics*, I, I.N. Sneddon, R. Hill, eds, North-Holland, Amsterdam: 329–398.

Cleja-Ţigoiu S. (1990). Large elasto-plastic deformations of materials with relaxed configurations-I.Constitutive assumptions, II. Role of the complementary plastic factor. *Int. J. Engng. Sci.*, 28:171–180, 273–284.

Cleja- Ţigoiu S. and Maugin G.A. (2000). Eshelby's stress tensors in finite elastoplasticity. *Acta Mechanica*, 139:231–249.

Cleja- Ţigoiu S. and Soós E. (1990). Elastoplastic models with relaxed configurations and internal state variables. *Appl. Mech. Rev.* 43:131–151.

Cleja-Ţigoiu S. (2002). Couple stresses and non-Riemannian plastic connection in finite elsto-plasticity. *ZAMP* 53:996–1013.

Cleja-Ţigoiu S. (2002). Small elastic strains in finite elasto-plastic materials with continuously distributed dislocations. *Theoretical and Applied Mechanics* 28–29:93–112.

Epstein M. and Maugin G. (2000). Thermomechanics of volumetric growth in uniform bodies. *Int. J. Plast.* 16:951–978.

Fleck N.A., Muller G.M., Ashby M.F. and Hutchinson J.W. (1994). Strain gradient plasticity: theory and experiment. *Acta Metall. Mater.* 42:475–487.

Fleck N.A. and Hutchinson J.W. (2001). A reformulation of strain gradient plasticity. *J. Mech. Phys. Solids* 49:2245–2271.

Forest S., Cailletand G. and Sievert R. (1997). A Cosserat theory for elastoviscoplastic single crystals at finite deformation. *Arch. Mech.* 49(4):705–736.

Gurtin M.E. (2000). On the plasticity of single crystal: free energy, microforces, plastic-strain gradients. *J. Mech. Phys. Solids* 48:989–1036.

Hwang K.C., Jiang H., Huang Y., Gao H. and Hu N. (2002). A finite deformation theory of strain gradient plasticity. *J. Mech. Phys. Solids* 50:81–99.

Kondo K. and Yuki M. (1958). On the current viewpoints of non-Riemannian plasticity theory. *In RAAG memoirs of the unifying study of basic problems in engng. and physical sciences by means of geometry*, II (D):202–226.

Le K.C. and Stumpf H. (1996). A model of elastoplastic bodies with continuously distributed dislocations. *Int. J. Plast.* 12(5):611–627.

Le K.C. and Stumpf H. (1996). Nonlinear continuum with dislocations. *Int. J. Engng. Sci.,* 34:339–358.

Mandel J. (1972). *Plasticité classique et viscoplasticité.* CISM- Udine, Springer- Verlag, Vienna, New-York.

Naghdi P.M. and Srinivasa A.R. (1994). Characterization of dislocations and their influence on plastic deformation in single crystal. *Int. J. Engng. Sci.* 32(7):1157–1182.

Noll W. (1967). Materially uniform siomple bodies with inhomogeneities. *Arch. Rat. Mech. Anal.* 27:1–32.

Schouten J.A. (1954). *Ricci-Calculus.* Springer-Verlag/ Berlin. Gottingn. Heidelberg.

Shizawa K. and Zbib H.M. (1999). A thermodynamical theory of gradient elastoplasticity with dislocation density tensor. I. Fundamentals. *Int. J. Plast.* 15:899–938.

Steinmann P. (1996). Views on multiplicative elastoplasticity and continuum theory of dislocations. *Int. J. Engng. Sci.,* 34:1717–1735.

Svendsen B. (2002). Continuum thermodynamic models for crystal plasticity including the effects of geometrically-necessary dislocations *J. Mech.Phys.Solids.* 50:1297–1329.

Teodosiu C. (1970). A dynamic theory of dislocations and its applications to the theory of the elastic-plastic continuum. In *Fundamental Aspects of Dislocation Theory*, J.A. Simmons, R. de Witt, and R. Bullough, eds., Nat. Bur. Stand. (U.S.), Spec. Publ. 317, II,:837–876.

Wang C.C. (1967). On the geometric structure of simple bodies, a Mathematical foundation for the theory of continuous distributions of dislocations. *Arch. Rat. Mech. Anal.* 27:33–94.

Configurational Mechanics, Kalpakides & Maugin (eds)
© 2004, Taylor & Francis Group, London, ISBN 90 5809 667 X

Branching dynamics of brittle fracture

M. Adda-Bedia & M. Ben Amar
Laboratoire de Physique Statistique de l'Ecole Normale Supérieure, France

ABSTRACT: We present calculations of the static stress intensity factors of a branched crack for both antiplane and inplane configurations. The result s are applied to the problem of dynamic crack branching, using an approximated form of the dynamic stress fields. We show that the finite waves speeds in the material cause a time delay in the interaction between the branches. The time delay increases with the propagation velocity, v_b, of the two branches, and causes each branch to behave as if the other branch was shorter. Despite these effects, we show that for any branching angle, the maximum energy release rate is always attained when $v_b = 0$. Following Eshelby, we show that the branching of a single propagating crack is energetically possible when its speed v exceeds a threshold value. Our calculations predict a relatively large branching angles, which is consistent with experimental observations. In addition, we show that an increasing fracture energy with the velocity results in a decrease in the critical velocity at which branching is energetically possible. Finally, we show that the principle of local symmetry and the principle of maximum energy release rate predict two different crack paths.

1 INTRODUCTION

A well-known phenomenon in brittle and semi-brittle crack propagation is the bifurcation of the crack tip into two or more branches (Boudet et al., 1996; Boudet and Ciliberto, 2000; Fineberg et al., 1992; Gross, 1993; Lawn, 1993; Ravi-Chandar and Knauss, 1984; Sharon et al., 1995; Sharon and Fineberg, 1996, 1999). Recent experiments in PMMA and glass plates (Sharon et al., 1995; Sharon and Fineberg, 1996, 1999) have clearly established that this phenomenon results from a dynamic instability of a single propagating crack. The instability occurs when the crack speed v exceeds a critical velocity v_c, which does not depend on the applied tractions and the geometry of the plate. Above v_c, a single crack is no longer stable, instead, a repetitive process of micro-branching occurs, which changes the crack dynamics. The acoustic emission from the crack increases (Boudet et al., 1996; Boudet and Ciliberto, 2000; Gross, 1993), the crack velocity develops strong oscillations and a pattern (which is correlated with the velocity oscillations) is created on the fracture surface (Fineberg et al., 1992). Crack branching has also been observed in simulations of dynamic crack propagation using molecular dynamics (Abraham et al., 1994) and finite element calculations, numerical resolution of constitutive equations on a lattice (Xu and Needleman, 1994), and by modeling the elastic medium as a two-dimensional lattice of coupled springs (Marder and Gross, 1995).

In 1951, Yoffe (Yoffe, 1951) observed that for crack speeds less than a critical velocity v_y, the transverse tensile stress in the vicinity of a crack tip attains its maximum along the direction of crack growth. For crack speeds greater than v_y, this component of the stress develops a maximum in two other symmetric directions. Yoffe suggested that this modification of the local singular stress field could account for the observation that rapidly growing cracks in brittle materials bifurcate into branched cracks. Since then, the origin of the micro-branching instability has been discussed by several authors (Adda-Bedia et al., 1996). However, all the attempts to explain this transition have led to critical speeds much larger than the experimental ones. A possible cause for the failure of theoretical predictions of this kind is that they were focused on the stress distribution around the tip of the single straight crack, prior to branching. These analysis clearly indicate that the stress field around the crack tip is deformed at high velocities, however, this does not set a crack growth or branching criterion.

As in the single crack case, a growth criterion for a branched crack must be based on the equality between the energy flux into the two propagating tips and the surface energy which is added as a result of this propagation (Griffith, 1920). Eshelby proposed this approach in 1970 (Eshelby, 1970; Rice et al., 1994), posing the question of how large must the single crack speed be so that there will be enough energy available to form two slow cracks instead of the single fast one. Since after branching twice as much surface area is being created, the branches would not advance unless their velocity is lower than that of the single crack. The simplest branched configuration that Eshelby analyzed is the idealized limiting case where the branches subtend a vanishingly small angle and both prolong the original crack plane. In this case, the energy release rate of two branches propagating at velocity v_b, equals twice that of a single crack moving at velocity v, and it is maximal when the branches velocity is $v_b = 0^+$. For the mode III case[1], Eshelby reported $v = 0.6c_s$ as the minimum speed which could allow branching or surface roughening (Eshelby, 1970). This result is clearly a rough estimate, as it is based on the non-realistic assumption of zero branching angle. The resultant zero velocity after branching is non-realistic as well. Still, the zero angle assumption was necessary as no solution was available for two branches with an arbitrary angle between them.

In this work, we provide exact calculations of the stress intensity factors of two static branches with an arbitrary branching angle. Using this solution, and introducing an approximate form of both the kinetic energy and the delayed stress fields due to the finite wave speed in the material, we calculate the stress intensity factors of the dynamic branching problem. Using the energy release rate of the dynamic branches, we calculate the branching velocity threshold and the branching angle in the light of Eshelby's hypothesis.

The paper is presented as follows. In Section II, we present the general elastostatic problem of a uniaxially stretched sheet, containing a crack of finite length and $(n - 1)$ side-branches of infinitely small lengths. In Section III, we present the solution for the case $n = 3$. In Section IV, we construct an approximate solution for the dynamic branching case by taking into account the finite waves velocities in the material, which cause a delay in the interaction between the two propagating branches. Due to this delay mechanism, each propagating branch interacts with an effectively shorter one. In Section V, we combine these results with Eshelby's approach to the problem of dynamic crack branching criterion. We show that the maximum energy release rate is attained when the branches propagate at a non zero branching angle, but at vanishing velocity. The critical velocity for branching of the initial single crack does not vary a lot when we apply to path selection the principal of local symmetry, although the branching angle increases. As this branching criterion is based on an energy balance, it is influenced by the velocity dependence of the fracture energy. The branching velocity is predicted to decrease when the fracture energy increases with velocity. In the last section, we compare our results to experimental data, and discuss their relevance in relation to other models of dynamic crack branching.

2 THE GENERAL ELASTOSTATIC PROBLEM

Let us start by giving the general solution of the elastostatic problem of an infinite, uniaxially stretched sheet with a crack contour consisting of an arbitrary number n of straight rays of different lengths and different directions, emerging from a common origin. The solution for any stress conditions at infinity can be obtained by superposition. In particular, the case of a main crack with $(n - 1)$ side-branches of infinitely small lengths is studied. The complex stress function method is used and the stress intensity factors are derived.

[1]At that time, the Freund–Kostrov solution for unsteady crack motion in the tensile mode was not available.

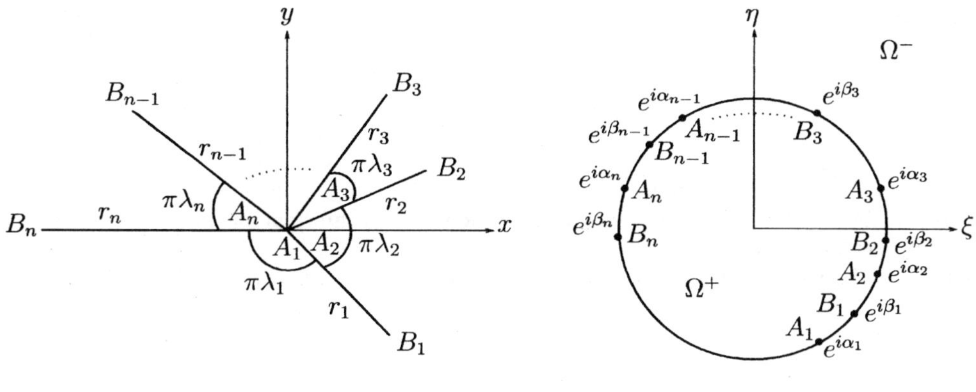

z–plane **ζ–plane**

Figure 1. Conformal mapping of a star shape crack in the z-plane onto the exterior of the unit circle in the ζ-plane.

In the following, the resolution of the mode III loading is given in details, while the mixed mode I–II loading is briefly presented. Detailed analysis of a similar problem can be found in (Amestoy and Leblond, 1992).

2.1 *Mapping function*

The mapping of the exterior of a star shape crack in the z-plane, $z = x + iy$, onto the exterior of the unit circle in the ζ-plane, $\zeta = \xi + i\eta$, is considered. The conformal mapping of a contour consisting of n rays from a common origin to the unit circle, according to Fig. 1 is given by the function (Andersson, 1969)

$$z = \omega(\zeta) = A\exp[i\phi]\,\zeta^{-1}\prod_{k=1}^{n}\left(\zeta - e^{i\alpha_k}\right)^{\lambda_k}\;; \tag{1}$$

where A and α_k, ($k = 1, \ldots, n$), are real constants. As can be seen from Fig. 1,

$$\sum_{k=1}^{n}\lambda_k = 2\,.$$

The values $\zeta_k = e^{i\beta_k}$, ($k = 1, \ldots, n$), corresponding to the tips B_k of the star shape crack in the z-plane, are the maxima of $\omega(\zeta)$. They satisfy the conditions

$$\sum_{k=1}^{n}\lambda_k\cot\left(\frac{\beta_j-\alpha_k}{2}\right) = 0\;; \tag{2}$$

with,

$$\alpha_1 \le \beta_1 \le \alpha_2 \le \ldots \le \alpha_n \le \beta_n \le 2\pi + \alpha_1\,.$$

The lengths of the rays are then expressed in the form

$$r_j = 4A\prod_{k=1}^{n}\left|\sin\left(\frac{\beta_j-\alpha_k}{2}\right)\right|^{\lambda_k}\,. \tag{3}$$

The x-axis is taken to be parallel to the nth crack, and the argument of z is defined in the interval $[-\pi, \pi]$. Then, the phase ϕ satisfies the condition

$$\phi = -\frac{1}{2}\sum_{k=1}^{n} \lambda_k \alpha_k \, . \tag{4}$$

Equations (2, 3, 4), in addition to an additional arbitrary condition due to the invariance under a global rotation in the ζ-plane, give a unique solution of the conformal mapping parameters A, ϕ, α_k and β_k as functions of r_k and λ_k.

In the vicinity of the tips B_k, the mapping function $z = \omega(\zeta)$ satisfies the following behavior

$$z - z_k \equiv \omega(\zeta) - \omega(\zeta_k) = \frac{1}{2}w''(\zeta_k)(\zeta - \zeta_k)^2 \, ; \tag{5}$$

$$\omega'(\zeta) = w''(\zeta_k)(\zeta - \zeta_k) = \sqrt{2w''(\zeta_k)(z - z_k)} \, ; \tag{6}$$

where,

$$\omega''(\zeta_k) = \frac{r_k}{4} e^{i(\delta_k - 2\beta_k)} \sum_{j=1}^{n} \frac{\lambda_j}{\sin^2\left(\frac{\beta_k - \alpha_j}{2}\right)} \, ; \tag{7}$$

and

$$\delta_k = \pi \sum_{j=1}^{k} \lambda_j - \pi \, . \tag{8}$$

In the following, we will be mostly interested in the limiting case where all the cracks, except the nth one, are of infinitely small lengths. For this case, one can define in the z-plane

$$r_n = L \, ; \; r_k = \frac{L}{4} l_k \epsilon \;\; (k = 1, \ldots, n-1) \, ;$$

with, $\epsilon \to 0$. Consequently, in the ζ-plane one has

$$\beta_n = \pi \, ; \; \beta_k = b_k \sqrt{\epsilon} \;\; (k = 1, \ldots, n-1) \, ; \; \alpha_k = a_k \sqrt{\epsilon} \;\; (k = 1, \ldots, n);$$

with,

$$a_1 \leq b_1 \leq a_2 \ldots \leq b_{n-1} \leq a_n \, .$$

To leading order in ϵ, Equations (2, 3, 4) thus give

$$A = \frac{L}{4} \, ; \tag{9}$$

$$\phi = 0 \, ; \tag{10}$$

$$\sum_{j=1}^{n} \frac{\lambda_j}{b_k - a_j} = 0 \, ; \tag{11}$$

$$\prod_{j=1}^{n} |b_k - a_j|^{\lambda_j} = l_k \, ; \tag{12}$$

for $k = 1, \ldots, n-1$. Finally, Eq. (7) reduces to

$$\omega''(\zeta_k) = \frac{L}{4} l_k e^{i\delta_k} \sum_{j=1}^{n} \frac{\lambda_j}{(b_k - a_j)^2} \; ; \qquad k = 1, \ldots, n-1 \, . \tag{13}$$

2.2 *The mode III problem*

In anti-plane strain deformation, the displacement $u_3(x, y)$ normal to the xy plane satisfies Laplace's equation

$$\Delta u_3 = 0 \, . \tag{14}$$

A traction free boundary conditions are taken on crack surfaces, and the infinite elastic body is supposed to be loaded by an external shear stress σ_{23}^{∞}. These conditions are written

$$\vec{n} \cdot \vec{\nabla} u_3 = 0 \quad , \qquad \text{on the crack surfaces;} \tag{15}$$

$$\sigma_{13} - i\sigma_{23} = -i\sigma_{23}^{\infty} \quad , \qquad \text{at infinity.} \tag{16}$$

The displacement and stress fields can be expressed by means of a complex function $\Phi(z)$, which is holomorphic outside the crack region. In the z-plane, one has

$$u_3 = \frac{1}{2\mu} \left[\Phi(z) + \overline{\Phi(z)} \right] \; ; \tag{17}$$

$$\sigma_{13} - i\sigma_{23} = \Phi'(z) \; ; \tag{18}$$

where μ is the Lamé shear coefficient and the overline indicates the complex conjugate. Note that $\overline{\Phi(z)}$ is a complex function which is holomorphic inside the crack region. The problem can be easily solved in the ζ-plane. Effectively, the boundary conditions (15, 16) read

$$\Phi(\zeta) - \overline{\Phi(\zeta)} = 0 \quad , \qquad \text{when} \qquad |\zeta| \to 1^+ \; ; \tag{19}$$

$$\Phi'(\zeta) = -i\sigma_{23}^{\infty} A e^{i\phi} \quad , \qquad \text{when} \qquad |\zeta| \to \infty \, . \tag{20}$$

The solution is readily obtained if $\Phi(\zeta)$ is given by (Sih, 1965)

$$\Phi(\zeta) = -i\sigma_{23}^{\infty} A \left[e^{i\phi}\zeta - e^{-i\phi}\zeta^{-1} \right] . \tag{21}$$

The stress intensity factors[2] $K_3^{(k)}$ at each tip B_k, ($k = 1, \ldots, n-1$), are defined by

$$K_3^{(k)} = i \lim_{z \to z_k} \sqrt{2\pi e^{i\delta_k}(z - z_k)} (\sigma_{13} - i\sigma_{23}) \, . \tag{22}$$

Using the behavior of $\Phi(\zeta)$ when $\zeta \to \zeta_k$ and relating it to the square root singularity of the stress field leads to the following result

$$K_3^{(k)} = i\Phi'(\zeta_k) \sqrt{\frac{\pi e^{i\delta_k}}{w''(\zeta_k)}} \, . \tag{23}$$

[2]For convenience, we will use in the following arabic numbers (1, 2 and 3) for the subscripts that label the loading modes instead of roman numbers.

One can easily verify that this quantity is real. For the limiting case where all the cracks, except the *nth* one, are of infinitely small lengths, one has

$$K_3^{(k)} = F_{33}^{(k)} K_3^{(0)} \ , (k = 1, \ldots, n-1) \ ; \tag{24}$$

where $K_3^{(0)} = \sigma_{23}^{\infty}\sqrt{\pi L/2}$, is the stress intensity factor of the main crack of length L in the absence of the $n-1$ branches. The general form of F_{33} is given by

$$F_{33}^{(k)} = \left[\frac{l_k}{2} \sum_{j=1}^{n} \frac{\lambda_j}{(b_k - a_j)^2} \right]^{-1/2} \ ; \tag{25}$$

which depends only of the conformal mapping parameters. This result is universal in the sense that it is independent of the applied loading. The discontinuity introduced by the vertices A_j is not intuitive, since the stress intensity factor is not given by a simple angular contribution.

2.3 *The mixed mode I–II problem*

In the following, the branching problem in plane situation is presented briefly. Effectively, the approach is analogous to the kinked crack problem which has been studied previously (Amestoy and Leblond, 1992). More details of the analysis can be found in (Amestoy and Leblond, 1992) and references therein.

According to Muskhelishvili (Muskhelishvili, 1953), the stresses at a point $z = x + iy = \omega(\zeta)$ can be expressed, in the ζ-plane, by the complex stress functions $\Phi(\zeta)$ and $\Psi(\zeta)$. For the case where a traction free boundary conditions are taken on crack surfaces and where the loading is given by external stresses σ_{11}^{∞}, σ_{22}^{∞}, and σ_{12}^{∞}, the complex stress functions satisfy the following conditions

$$\Phi(\zeta) + \frac{\omega(\zeta)}{\omega'(\zeta)}\overline{\Phi'(\zeta)} + \overline{\Psi(\zeta)} = Cte \quad , \qquad \text{when} \qquad |\zeta| \to 1^+ \ ; \tag{26}$$

$$\Phi(\zeta) = \Gamma A e^{i\phi} \quad , \qquad \text{when} \qquad |\zeta| \to \infty \ ; \tag{27}$$

$$\Psi(\zeta) = \Gamma' A e^{i\phi} \quad , \qquad \text{when} \qquad |\zeta| \to \infty \ ; \tag{28}$$

where $\Gamma = (\sigma_{11}^{\infty} + \sigma_{22}^{\infty})/4$ and $\Gamma' = (\sigma_{22}^{\infty} - \sigma_{11}^{\infty})/2 + i\sigma_{12}^{\infty}$.

We focus on the asymptotic case where all the cracks, except the *nth* one, are of infinitely small lengths. Let us perform a second conformal mapping given by

$$\zeta = \exp i\sqrt{\epsilon}Z \ ; \tag{29}$$

which maps the region $|\zeta| > 1$ onto the domain $\Im Z < 0$ (see Fig. 2).

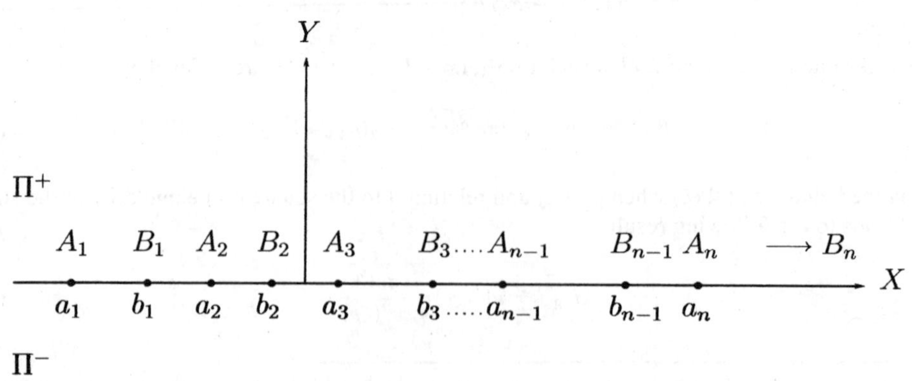

Figure 2. The Z-plane corresponding to the conformal mapping $\zeta = \exp i.\sqrt{\epsilon}\ Z$

Following the same steps as in (Amestoy and Leblond, 1992), one arrives to the result that the problem reduces to solve the integral equation

$$U(Z) = K^{(0)} + \sum_{k=1}^{n-1}(1 - e^{2i\delta_k}) \int_{a_k}^{a_{k+1}} q(t) \frac{\overline{U(t)}}{(t-Z)^2} \frac{dt}{4i\pi} ; \tag{30}$$

where

$$U(Z) = \sqrt{\frac{8\pi}{L}}\Phi'(Z) ; \tag{31}$$

is a holomorphic function in the domain $\Im Z < 0$. The function $q(t)$ is given by

$$q(t) = \frac{\prod_{j=1}^{n}(t - a_j)}{\prod_{j=1}^{n-1}(t - b_j + i\epsilon)} . \tag{32}$$

The term $K^{(0)}$ is the complex stress intensity factor of the main crack of length L in the absence of the $n - 1$ branches. It is given by

$$K^{(0)} = K_1^{(0)} - iK_2^{(0)} = (\sigma_{22}^{\infty} - i\sigma_{12}^{\infty})\sqrt{\frac{\pi L}{2}} \tag{33}$$

On the other hand, it can be shown (Amestoy and Leblond, 1992) that the complex stress intensity factors $K^{(k)}$ at each tip B_k, ($k = 1, \ldots, n - 1$), are given by

$$K^{(k)} = F_{33}^{(k)} e^{-i\delta_k} U(b_k) \tag{34}$$

which depends on the conformal mapping parameters and linearly of the stress intensity factors $K_1^{(0)}$ and $K_2^{(0)}$. This result is also universal in the sense that it is independent of the applied loading. Once the integral equation (30) is solved the stress intensity factors $K^{(k)}$ are uniquely determined.

Finally, the main results of this section can be summarized by the matricial relation

$$\begin{pmatrix} K_1^{(k)} \\ K_2^{(k)} \\ K_3^{(k)} \end{pmatrix} = \begin{pmatrix} F_{11}^{(k)} & F_{12}^{(k)} & 0 \\ F_{21}^{(k)} & F_{22}^{(k)} & 0 \\ 0 & 0 & F_{33}^{(k)} \end{pmatrix} \begin{pmatrix} K_1^{(0)} \\ K_2^{(0)} \\ K_3^{(0)} \end{pmatrix} ; \tag{35}$$

at each crack tip B_k; $k = 1, \ldots, n - 1$; of the infinitely small branches.

3 THE THREE-BRANCHES STATIC PROBLEM

In this section, we will focus on the case of a main crack of length L with two symmetric side branches of lengths r_1 and r_2, as shown in Fig. 3. The lengths r_1 and r_2 satisfy the conditions $r_1/L \to 0$, $r_2/L \to 0$ with $r_1/r_2 (\equiv l_1/l_2)$ finite. The angle between the two side branches is denoted by $2\pi\lambda$. Without loss of generality, we will take $l_1 \leq l_2$. We have

$$\lambda_1 = 1 - \lambda , \quad \lambda_2 = 2\lambda , \quad \lambda_3 = 1 - \lambda . \tag{36}$$

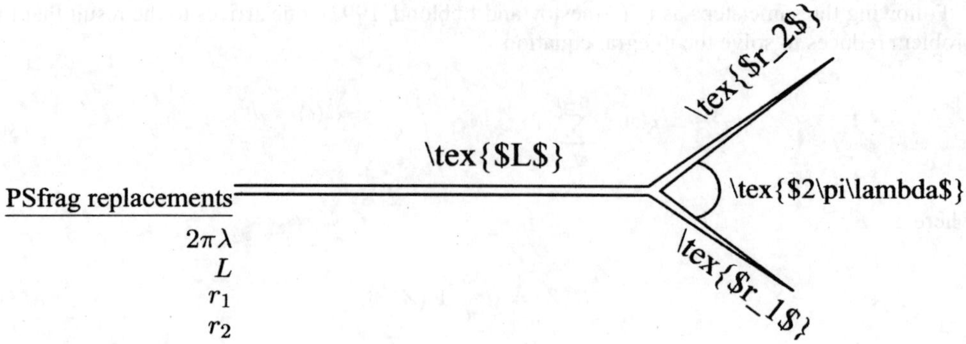

Figure 3. Schematic representation of the three-branches static problem.

In addition to Equations (11, 12), we have two supplementary free conditions, due to the invariance by translation $(a_k, b_k) \rightarrow (a_k + \delta, b_k + \delta)$ and to the invariance under the scaling change $(a_k, b_k, l_k) \rightarrow (\delta a_k, \delta b_k, \delta^2 l_k)$. Therefore, we can choose

$$a_1 = -1\,, \quad a_2 = -2a\,, \quad a_3 = 1 \tag{37}$$

with $0 \leq 2a \leq 1$. Consequently, Equations (11,12) reduce to

$$b_1 = -(1-\lambda)a - \sqrt{\lambda + (1-\lambda)^2 a^2}\,; \tag{38}$$

$$b_2 = -(1-\lambda)a + \sqrt{\lambda + (1-\lambda)^2 a^2}\,; \tag{39}$$

$$l_1 = \left(1 - b_1^2\right)^{(1-\lambda)} (-2a - b_1)^{2\lambda}\,; \tag{40}$$

$$l_2 = \left(1 - b_2^2\right)^{(1-\lambda)} (2a + b_2)^{2\lambda}\,. \tag{41}$$

It is seen that once the constant a is fixed, the remaining conformal mapping parameters can be determined. For given λ and l_1/l_2, the numerical problem consists on solving the following nonlinear equation

$$\frac{l_1}{l_2} = \left(\frac{1 - b_1^2}{1 - b_2^2}\right)^{(1-\lambda)} \left(\frac{-2a - b_1}{2a + b_2}\right)^{2\lambda}\,; \tag{42}$$

which is straightforward. For the fully symmetric case where $l_1 = l_2$, one finds

$$a = 0, \quad b_1 = -\sqrt{\lambda}, \quad b_2 = \sqrt{\lambda}, \quad l_1 = l_2 = \lambda^{\lambda}(1-\lambda)^{(1-\lambda)} \tag{43}$$

3.1 *The mode III problem*

We are interested on calculating $K_3^{(2)}$, the stress intensity factor just after branching at the upper crack tip. In order to simplify notations, we will drop in the following the upper indice labeling the branch number. The stress intensity factor depends on λ and on r_1/r_2 through the universal function F_{33} as follows

$$\frac{K_3}{K_3^{(0)}} = F_{33}(\lambda, l_1/l_2) = \left[l_2\left(\frac{(1-\lambda)(1+b_2^2)}{(1-b_2^2)^2} + \frac{\lambda}{(2a+b_2)^2}\right)\right]^{-1/2}. \tag{44}$$

If $l_1 = l_2$, one easily finds that

$$F_{33}(\lambda) = \frac{1}{\sqrt{2}}\left[\frac{1-\lambda}{\lambda}\right]^{\lambda/2} . \tag{45}$$

It can be shown that for this particular case, $F_{33}(\lambda)$ shows a maximum for a given value of, $\lambda \neq 0$. This behavior is different from the kinked crack case, where the corresponding function F_{33} is monotonic (Sih, 1965).

3.2 *The mixed mode I–II problem*

For the three-branches case depicted in Fig. 3, Eq. (30) reduces to

$$U(Z) = K^{(0)} + (1 - e^{-2i\pi\lambda}) \int_{-1}^{-2a} q(t)\frac{\overline{U(t)}}{(t-Z)^2}\frac{dt}{4i\pi} + (1 - e^{2i\pi\lambda}) \int_{-2a}^{1} q(t)\frac{\overline{U(t)}}{(t-Z)^2}\frac{dt}{4i\pi} ; \tag{46}$$

with,

$$q(t) = \frac{(t^2-1)(t+2a)}{(t-b_1+i\epsilon)(t-b_2+i\epsilon)} . \tag{47}$$

The complex stress intensity factor just after branching at the upper crack tip is given by

$$K = F_{33}(\lambda, l_1/l_2)e^{-i\pi\lambda}U(b_2) . \tag{48}$$

In order to solve the integral equation (46), a useful decomposition of $U(Z)$ is given by

$$U(Z) = K_1^{(0)}U_1(Z) - iK_2^{(0)}U_2(Z) . \tag{49}$$

Eq. (46) is now decomposed into two independent equations

$$U_{1,2}(Z) = 1 \pm (1 - e^{-2i\pi\lambda}) \int_{-1}^{-2a} q(t)\frac{\overline{U_{1,2}(t)}}{(t-Z)^2}\frac{dt}{4i\pi} \pm (1 - e^{2i\pi\lambda}) \int_{-2a}^{1} q(t)\frac{\overline{U_{1,2}(t)}}{(t-Z)^2}\frac{dt}{4i\pi} ; \tag{50}$$

Eq. (50) can be easily computed numerically using an iterative method similar to that used in determining the corresponding functions for the kinked crack problem. The computation of the functions $U_1(Z)$ and $U_2(Z)$ can be performed on any curve belonging to the lower half Z-plane, and that includes the points a_1, a_2 and a_3. The details of this method can be found in (Amestoy and Leblond, 1992).

The remaining elments of the matrix in Eq. (35) can be now written under the form

$$F_{11}(\lambda, l_1/l_2) = F_{33}\Re\left[e^{-i\pi\lambda}U_1(b_2)\right] ; \tag{51}$$

$$F_{12}(\lambda, l_1/l_2) = F_{33}\Im\left[e^{-i\pi\lambda}U_2(b_2)\right] ; \tag{52}$$

$$F_{21}(\lambda, l_1/l_2) = -F_{33}\Im\left[e^{-i\pi\lambda}U_1(b_2)\right] ; \tag{53}$$

$$F_{22}(\lambda, l_1/l_2) = F_{33}\Re\left[e^{-i\pi\lambda}U_2(b_2)\right] . \tag{54}$$

They are determined once the computations of the functions $U_1(Z)$ and $U_2(Z)$ are performed.

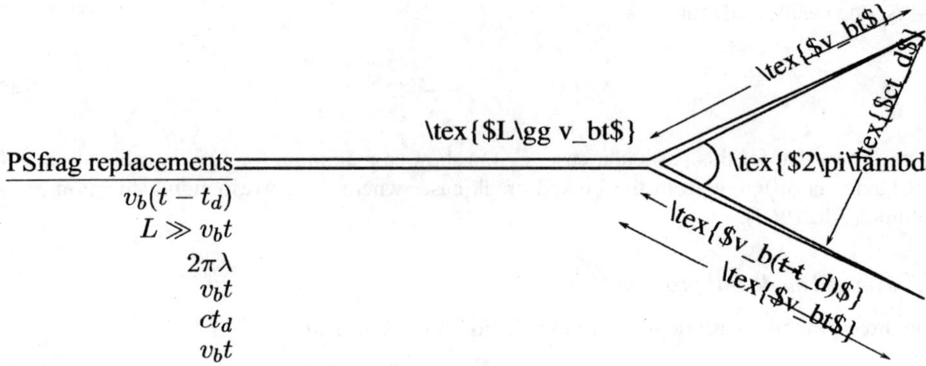

Figure 4. Schematic representation of the dynamic branching problem.

4 DYNAMIC STRESS INTENSITY FACTORS AFTER BRANCHING

A general solution for the dynamical branching problem is not available. The only known solutions concern the case of a semi-infinite crack, initially at rest, that starts to propagate by kinking or branching under the action of stress wave loadings applied on the crack faces (Achenbach et al., 1984; Burgers, 1983; Dempsey et al., 1982). Also, it is well known (Freund, 1990; Kostrov, 1975) that the dynamic stress intensity factors, $K_l(t)$, of a straight crack are related to the static stress intensity factors, $K_{0l}(t)$, of the same configuration by

$$K_l(t) = k_l(v) K_{0l}(t) \; ; \tag{55}$$

where $k_l(v)$, $l = 1, 2, 3$, are known universal functions of the instantaneous crack tip speed (Freund, 1990; Kostrov, 1975). Their explicit forms can be found in (Freund, 1990).

The process of crack branching can be seen as follow. A semi-infinite straight crack that propagates at a speed $v(t)$ for $t < 0$ suddenly stops at $t \to 0^-$. At $t \to 0^+$, the crack branches locally with a branching angle equal to $\lambda\pi$ (see Fig. 4). For $t > 0$, each branch propagates at a velocity $v_b(t)$, following the direction $\pm\lambda\pi$. Since there is no time scale in this problem, and consequently no length scale, the dynamic stress intensity factors just after branching, $K_{bl}(t)$ for $t \to 0^+$, at each crack tip can always be written under the form of a universal function of the velocities and branching angle times the rest stress intensity factors before branching, $K_{0l}(t)$ for $t \to 0^-$

$$K_{bl} = \sum_m H_{lm}(\lambda, v, v_b) K_{0m} \; . \tag{56}$$

As in the quasistatic case (Leblond, 1992), the matrix H is universal in the sense that it does not depend neither on loading configuration nor on the geometry of the body. Moreover, H should satisfy the following property

$$\lim_{v_b \to 0} H_{lm}(\lambda, v, v_b) \;\; = \;\; F_{lm}(\lambda) \; ; \tag{57}$$

4.1 The mode III loading

For the mode III case, when the initial crack stops, a static stress distribution is restored behind a wave front that propagates from the crack tip at the shear wave speed (Eshelby, 1969). Thus for $t > 0$, the propagation of the branches occurs within a stationary stress field. The branching problem is then equivalent to solve the problem of two symmetric branches reminiscent from a preexisting stationary straight crack, that propagate in the directions $\pm\lambda\pi$, with a velocity v_b, and subjected to traction distributions on the new broken surfaces that negate the stationary stress intensity factor field induced by the arrest of the original single crack. Consequently, H_{33} must be independent of the velocity prior to branching

$$H_{33}(\lambda, v, v_b) \equiv H_{33}(\lambda, v_b/c_s) \,. \tag{58}$$

The finite wave velocity in the material induces a relativistic effect that causes a delay in the interaction between the two propagating branches. Stress variations due to crack propagation in the direction $-\pi\lambda$ arrive with a delay time at the crack faces of the branch propagating in the $\pi\lambda$ direction, and *vice-versa*. Therefore, the tip of a branch of length $l_2 = v_b t$ propagates under the action of tractions induced by an effectively shorter branch. The "shorter" branch length is $l_1 = v_b(t - t_d)$, where t_d is the propagation time of a cylindrical wave of radius $d = c_s t_d$ (see Fig. 4). This effect must increase the stress intensity at the branches tips and, thus, must be taken into account when addressing the problem of dynamic branching. In the following, the main hypothesis we make is that the dynamic stress intensity factor just after branching depends on the velocity v_b mainly through the usual function $k_3(v_b)$ and through this retardation effect. While the first part of the velocity dependence has already been introduced by Eshelby (Eshelby, 1970), the second effect has not been taken into account previously. By splitting the influence of crack velocity into these two different parts, we suppose that the *rest* stress intensity factor just after branching is well approximated by the *static* stress intensity factor, at the tip of the longer branch, of two-branches configuration satisfying $l_1 \to 0$ and $l_2 \to 0$. Using simple geometry considerations, one can get the ratio between the effective lengths of the two branches

$$\frac{d}{c_s} = \frac{l_2 - l_1}{v_b} \,; \tag{59}$$

$$d^2 = l_1^2 + l_2^2 - 2 l_1 l_2 \cos 2\lambda\pi \,. \tag{60}$$

Thus,

$$\frac{l_1}{l_2} \equiv l_{12}(\lambda, v_b/c_s) = 1 - \frac{2(v_b/c_s)\sin\lambda\pi}{1-(v_b/c_s)^2}\sqrt{1-(v_b/c_s)^2\cos^2\lambda\pi} \,. \tag{61}$$

Note that this approach is valid for branching angles and crack speeds satisfying $0 \le \lambda \le 1/2$ and $v_b \le (\sqrt{2} - 1)c_s$, which are within the ranges of interest. As one can see from Fig. 5, at non zero branching angle and branches velocity, l_2 can be considerably larger than l_1. The difference between the effective lengths increases with the branches velocity and with the branching angle.

Using the previous arguments, the matrix element H_{33} can be approximated by

$$H_{33}(\lambda, v_b/c_s) = k_3(v_b) F_{33}(\lambda, l_{12}(\lambda, v_b/c_s)) \,. \tag{62}$$

This form satisfies the following asymptotic behavior

$$H_{33} \to F_{33}(\lambda) \,, \qquad \text{for} \qquad v_b \to 0 \,; \tag{63}$$

$$H_{33} \to \frac{1}{\sqrt{2}} k_3(v_b) \,, \qquad \text{for} \qquad \lambda \to 0 \,. \tag{64}$$

Thus, one recovers the limiting cases given by Eq. (57) and the $\lambda = 0$ result of (Eshelby, 1970; Kostrov, 1975). In Fig. 6, the function H_{33} is plotted for different values of the crack tip velocity after branching. It is seen that the delay mechanism modifies the magnitude of the function H_{33}. At small branching angles, the matrix element H_{33} increases with the branching angle from the value corresponding to $\lambda = 0$. However, for any fixed branching angle, the magnitude of H_{33} is found to be always maximum when $v_b = 0$.

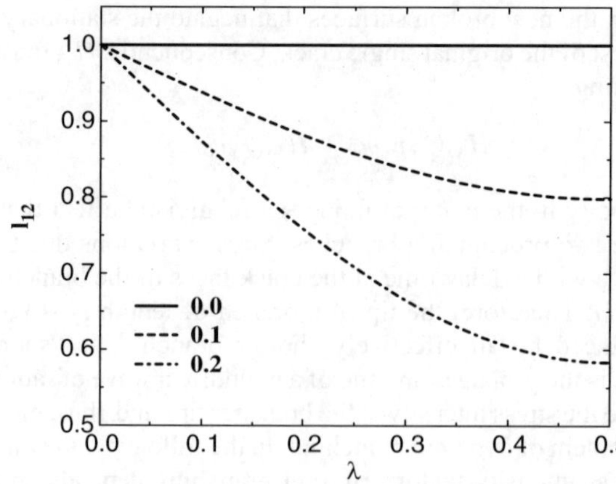

Figure 5. Plot of the function $l_{12}(\lambda, v_b/c)$ as a function of the branching angle $\lambda\pi$, for different values of v_b/c.

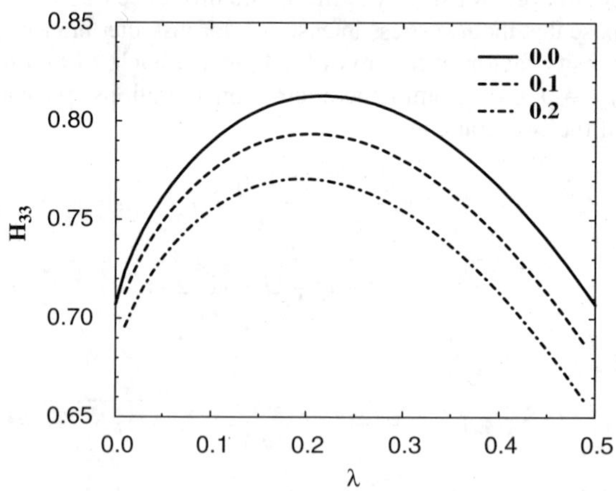

Figure 6. Plot of the function $H_{33}(\lambda, l_{12}(\lambda, v_b/c_s))$ as a function of λ, for different values of v_b/c_s.

4.2 *The mixed mode I–II loading*

For the present case, when the initial crack stops, static stress distributions are restored in the crack plane only (Freund, 1990): ahead of the crack tip behind a wave front that propagates at the shear wave speed, and rear the crack tip behind a wave front that propagates at the Rayleigh wave speed. On the contrary, at any point outside the crack plane the stress distributions relax continuously and reach the static ones for $t \to \infty$ only (Freund, 1990; Kostrov, 1975). Thus in general, the propagation of the branches does not occur within stationary stress fields. However, if one restricts to small branching angles and low branching velocities, the branching problem is well approximated by the one with boundary conditions on the new broken surfaces that negate traction distributions given by stationary stress intensity factor fields induced by the arrest of the original single crack. Then, within zeroth order approximations, one can take H_{lm} as independent of the velocity prior to branching.

Therefore, if one considers small branching angles and low branching speeds, one can still apply the approximations used for the mode III case. At leading order, the dynamic stress intensity factors just after branching at each crack tip can be written under the form of the universal function of the velocity $k_l(v_b)$ times a *rest* stress intensity factor given by the *static* stress intensity factor, at the tip of the longer branch, of two-branches configuration satisfying $l_1 \to 0$ and $l_2 \to 0$, and

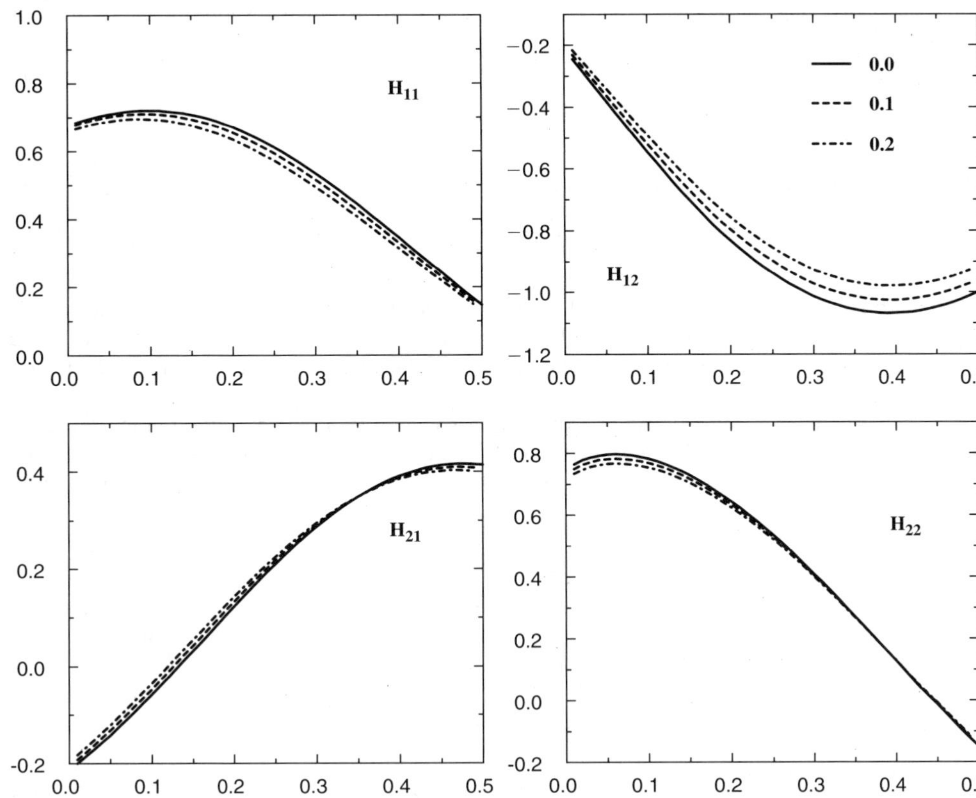

Figure 7. Plot of the elements of the matrix $H_{lm}(\lambda, l_{12}(\lambda, v_b/c_d))$ as a function of λ, for $(c_d/c_s)^2 = 3$ and for different values of v_b/c_s.

$$\frac{l_1}{l_2} = l_{12}(\lambda, v_b/c_d) \; ; \tag{65}$$

where l_{12} is still given by Eq. (61), with c_s replaced by c_d since the dilatational waves propagate faster.

Thus, using the previous arguments, the matrix elements H_{lm}, for l and m equal 1 or 2, can be approximated by

$$H_{lm} = k_l(v_b) F_{lm}(\lambda, l_{12}(\lambda, v_b/c_d)) \tag{66}$$

which should be an accurate approximation for small branching angles and low branching velocities. In Fig. 7, the functions H_{lm} are plotted for different values of the crack tip velocity after branching. It is seen that the delay mechanism modifies slightly the magnitude of the functions H_{lm} from their values at $v_b = 0$.

5 DYNAMIC BRANCHING INSTABILITY

Similarly to the single crack case, we will base our growth criterion for a branched crack on the equality between the energy flux into each propagating tip and the surface energy which is added as a result of this propagation (Griffith, 1920). The dynamic energy release rate is defined as the rate of mechanical energy flow out of the body and into the crack tip per unit crack advance. It is well established that the energy release rate G for a straight moving single crack is given by (Freund, 1990; Kostrov, 1975)

$$G = \frac{1}{2\mu}\sum_{l} A_l(v)K_l^2 = \frac{1}{2\mu}\sum_{l} g_l(v)K_{0l}^2\,, \tag{67}$$

where μ is the Lamé shear coefficient, and

$$g_l(v) = A_l(v)k_l^2(v)\,. \tag{68}$$

The functions $A_l(v)$ and $g_l(v)$ are universal in the sense that they do not depend on the details of the applied loading or the configuration of the body being analyzed. They do depend on the local *instantaneous* speed of the crack tip and on the properties of the material only. The universal functions $g_l(v)$ are plotted in Fig. 8.

The dynamic energy release rate is a quantity associated to a single moving crack tip. Thus, after branching one has to determine it for each crack tip. Due to the symmetry of the branching configuration, the energy release rate just after branching G' for each crack tip is still given by

$$G_b = \frac{1}{2\mu}\sum_{l} A_l(v_b)K_{bl}^2\,. \tag{69}$$

When the main crack before branching is propagating under a pure mode III loading, and using the results of the previous section, this reduces to

$$G_b = \frac{1}{2\mu} g_3(v_b)F_{33}^2(\lambda, l_{12}(\lambda, v_b/c_s))K_{03}^2\,, \tag{70}$$

where K_{03} is the mode III rest stress intensity factor just before branching. Equivalently, if the loading configuration before branching is a pure opening one, the energy release rate just after branching G' at each crack tip is given by

$$G_b = \frac{1}{2\mu}\left[g_1(v_b)F_{11}^2(\lambda, l_{12}(\lambda, v_b/c_d)) + g_2(v_b)F_{21}^2(\lambda, l_{12}(\lambda, v_b/c_d))\right] K_{01}^2 \tag{71}$$

where K_{01} is the rest mode I stress intensity factor just before branching.

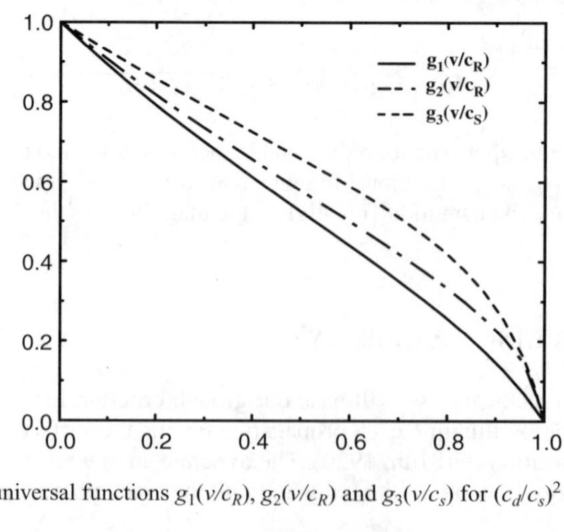

Figure 8. Plot of the universal functions $g_1(v/c_R)$, $g_2(v/c_R)$ and $g_3(v/c_s)$ for $(c_d/c_s)^2 = 3$.

In the study of crack growth process in materials which fail in a purely brittle manner, or for which any inelastic crack tip zone is contained within a surrounding elastic zone, the most commonly used crack growth criteria for defining the *equation of motion* of the crack tip are the generalization of Griffith's critical energy release rate criterion. According to the generalized Griffith criterion (Griffith, 1920), the crack must grow in such a way that the energy release rate is always equal to a newly defined quantity: the dynamic fracture energy of the material, Γ. The energy release rate is a property of the local mechanical fields. The dynamic fracture energy, on the other hand, represents the resistance of the material to crack advance; it is assumed to be a property of the material and its value may depend on the instantaneous crack tip speed (Boudet et al., 1996; Sharon and Fineberg, 1999). This growth criterion should be applied for the crack tips before and after branching. Especially, at the onset of branching the conditions

$$G \equiv \Gamma(v) ; \tag{72}$$

$$G_b \equiv \Gamma(v_b) , \tag{73}$$

must be satisfied. This insures that each crack tip is propagating according to the Griffith criterion. Therefore, the growth criterion introduces an intrinsic relation between the energy release rates just before and just after branching, which reads

$$G_b \equiv \frac{\Gamma(v_b)}{\Gamma(v)} G . \tag{74}$$

Eq. (74) is a necessary condition for the existence of a branching configuration. Otherwise, the single crack tip propagation should be maintained. This equation, however, is not a sufficient condition for determining the branching threshold parameters, which must be deduced from some other criteria. For instance, the maximum energy release rate criterion (Erdogan and Sih, 1963) or the principal of local symmetry (Gol'dstein and Salganik, 1974). These two criteria have been applied essentially to problems related to the selection of single quasistatic crack paths, and give almost similar numerical predictions. However, it has been shown that the principal of local symmetry is more coherent and it is now widely admitted as the second additional equation of motion (Adda-Bedia et al., 1999; Leblond, 1992). In the following, we will use these two criteria to predict the branching instability and compare them with the experimental results.

5.1 *The maximum energy release rate (MERR) criterion*

Eshelby (Eshelby, 1970) posed the question of how large must be the single crack velocity v, so that by decelerating, there is enough energy available to form two cracks propagating at a velocity v_b. He analyzed the idealized mode III branched configuration when the branches subtend a vanishing angle and both prolong the original crack plane. Thus, he found that the least v before deceleration corresponds to $v_b = 0^+$ just afterwards, and he reported $v_c = 0.6c_s$ as the minimum speed which could allow branching. This approach can be interpreted as being equivalent to the maximum energy release rate criterion.

As one can see, non zero branching angle and branches velocity, lead to a finite travel time of stress waves from one tip to the other. This time delay in the interaction increases with the branches velocity and with the branching angle. As this delay increases, each of the cracks is screened by a shorter competitor. However, Fig. 9 shows that the introduction of the delay mechanism results in decreasing the energy release rate with the branches velocity. Therefore, our approach leads to the result that the energy release rate after branching is always the largest when $v_b = 0^+$. Thus the maximum energy release rate after branching depends on the branching angle only.

Let us start by the case of constant fracture energy and focus on the inplane crack propagation. The analysis of the mode III case being equivalent. Using Eq. (74), one deduces that the critical velocity before branching must be a solution of

$$g_1(v) = F_{11}^2(\lambda) + F_{21}^2(\lambda) . \tag{75}$$

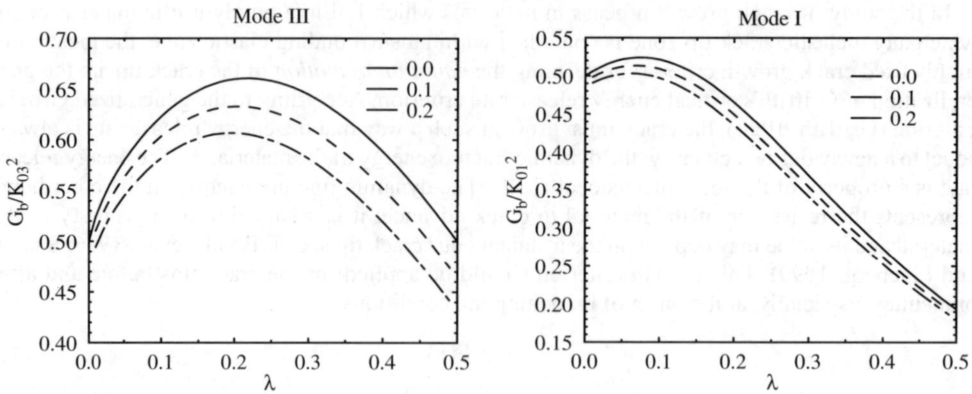

Figure 9. Plot of the energy release rate just after branching, for $(c_d/c_s)^2 = 3$ and for different values of the branching velocity v_b/c_s.

Fig. 8 shows that $g_1(v)$ is a decreasing function of the velocity that satisfies $g_1(0) = 1$, and $g_1(c_R) = 0$. On the other hand, the right hand side of Eq. (75) displays a maximum, whose value is less than unity (see Fig. 9-b). Therefore, this equation is not always satisfied, and branched solutions exist if the velocity v exceeds a critical velocity v_c only. This threshold value is given by the least velocity v for which Eq. (75) admits a solution. It corresponds to a branched solution with a nonvanishing branching angle, $\lambda_c = 0.07$, given by the maximum of $F_{11}^2(\lambda) + F_{21}^2(\lambda)$, and the corresponding value of $v_c \approx 0.5c_R$ is found to be below the critical velocity which can be deduced from Eshelby's approach (see Table 1).

5.2 *The principle of local symmetry (PLS)*

While the MERR criterion states that the crack should follow a direction of maximum energy release rate, the principal of local symmetry states that the path taken by a crack in a brittle homogeneous isotropic material is the one for which the local stress field at the tip is of mode I type (Gol'dstein and Salganik, 1974). It is not obvious that the crack propagation still satisfies this criterion in the dynamic case. However, there is an argument in favor of this scenario (Adda-Bedia et al., 1999). The energy release rate can be seen as the component, F_1, of the configurational force along the direction of crack motion. The Griffith energy criterion can thus be reinterpreted as material forces balance between F_1 and a resistance force to crack advance per unit length of the crack front: $F_1 = \Gamma$. However, this equation of motion is not sufficient to determine the trajectory of a crack that is allowed to deviate from straight line motion. If one assumes that configurational forces balance hold at the crack tip, one should also impose that the component of the material force perpendicular to the direction of crack propagation must vanish (Adda-Bedia et al., 1999). Since this quantity is proportional to K_2, it results that the crack propagation occurs in such a way as to keep a purely opening mode at its tip.

In the case of PLS, at any stage of crack propagation two equations of motion must be fulfilled. When a main crack subjected to a pure opening loading branches into two symmetric cracks, these equations are satisfied before branching once the Griffith criterion is satisfied. On the other hand, just after branching, these two conditions can be written as

$$g_1(v) = F_{11}^2(\lambda)\,; \tag{76}$$

$$F_{21}(\lambda) = 0\,. \tag{77}$$

Now, the branching angle, λ_c, is selected in a different way from the MERR criterion, while the arguments for determining the critical velocity, v_c, are the same. Thus, the result of application of these equations gives also a selected branching angle and a critical speed for branching. From

Table 1. Branching thresholds v_c and λ_c for different branching criteria and loading modes. The values in parenthesis correspond to thresholds using the Eshelby's approach. For the inplane case, the thresholds are given using a Poisson ratio such that $(c_d/c_s)^2 = 3$.

	MERR, mode III	MERR, mode I	PLS
v_c	$0.392\ c_s\ (0.6\ c_s)$	$0.503\ c_R\ (0.53\ c_R)$	$0.518\ c_R$
λ_c	0.22 (0)	0.07 (0)	0.13

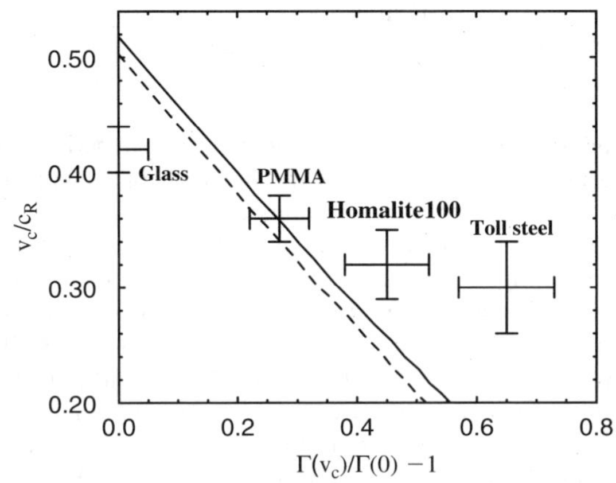

Figure 10. Plot of the critical velocity v_c/c_R for non constant fracture energy $(c_d/c_s)^2 = 3$. The solid line corresponds to the values deduced from PLS and the dashed line to those deduced from the MERR criterion. The experimental values reported are estimates taken from (Sharon and Fineberg, 1999) for Glass and PMMA, (Dally, 1979) for Homalite 100 and (Anthony et al., 1970; Congleton, 1973) for Toll steel.

Table 1, it is seen that the critical branching velocity $v_c \approx 0.52c_R$ does not differ too much from the one deduced by using MERR criterion, while the critical branching angle is twice larger, $\lambda_c = 0.13$.

5.3 *Effect of velocity dependent fracture energy*

We now turn to look at the effect of the variation of the fracture energy, $\Gamma(v)$, with the crack velocity, on the results of the latter calculations. As the two branching criteria are based to some level on energy balance, their results should be influenced by the behavior of $\Gamma(v)$. When taking into account this new degree of freedom, Eq. (75), related to MERR criterion, becomes

$$\frac{\Gamma(0)}{\Gamma(v)}\, g_1(v) = F_{11}^2(\lambda) + F_{21}^2(\lambda)\ ; \tag{78}$$

and in the case of PLS, Eq. (76) becomes

$$\frac{\Gamma(0)}{\Gamma(v)}\, g_1(v) = F_{11}^2(\lambda)\ , \tag{79}$$

while Eq. (77) is not modified. In general, $\Gamma(v)$ is an increasing function of the velocity (Boudet et al., 1996; Sharon and Fineberg, 1999). Therefore, the left hand side of Equations (78, 79) decreases faster than in the constant fracture energy case, and the energy balance can thus be achieved at lower velocity, while the critical branching angles remain the same. Although $\Gamma(v)$ can be a non-linearly dependent function of the crack tip speed, it is only the amount of $\Gamma(v_c)/\Gamma(0)$ which is of importance in determining the critical crack tip velocity v_c. In Fig. 10, this quantity is plotted for different values of $\Gamma(v_c)/\Gamma(0)$, when either MERR criterion or PLS is applied.

6 DISCUSSION

We performed exact calculations of the stress intensity factors of two static branches with an arbitrary branching angle. We introduced the effect of delayed stress fields due to the finite wave speed in the material, and derived approximate forms of the stress intensity factors of the dynamic branching problem. Given the energy release rate of the dynamic branches, we found the critical crack velocity, at which a single mode I crack can bifurcate into two branches. The branches' velocity is always equal to zero at the onset of the instability. The critical velocity v_c is weekly dependent on the Poisson ratio of the material, and is only slightly modified by changing the crack propagation criterion from the maximum energy release rate criterion to the principle of local symmetry (see Table 1 and Fig. 10). Still, the branching angle in the case of PLS is twice as large as the one that results from the application of MERR criterion; 23.4 deg. instead of 12.6 deg. Finally, for velocity dependent fracture energy, the critical velocity for branching v_c decreases with increasing $\Gamma(v_c)/\Gamma(0)$.

We start our discussion by comparing our results with existing experimental data of critical speeds for branching. Looking at Fig. 10, we see that the decrease of v_c with increasing $\Gamma(v_c)/\Gamma(0)$ is found in experiments as well. The experimental data, however, show a weaker dependence of v_c on $\Gamma(v_c)/\Gamma(0)$. This difference may result from the approximate forms we used for the dynamic stress intensity factors after branching. Moreover, the result of zero branch velocity at the onset of the instability is non-realistic and clearly places too much importance on the increasing form of $\Gamma(v)$. One can see that when $v_b > 0$, we should replace $\Gamma(0)$ in Equations (78, 79) with $\Gamma(v_b)$. This will result in a more moderate dependence of v_c on $\Gamma(v_c)/\Gamma(0)$. We would like to point out that in order to find a dependence of v_c on $\Gamma(v)$, one must use the energy balance approach. It can not be included in calculations based on analysis of the stress field around a single crack tip.

We now turn to the question of path selection of the branches. Unlike the quasistatic case, the MERR criterion and the PLS result in highly different branching angles. Looking at Fig. 9, we see that the insensitivity of v_c to the type of applied criterion results from the flat maximum of the energy release rate G' with respect to the branching angle. Moreover, the flatness of G' at low branching angles is an indication in favor of the PLS versus the MERR criterion. If the MERR criterion were the relevant one, there would not be a clear selection of the branching angle and we would expect to find branching for a wide range of angles. This prediction contradicts experimental observations (Sharon et al., 1995), which indicated a clear selection of the branching angle. On the other hand, $F_{21}(\lambda)$ is steep around its zero value and thus, the PLS would result in distinct selection of the branching angle. The much larger branching angle, which is given by the PLS criterion, results in an incremental decrease in the value of G' and thus, a small change in the velocities involved in the branching process. Applying the PLS, we found that the selected branching angle is equal to 23.4 deg, independent of Poisson ratio. This branching angle is larger than the measured angles in PMMA and Homalite 100 (Cotterell, 1965; Ramulu and Kobayashy, 1985), however these measurements were performed on typical scale of 100–300 microns. Sharon and Fineberg (Sharon et al., 1995) reported that the micro-branches which appear just after the onset of the branching instability, are not straight. Instead, the "branching angle" increases as we approach the branching point. Measuring the tangent to the branch profile at 5 microns from the branching point, the authors reported branching angles of 30 deg, for glass and PMMA. These angles are not significantly larger than the predictions of our model, based on PLS, but nearly three times larger than the predictions when using MERR. Our observations, thus, suggest that the relevant criterion for branching must be the PLS.

Finally, we have to check the relevance of our results to the dynamic branching problem and compare our work with other approaches to the problem. Looking at previous theoretical works on the subject within the continuum frame work, we believe that despite being based on some approximations, the calculations we presented here have some advantages. Many attempts to predict the critical velocity for branching in the framework of linear elasticity were focused on the properties of the stress fields around the tip of a single fast crack. These works indicated that above a critical velocity a component of the singular stress field, which was regarded as the relevant one for path selection, attains a maximum off the original direction of propagation. It was suggested that in these conditions the single straight crack solution becomes unstable. Still, none of these calculations dealt with the resultant state of the system, that of two branches. This state must be analyzed in the same way that a single crack state is analyzed: One must verify that the new "two-branches" state is possible energetically and the growth criterion must be that of equality between the energy release rate and the fracture energy. Eshelby's approach targets these points, applying this growth criterion to the two-branches state.

More recent works challenge the dynamic branching problem, using simulations and calculations on the molecular scale (Abraham et al., 1994; Marder and Gross, 1995). One can argue that since the first stages of the branching process occur within the process zone, branching criteria based on continuum models are meaningless. It might be that the branching criterion must be based on processes at the molecular scale. Indeed, our calculations do not target the question of when a tip of a crack will bifurcate. Instead, we find the conditions which are necessary for the propagation of two branches. It might be that calculations at molecular scale will show that even at lower velocities the tip is unstable and the material tends to be opened off the original direction. However, our calculations show that these flows will not be able to grow to the continuum scale, and instead become branches with singular stress fields at their tips. Instead they might affect the fracture energy of the *single crack* and its fracture surface topology. As experimental data show (Sharon et al., 1995; Sharon and Fineberg, 1996), the change in the dynamics of the crack occurs simultaneously with the appearance of micro-branches of 10 to 100 microns length. This is certainly a continuum scale. Thus, deviations from straight crack which occur before the appearance of "continuum scale branches" do not change drastically the crack's dynamics and can be included within the process zone.

As a conclusion, we believe that the approach we presented here may serve as a useful tool in understanding the process of dynamic crack branching. In its current form it includes some approximations that obviously affect its accuracy. We believe, however, that once the exact branching problem is solved, this approach, which follows from Eshelby, will lead to the exact branching thresholds.

ACKNOWLEDGMENTS

We are grateful to J.R. Rice and E. Sharon for useful discussions and enlightening comments on the present work. Laboratoire de Physique Statistique is "associé au CNRS (UMR 8550) et aux Universités Paris VI et Paris VII".

REFERENCES

Abraham, F. F., Brodbeck, D., Rafey, R. A. and Rudge, W. E. (1994) Instability dynamics of fracture, a computer simulation investigation. *Phys. Rev. Lett.* 73: 272–275.

Achenbach, J. D., Kuo, M. K. and Dempsey, J. P. (1984) Mode III and mixed mode I–II crack kinking under stress wave loading. *Int. J. Solids Structures* 20: 395–410.

Adda-Bedia, M., Ben Amar, M. and Pomeau, Y. (1996) Morphological instabilities of dynamic fracture in brittle solids. *Phys. Rev. E* 54: 5774–5779.

Adda-Bedia, M., Arias, R., Ben Amar, M. and Lund, F. (1999) Generalized Griffith criterion for dynamic fracture and the stability of crack motion at high velocities. *Phys. Rev. E* 60: 2366–2376.

Amestoy, M. and Leblond J. B. (1992) Crack paths in plane situations-II. Detailed form of the expansion of the stress intensity factors. *Int. J. Solids Structures* 29: 465–501.

Andersson, H. (1969) Stress intensity factors at the tips of a star-shaped contour in an infinite tensile sheet. *J. Mech. Phys. Solids* 17: 405–417.

Anthony, S. R., Chubb, J. P. and Congleton, J. (1970) The Crack-Branching Velocity. *Phil. Mag.* 22: 1201–1216.

Boudet, J. F., Ciliberto, S. and Steinberg, V. (1996) Dynamics of crack propagation in brittle materials. *J. Phys. II France* 6: 1493–1516.

Boudet, J. F. and Ciliberto, S. (2000) Interaction of sound with fast crack propagation: an equation of motion for the crack tip. *Physica D* 142: 317–345.

Burgers, P. (1983) Dynamic kinking of a crack in plane strain. *Int. J. Solids Structures* 19: 735–752.
Congleton, J. (1973) Practical applications of crack-branching measurements *Proceedings of the International Conference on Dynamic Crack Propagation*, Noordhoff International Publishing: 427–438.
Cotterell, B. (1965) On brittle fracture paths. *Int. J. Fracture Mech.* 1: 96–103.
Dally, J. W. (1979) Dynamic photoelastic studies of fracture. *Experimental Mechanics* 19: 349–361.
Dempsey, J. P., Kuo, M. K. and Achenbach, J. D. (1982) Mode III kinking under stress wave loading. *Wave Motion* 4: 181–190.
Erdogan, G. and Sih, G. C. (1963) On the crack extension in plates under plane loading and transverse shear. *J. Basic Engng.* 85: 519–527.
Eshelby, J. D. (1969) The elastic field of a crack extending nonuniformly under general anti-plane loading. *J. Mech. Phys. Solids* 17: 177–199.
Eshelby, J. D. (1970) Energy relations and the energy-momentum tensor in continuum mechanics. *Inelastic Behaviour of Solids* (ed. Kanninen, M. F., Adler, W. F., Rosenfield, A. R. and Jaffee, R. I.), McGraw-Hill, New York: 77–115.
Fineberg, J., Gross, S. P., Marder, M. and Swinney, H. L. (1992) Instability in the propagation of fast cracks. *Phys. Rev. B* 45: 5146–5154.
Freund, L. B. (1990) *Dynamic Fracture Mechanics*, New York: Cambridge University Press.
Gol'dstein, R. V. and Salganik, R. L. (1974) Brittle fracture of solids with arbitrary cracks. *Int. J. Fracture* 10: 507–523.
Griffith, A. A. (1920) The phenomenon of rupture and flow in solids. *Phil. Trans. R. Soc. (London)* A221: 163–198.
Gross, S. P., Fineberg, J., Marder, M., McCormick, W. D. and Swinney, H. L. (1993) Acoustic emission in dynamic fracture. *Phys. Rev. Lett.* 71: 3162–3165.
Kostrov, B. V. (1975) On the crack propagation with variable velocity. *Int. J. Fracture* 11: 47–56.
Lawn, B. (1993) *Fracture of brittle solids*, New York: Cambridge University Press.
Leblond, J. B. (1992) Crack paths in plane situations-I. Detailed form of the expansion of the stress intensity factors. *Int. J. Solids Structures* 29: 465–501.
Marder, M. and Gross, S. P. (1995) Origin of crack tip instabilities. *J. Mech. Phys. Solids* 43: 1–48.
Muskhelishvili, N. I. (1953) *Some basic problems of the mathematical theory of elasticity*. (Noordhoff, Groningen).
Ramulu, M. and Kobayashy, A. S. (1985) Mechanics of crack curving and branching – a dynamic fracture analysis. *Int. J. Fracture* 27: 187–201.
Ravi-Chandar, K. and Knauss, W. G. (1984) An experimental investigation into dynamic fracture-III. On steady-state crack propagation and crack branching. *Int. J. Fracture* 26: 141–154.
Rice, J. R., Ben-Zion, Y. and Kim, K. S. (1994) Three-dimensional perturbation solution for a dynamic planar crack moving unsteadily in a model elastic solid. *J. Mech. Phys. Solids* 42: 813–843.
Sharon, E., Gross, S. P. and Fineberg, J. (1995) Local branching as a mechanism in dynamic fracture. *Phys. Rev. Lett.* 74: 5096–5099.
Sharon, E. and Fineberg, J. (1996) Microbranching instability and the dynamic fracture of brittle materials. *Phys. Rev. B* 54: 7128–7139.
Sharon, E. and Fineberg, J. (1999) Confirming the continuum theory of dynamic brittle fracture for fast cracks. *Nature* 399: 333–335.
Sih, G. C. (1965) Stress distribution near internal crack tips for longitudinal shear problems. *J. Appl. Mech.* 32: 51–58.
Xu, X. P. and Needleman, A. (1994) Numerical simulations of fast crack growth in brittle solids. *J. Mech. Phys. Solids* 42: 1397–1434.
Yoffe, E. H. (1951) The moving Griffith crack. *Phil. Mag.* 42: 739–750.

Configurational Mechanics, Kalpakides & Maugin (eds)
© 2004, Taylor & Francis Group, London, ISBN 90 5809 667 X

On the dynamic fracture for a curved crack

E.K. Agiasofitou & V.K. Kalpakides
Department of Mathematics, Division of Applied Mathematics and Mechanics, University of Ioannina, Greece

ABSTRACT: This work aims at the study of the dynamic fracture of an elastic material in the framework of the configurational mechanics. The analysis is based on the global balances for the physical and configurational fields. Thus, the concept of the balance law for an elastic fractured body, in Euclidean and material space, is treated in detail. In the spirit of modern continuum mechanics, a rigorous localization process is proposed. This procedure provides the equations in Euclidean and material space as well as the new contributions for the configurational forces and moments at the crack tip. In addition, it facilitates the derivation of the relationship between the energy release rate (or the rotational release rate) and the configurational force (or the configurational moment). The results are compared with the corresponding ones of fracture mechanics and some new interpretations are discussed.

1 INTRODUCTION

The propagation of a crack of any curvature in a deformable body is a complex phenomenon because apart from the dynamics of the elastic motion, the evolution of the crack must be accounted for, too. The evolution of the crack takes place, not in the physical space, but within the material body, that is, the material space. Thus, we believe that configurational mechanics (Maugin 1993, 1995; Gurtin 2000) should be the appropriate framework in which this problem can be efficiently studied.

To this scope, we start with the global balance laws as it is used to do in any other problem in continuum mechanics. More specifically, we consider an elastic body with a propagating crack in its interior and postulate the balances for all relevant fields, included the configurational ones, for any arbitrary part of the body (Agiasofitou and Kalpakides 2003). In the presence of the crack, this procedure becomes much more complicated because of two reasons. Firstly, the involved fields are not continuous across the crack, even more, they may have a singularity at the crack tip and second, the underlying kinematics is more complicated due to the presence of the separate crack kinematics. In particular, the singularities at the crack tip make necessary to reformulate the transport and divergence theorems, which are indispensable for any localization process.

In literature, such a view can be found in the work of (Steinmann 2000) who presented balance laws in both the physical and material space for elastostatics of a smooth elastic body. Also, reports to equations, which can be considered as balance laws for a fractured body, appeared in (Maugin 1993, 1995; Dascalu and Maugin 1995; Gurtin and Podio-Guidugli 1996; Gurtin 2000; Kienzler and Herrmann 2000) but, to the best of our knowledge, up to now there is not a complete and consistent analysis in the spirit of modern continuum mechanics.

The global view adopted in this paper can shed light on the relationship between the configurational fields at the crack tip and the energy release rates as well as the connection between the first ones and the J and L integrals. For instance, starting from the pseudomomentum and energy equations, one can establish a connection between the energy release rate and the configurational force at the crack tip. This quantity is referred to by Maugin as global material force and it is directly related to the J-integral (Maugin 1993; Dascalu and Maugin 1995). One of our goal in this paper is to explore an analogous relation starting from the material angular momentum and energy equations. In this case, it is expected a connection between the configurational moment at the crack tip and the rotational energy release rate. Such a relation has been provided by (Maugin and Trimarco 1995) for the case where the defect is a disclination line.

Furthermore, (Golebiewska Herrmann and Herrmann 1981) considered the case of a stationary crack which rotates and they computed the rotational energy release rate. Also, (Eischen and Herrmann 1987) tried to connect the conservation (and balance) laws with the energy release rates and the J, L and M integrals. In these works, a straight stationary crack is considered and the rotational energy release rate emerges by a virtual rotation of the crack around its center. Although this is a very successful and meaningful Manipulation (in the sense that the associated conservation law is coming from the invariance of the action functional under the group of rotations), it can not be related to a real situation of a propagating crack.

Looking for a more physical interpretation, the propagation of a crack along a curve of arbitrary curvature is considered in such a way that the linear and the angular velocity of the crack tip to be inserted. The balance laws are postulated and from the localization process the configurational fields at the crack tip naturally arise. Finally, these quantities are correlated with the energy release rates and the J and L integrals of fracture mechanics.

Although the crack propagation in a deformable body is a dissipative phenomenon, in this paper no mention is made to the second law of thermodynamics and to the subsequent discussion about constitutive relations.

In Section 2, some preliminaries concerning the proper kinematics for a cracked elastic body are presented. In Section 3, an abstract balance law is postulated, the conditions under which it is meaningful are examined and its consequences are extracted rigorously. The application of this procedure to the physical and configurational fields, related to the problem under study, is made in Sections 4 and 5, respectively. Finally, in Section 6, the obtained results are used to derive the relations between the energy release rates and the configurational fields at the crack tip.

2 PRELIMINARIES

Let $\mathcal{B}_R$ be the reference configuration containing a crack which is described by a smooth, non-intersecting curve C_R with the one end point to lie on the boundary of the body and the other one to be the crack tip, $\mathbf{Z}_0$. We consider that the crack evolves, not necessarily in straight direction, following the "motion" of the crack tip within the body. Thus at the time t, the crack is represented by a smooth curve $C(t)$ belonging to a material configuration $\mathcal{B}_t$, $t \in I \subset IR$, where I denotes a time interval. The only difference between the reference configuration $\mathcal{B}_R$ and the material configurations $\mathcal{B}_t$ lies in the different curve they contain. Certainly, it is required for $t_1 > t_2$ to imply $C(t_2) \subset C(t_1)$.

We focus now on the end point of the crack at time t, $\mathbf{Z}(t)$. We consider that $\mathbf{Z}(t)$ is a smooth, time dependent mapping, hence its derivative

$$\mathbf{V}(t) = \frac{d\mathbf{Z}}{dt} \tag{1}$$

provides the propagation velocity of the crack. Also, if we denote with $\mathbf{t}$ the tangent vector to the crack curve, we can write $\mathbf{V} = V\mathbf{t}$.

Taking the standard view of fracture mechanics, we consider a disc of radius ϵ centered at the crack tip $\mathbf{Z}(t)$ for any time t, denoted by $D_\epsilon(t)$:

$$D_\epsilon(t) = \{\mathbf{X} \in \mathcal{B}_t : |\mathbf{X} - \mathbf{Z}(t)| \leq \epsilon\}. \tag{2}$$

At the time t_0, the tip disc is given by:

$$D_{\epsilon_0} = \{\mathbf{Y} \in \mathcal{B}_R : |\mathbf{Y} - \mathbf{Z}_0| \leq \epsilon\}.$$

Notice here that $D_{\epsilon_0} \subset \mathcal{B}_R$ and $D_\epsilon(t) \subset \mathcal{B}_t$. Also, we will denote the part of the crack curve which lies on $D_\epsilon(t)$ with γ_D, i.e., $\gamma_D = D_\epsilon(t) \cap C(t)$.

Taking into account the crack tip evolution, we can establish a fictitious motion of the tip disc (Fig.1) in the following form

$$\mathbf{X} = X(\mathbf{Y}, t), \quad \mathbf{X} \in D_\epsilon(t), \quad \mathbf{Y} \in D_{\epsilon 0}, \ t \in I. \tag{3}$$

Without any loss of generality, we assume that this "motion" is a rigid one (Gurtin 1981) and particularly, it is a simple translation which follows the crack tip evolution, that is

$$X(\mathbf{Y}, t) = \mathbf{Y} + \mathbf{Z}(t) - \mathbf{Z}_0, \ \text{ for all } \ \mathbf{Y} \in D_{\epsilon 0}. \tag{4}$$

It is obvious that every point of $D_{\epsilon 0}$ "moves" with the velocity of the crack tip, i.e.,

$$\mathbf{V}(\mathbf{Y}, t) = \frac{\partial X}{\partial t}(\mathbf{Y}, t) = \frac{d\mathbf{Z}}{dt} = \mathbf{V}(t), \ \text{ for all } \ \mathbf{Y} \in D_{\epsilon 0}. \tag{5}$$

Consider now the physical motion

$$\mathbf{x} = \chi(\mathbf{X}, t), \quad \mathbf{x} \in B_t, \quad \mathbf{X} \in \mathcal{B}_t, \ t \in I \subset I\!R, \tag{6}$$

which is twice-differentiable for all $(\mathbf{X}, t) \in (\mathcal{B}_t \backslash C(t)) \times I$. Also, it is continuous along the crack curve $C(t) \backslash \mathbf{Z}(t)$, as we assume that the crack faces are in perfect contact. We observe that the material points $\mathbf{X} \in D_\epsilon(t)$ depend on t via the mapping X, while the material points $\mathbf{X} \in \mathcal{B}_t \backslash D_\epsilon(t)$ do not depend on t. Consequently, we can compose the mappings X and χ for all $\mathbf{X} \in D_\epsilon(t)$ to interpret both the crack evolution and the motion of the body in the physical space (Fig. 2). Note that this composition holds only for those $\mathbf{X}$ that belong to $D_\epsilon(t)$ at the time t. As a result, we can write for all $\mathbf{X} \in D_\epsilon(t)$

$$\widetilde{\chi} = \chi \circ X, \quad \mathbf{x} = \widetilde{\chi}(\mathbf{Y}, t) = \chi(X(\mathbf{Y}, t), t), \mathbf{Y} \in D_{\epsilon 0}. \tag{7}$$

The partial derivative of $\widetilde{\chi}$ with respect to time will be denoted by $\mathring{\mathbf{x}}$ and the following chain differentiation will hold

$$\mathring{\mathbf{x}} = \frac{\partial \chi}{\partial \mathbf{X}}(\mathbf{X}, t) \frac{\partial X}{\partial t}(\mathbf{Y}, t) + \frac{\partial \chi}{\partial t}(\mathbf{X}, t), \quad \text{for all} \quad \mathbf{X} \in D_\epsilon(t) \setminus \gamma_D. \tag{8}$$

While for all $\mathbf{X} \in \mathcal{B}_t \backslash D_\epsilon(t)$ away from the crack, it holds

$$\mathring{\mathbf{x}} = \frac{\partial \chi}{\partial t}(\mathbf{X}, t), \quad \text{for all} \quad \mathbf{X} \in \mathcal{B}_t \setminus D_\epsilon(t).$$

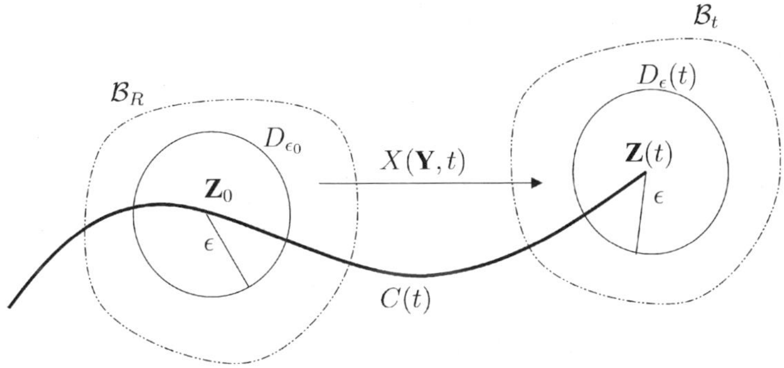

Figure 1. The motion of the tip disc.

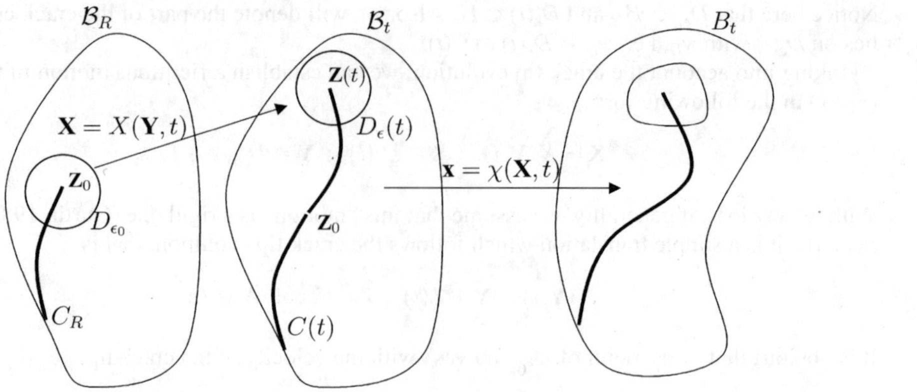

Figure 2. The total motion.

Denoting, as usually, with $\mathbf{F}(\mathbf{X}, t) = \partial\chi(\mathbf{X}, t)/\partial\mathbf{X}$ the deformation gradient and taking into account eq. (5), the equation (8) takes the form

$$\overset{\circ}{\mathbf{x}} = \mathbf{F}(\mathbf{X}, t)\mathbf{V}(t) + \dot{\mathbf{x}}(\mathbf{X}, t) =: \widetilde{\mathbf{V}}(\mathbf{X}, t), \quad \text{for all} \quad \mathbf{X} \in D_\epsilon(t) \setminus \gamma_D, \tag{9}$$

where $\dot{\mathbf{x}}(\mathbf{X}, t) = \partial\chi(\mathbf{X}, t)/\partial t$. Note that, from the above assumptions about the smoothness of x, we have that $\mathbf{F}(\mathbf{X}, t)$ and $\dot{\mathbf{x}}(\mathbf{X}, t)$ are continuous for $\mathbf{X} \in D_\epsilon(t) \backslash \gamma_D$. However, both $\mathbf{F}$ and $\dot{\mathbf{x}}$ are singular at the crack tip $\mathbf{Z}(t)$.

The quantity $\widetilde{\mathbf{V}}(\mathbf{X}, t)$ represents the velocity of the deformed tip disc accounting for the crack evolution velocity as well. Though $\widetilde{\mathbf{V}}(\mathbf{X}, t)$ is defined with the aid of the fields $\mathbf{F}$ and $\dot{\mathbf{x}}$ which are singular at the crack tip, we would like $\widetilde{\mathbf{V}}$ to be smooth at the crack tip. Thus, taking the view of (Gurtin 2000, Gurtin and Podio-Guidugli 1996), we assume the existence of a bounded, time-dependent function $\widetilde{\mathbf{U}}(t)$ such that

$$\lim_{\mathbf{X} \to \mathbf{Z}(t)} \widetilde{\mathbf{V}}(\mathbf{X}, t) = \widetilde{\mathbf{U}}(t), \ \text{ uniformly in } I. \tag{10}$$

Notice that the quantity $\widetilde{\mathbf{U}}(t)$ represents the velocity of the deformed crack tip.

3 AN ABSTRACT BALANCE LAW FOR A CRACKED BODY

Let Ω be any smooth domain of the body in the material configuration $\mathcal{B}_t$. If the crack tip $\mathbf{Z}(t)$ is an interior point of Ω, then there exists a radius ϵ such that $D_\epsilon(t) \subset \Omega$. In this case, we will denote with Ω_ϵ the subset of Ω which is defined as follows (Fig. 3),

$$\Omega_\epsilon(t) = \Omega \setminus D_\epsilon(t) \ \text{ or } \ \Omega = \Omega_\epsilon(t) \cup D_\epsilon(t). \tag{11}$$

Notice that $\partial\Omega_\epsilon = \partial\Omega \cup \partial D_\epsilon(t)$. Also, the parts of the crack $C(t)$ contained in Ω_ϵ and Ω will be denoted by γ_ϵ and γ_Ω, respectively , that is

$$\gamma_\epsilon = C(t) \cap \Omega_\epsilon(t), \quad \gamma_\Omega = C(t) \cap \Omega.$$

In standard continuum mechanics, one has the freedom to formulate a global balance law either in the reference configuration or in the current configuration. In the proposed framework, there are three distinct configurations (Fig. 2). We work on a material configuration $\mathcal{B}_t$, in which all the

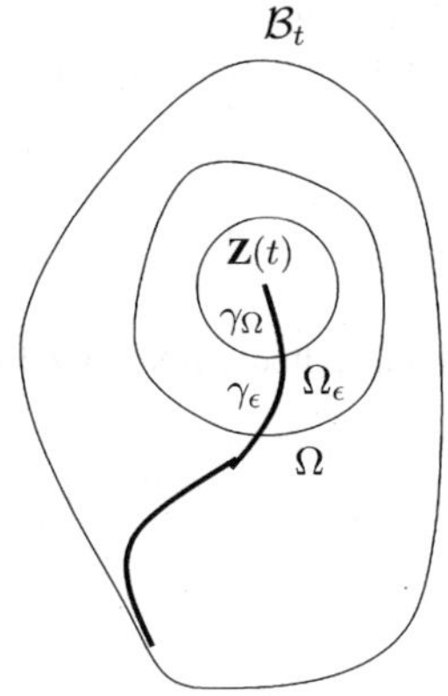

Figure 3. A domain Ω containing the crack tip.

relevant fields should be defined. Let $\phi(\mathbf{X}, t)$ be a scalar valued function defined in $\mathcal{B}_t$, representing some physical quantity, sufficiently smooth away from the crack tip and up to the crack $C(t)$ from either side, thus we let ϕ to have a singularity at the crack tip and to be discontinuous with finite jump along $C(t)\backslash\{\mathbf{Z}(t)\}$.

Taking the view of (Gurtin 2000), we will assume *the integrability of ϕ in the sense of Cauchy principal value*, i.e. for all $\Omega \in \mathcal{B}_t$

$$\int_\Omega \phi(\mathbf{X}, t)\, dA = \lim_{\epsilon \to 0} \int_{\Omega_\epsilon} \phi(\mathbf{X}, t)\, dA. \tag{12}$$

Analogously, it holds for the line integral of a vector valued function $\mathbf{g}(\mathbf{X}, t)$ along the curve γ_Ω in the sense

$$\int_{\gamma_\Omega} \mathbf{g}(\mathbf{X}, t) \cdot \mathbf{n}\, dl = \lim_{\epsilon \to 0} \int_{\gamma_\epsilon} \mathbf{g}(\mathbf{X}, t) \cdot \mathbf{n}\, dl, \tag{13}$$

where $\mathbf{n}$ is the unit normal to $C(t)$. Hereafter, when we refer to the integrability of any function over Ω and γ_Ω, it will be meant in the sense of eqs. (12) and (13).

Next, we consider a *global balance law for the quantity ϕ* of the form

$$\frac{d}{dt} \int_\Omega \phi(\mathbf{X}, t)\, dA = \int_{\partial\Omega} \mathbf{f}(\mathbf{X}, t) \cdot \mathbf{N}\, dS + \int_\Omega h(\mathbf{X}, t)\, dA + g(t), \tag{14}$$

where $\mathbf{N}$ is the outward unit normal to the boundary $\partial\Omega$ and $\mathbf{f}$ and h, are the flux and the source of ϕ, respectively. The time dependent function g represents the source of ϕ due to the crack evolution.

It is apparent that the integrability of ϕ is not enough to make eq. (14) meaningful. So, we must pose extra smoothness on the integrands. Denoting with $[\mathbf{f}]$ the jump of $\mathbf{f}$ across the crack, we assume the following conditions:

C1 $h, \ \phi$ are integrable over Ω.

C2 $\displaystyle \lim_{\epsilon \to 0} \int_{\Omega_\epsilon} \frac{\partial \phi}{\partial t}(\mathbf{X}, t)\, dA = \int_\Omega \frac{\partial \phi}{\partial t}(\mathbf{X}, t)\, dA,$ uniformly in I.

C3 $\displaystyle \int_{\partial D_\epsilon} \phi(\mathbf{X}, t)(\mathbf{V} \cdot \mathbf{N})\, dS$ converges uniformly in I as $\epsilon \to 0$.

C4 $\mathrm{Div}\, \mathbf{f}, \ [\mathbf{f}] \cdot \mathbf{n}$ are integrable over Ω and γ_Ω, respectively.

C5 $\displaystyle \int_{\partial D_\epsilon} \mathbf{f}(\mathbf{X}, t) \cdot \mathbf{N}\, dS$ converges to a time dependent function as $\epsilon \to 0$.

One can prove the following statement (Agiasofitou and Kalpakides 2003).

Assume that the Conditions 1, 2 and 3 hold. Then, $\int_\Omega \phi(\mathbf{X}, t)dA$ is a differentiable function of t. In addition, its derivative will be given by the relation,

$$\frac{d}{dt}\int_\Omega \phi(\mathbf{X},t)\,dA = \lim_{\epsilon\to 0}\left(\frac{d}{dt}\int_{\Omega_\epsilon}\phi(\mathbf{X},t)\,dA\right). \tag{15}$$

The transport theorem and the divergence theorem for any domain Ω_ϵ, can be written, respectively

$$\frac{d}{dt}\int_{\Omega_\epsilon}\phi(\mathbf{X},t)\,dA = \int_{\Omega_\epsilon}\frac{\partial\phi(\mathbf{X},t)}{\partial t}\,dA - \int_{\partial D_\epsilon}\phi(\mathbf{X},t)(\mathbf{V}\cdot\mathbf{N})\,dS \tag{16}$$

and

$$\int_{\Omega_\epsilon}\text{Div}\,\mathbf{f}(\mathbf{X},t)\,dA = \int_{\partial\Omega}\mathbf{f}(\mathbf{X},t)\cdot\mathbf{N}\,dS - \int_{\partial D_\epsilon}\mathbf{f}(\mathbf{X},t)\cdot\mathbf{N}\,dS + \int_{\gamma_\epsilon}[\mathbf{f}(\mathbf{X},t)]\cdot\mathbf{n}\,dl. \tag{17}$$

Using the Conditions 1–5 and the eqs. (16)–(17), one can prove the following versions for the transport theorem and divergence theorem appropriate for the problem under study (Agiasofitou and Kalpakides 2003)

$$\frac{d}{dt}\int_\Omega \phi(\mathbf{X},t)\,dA = \int_\Omega \frac{\partial\phi}{\partial t}(\mathbf{X},t)\,dA - \lim_{\epsilon\to 0}\int_{\partial D_\epsilon}\phi(\mathbf{X},t)(\mathbf{V}\cdot\mathbf{N})\,dS \tag{18}$$

and

$$\int_{\partial\Omega}\mathbf{f}\cdot\mathbf{N}\,dS = \int_\Omega \text{Div}\,\mathbf{f}\,dA + \lim_{\epsilon\to 0}\int_{\partial D_\epsilon}\mathbf{f}(\mathbf{X},t)\cdot\mathbf{N}\,dS - \int_{\gamma_\Omega}[\mathbf{f}(\mathbf{X},t)]\cdot\mathbf{n}\,dl. \tag{19}$$

Inserting eqs. (18) and (19) into eq. (14), we obtain

$$\begin{aligned}\int_\Omega\left(\frac{\partial\phi(\mathbf{X},t)}{\partial t} - \text{Div}\,\mathbf{f}(\mathbf{X},t) - h(\mathbf{X},t)\right)dA + \int_{\gamma_\Omega}[\mathbf{f}(\mathbf{X},t)]\cdot\mathbf{n}\,dl\, - \\ \lim_{\epsilon\to 0}\int_{\partial D_\epsilon}(\phi(\mathbf{X},t)(\mathbf{V}\cdot\mathbf{N}) + \mathbf{f}(\mathbf{X},t)\cdot\mathbf{N})\,dS - g(t) = 0,\end{aligned} \tag{20}$$

for all Ω containing the crack tip.

We remark that in the case where Ω does not contain the crack tip and any part of the crack, eq. (20) takes the simpler form

$$\int_\Omega\left(\frac{\partial\phi(\mathbf{X},t)}{\partial t} - \text{Div}\,\mathbf{f}(\mathbf{X},t) - h(\mathbf{X},t)\right)dA = 0. \tag{21}$$

Thus, due to the arbitrariness of Ω, we conclude that

$$\frac{\partial\phi(\mathbf{X},t)}{\partial t} - \text{Div}\,\mathbf{f}(\mathbf{X},t) - h(\mathbf{X},t) = 0, \text{ for all } t\in I,\ \mathbf{X}\in\mathcal{B}_t\setminus C(t). \tag{22}$$

Similarly, we can consider Ω containing a part of the crack apart from the crack tip. In this case, the global balance law (eq. (20)) takes the form

$$\int_\Omega\left(\frac{\partial\phi(\mathbf{X},t)}{\partial t} - \text{Div}\,\mathbf{f}(\mathbf{X},t) - h(\mathbf{X},t)\right)dA + \int_{\gamma_\Omega}[\mathbf{f}(\mathbf{X},t)]\cdot\mathbf{n}\,dl = 0. \tag{23}$$

However, $\int_\Omega (\partial\phi(\mathbf{X}, t)/\partial t - \text{Div } \mathbf{f}(\mathbf{X}, t) - h(\mathbf{X}, t))dA = 0$, because its integrand is zero almost everywhere due to eq. (22), i.e., it is zero everywhere apart from the crack line γ_Ω, which is a set of measure zero in Ω. Consequently, eq. (23) gives

$$\int_{\gamma_\Omega} [\mathbf{f}(\mathbf{X}, t)] \cdot \mathbf{n} \, dl = 0, \tag{24}$$

for all γ_Ω which do not contain the crack tip. Thus, we obtain

$$[\mathbf{f}(\mathbf{X}, t)] \cdot \mathbf{n} = 0, \quad \text{for all } \ t \in I, \ \mathbf{X} \in C(t) \setminus \{\mathbf{Z}(t)\}. \tag{25}$$

In the same line of argument, we consider arbitrary Ω which contains the whole crack. In this case, we must use the complete form of eq. (20). Taking into account the results provided by eqs. (22) and (25), we remark that the integrands of the first two terms of eq. (20) vanish almost everywhere in any Ω_ϵ and any γ_ϵ, respectively, thus we can write

$$\int_{\Omega_\epsilon} \left(\frac{\partial\phi(\mathbf{X}, t)}{\partial t} - \text{Div } \mathbf{f}(\mathbf{X}, t) - h(\mathbf{X}, t) \right) dA = 0,$$
$$\int_{\gamma_\epsilon} [\mathbf{f}(\mathbf{X}, t)] \cdot \mathbf{n} \, dl = 0,$$

for all $\epsilon > 0$.

Thus, recalling the sense of integrability given by eqs. (12) and (13), we conclude that

$$\int_{\Omega} \left(\frac{\partial\phi(\mathbf{X}, t)}{\partial t} - \text{Div } \mathbf{f}(\mathbf{X}, t) - h(\mathbf{X}, t) \right) dA = 0, \tag{26}$$

$$\int_{\gamma_\Omega} [\mathbf{f}(\mathbf{X}, t)] \cdot \mathbf{n} \, dl = 0, \tag{27}$$

for all Ω and γ_Ω, even they contain the crack tip. Finally, we obtain the localization of the balance law at the crack tip as follows:

$$g(t) = -\lim_{\epsilon \to 0} \int_{\partial D_\epsilon} (\phi(\mathbf{X}, t)(\mathbf{V} \cdot \mathbf{N}) + \mathbf{f}(\mathbf{X}, t) \cdot \mathbf{N}) dS, \quad \text{for all } \ t \in I. \tag{28}$$

To sum up, the requirement that the balance law (14) holds for all $\Omega \in \mathcal{B}_t$ implies the local equations (22), (25) and (28), that is,

$$\begin{aligned} &\frac{\partial\phi}{\partial t} - \text{Div } \mathbf{f} - h = 0, \qquad \text{for all } \ t \in I, \ \mathbf{X} \in \mathcal{B}_t \setminus C(t), \\ &[\mathbf{f}] \cdot \mathbf{n} = 0, \qquad \text{for all } \ t \in I, \ \mathbf{X} \in C(t) \setminus \{\mathbf{Z}(t)\}, \\ &g(t) = -\lim_{\epsilon \to 0} \int_{\partial D_\epsilon} (\phi(\mathbf{V} \cdot \mathbf{N}) + \mathbf{f} \cdot \mathbf{N}) dS, \qquad \text{for all } \ t \in I. \end{aligned} \tag{29}$$

4 BALANCE LAWS IN THE PHYSICAL SPACE

Throughout this and the next section, we assume that each field inserted in a global balance law at the position of the abstract functions Φ, **f** and h will enjoy the corresponding smoothness specified in the previous section.

4.1 *The balances of mass, momentum and angular momentum*

We assume that there are no sources of mass, momentum and angular momentum, due to the crack evolution. Thus, we accept that, apart from the energy, the crack evolution does not intervene directly in the balance of the physical fields. Nevertheless, we expect a new relation at the crack tip due to the singularities of the physical fields. We denote with ρ and $\mathbf{T}$ the mass density in the material configuration and the Piola-Kirchhoff stress tensor, respectively. Also, the position vector of $\mathbf{x}$ is denoted with $\mathbf{r} = \mathbf{x} - \mathbf{0}$. As in the standard continuum mechanics, it is postulated that the mass, the momentum and the angular momentum fulfill the following relations

$$\frac{d}{dt}\int_{\Omega} \rho(\mathbf{X}, t)\, dA = 0. \tag{30}$$

$$\frac{d}{dt}\int_{\Omega} \rho\dot{\mathbf{x}}\, dA = \int_{\partial\Omega} \mathbf{TN}\, dS, \tag{31}$$

$$\frac{d}{dt}\int_{\Omega} \mathbf{r} \times \rho\dot{\mathbf{x}}\, dA = \int_{\partial\Omega} (\mathbf{r} \times \mathbf{T})\mathbf{N}\, dS, \tag{32}$$

for every part Ω of $\mathcal{B}_t$ and for every t in some interval I.

The local form of the balances (30–32) outside the crack are extracted from eq. $(29)_1$

$$\frac{\partial \rho(\mathbf{X}, t)}{\partial t} = 0 \Rightarrow \rho = \rho(\mathbf{X}), \tag{33}$$

$$\frac{\partial}{\partial t}(\rho\dot{\mathbf{x}}) - \mathrm{Div}\mathbf{T} = 0, \tag{34}$$

$$\frac{\partial}{\partial t}(\mathbf{r} \times \rho\dot{\mathbf{x}}) - \mathrm{Div}(\mathbf{r} \times \mathbf{T}) = 0, \tag{35}$$

for all $t \in I$, $\mathbf{X} \in \mathcal{B}_t \backslash C(t)$.

Moreover, the localization process gives the following jump conditions (see eq. $(29)_2$) along the crack curve

$$[\mathbf{T}]\mathbf{n} = 0. \tag{36}$$

$$[\mathbf{r} \times \mathbf{T}]\mathbf{n} = 0. \tag{37}$$

for all $t \in I$, $\mathbf{X} \in C(t)\backslash\{\mathbf{Z}(t)\}$. The above local equations and jump conditions do not differ from the corresponding ones holding for any smooth elastic body with a material surface of discontinuity within it. Recalling that the motion $\chi(\mathbf{X}, t)$ is continuous along the crack $C(t)$ (hence, $\mathbf{r}$ is continuous as well), we easily conclude that the condition (37) follows from the jump condition (36).

The new results of the proposed approach concern the relations holding at the crack tip are derived from$(29)_3$ as follows:

$$\lim_{\epsilon \to 0} \int_{\partial D_\epsilon} \rho(\mathbf{V} \cdot \mathbf{N})\, dS = 0, \tag{38}$$

$$\lim_{\epsilon \to 0} \int_{\partial D_\epsilon} (\rho\dot{\mathbf{x}}(\mathbf{V} \cdot \mathbf{N}) + \mathbf{TN})\, dS = 0, \tag{39}$$

$$\lim_{\epsilon \to 0} \int_{\partial D_\epsilon} \mathbf{r} \times (\rho\dot{\mathbf{x}}(\mathbf{V} \cdot \mathbf{N}) + \mathbf{TN})\, dS = 0. \tag{40}$$

for all $t \in I$.

Equation (38) shows that the rate of mass flow through ∂D_ϵ vanishes, when the boundary shrinks onto the crack tip. Eqs. (39) and (40) represent the balance of linear momentum and angular momentum at the crack tip, respectively. Adopting the *standard momentum condition* of (Gurtin 2000), that is,

$$\lim_{\epsilon\to 0}\int_{\partial D_\epsilon} \rho\dot{\mathbf{x}}\otimes\mathbf{N}\,dS = 0,$$

we take from eq. (39)

$$\lim_{\epsilon\to 0}\int_{\partial D_\epsilon} \rho\dot{\mathbf{x}}(\mathbf{V}\cdot\mathbf{N})\,dS = 0,\quad \lim_{\epsilon\to 0}\int_{\partial D_\epsilon} \mathbf{TN}\,dS = 0.$$

4.2 *The balance of energy*

Unlike the balances for mass, linear momentum and angular momentum, we assume that the balance of energy is directly influenced by the crack growth. This is quite reasonable because the crack propagation is a dissipative phenomenon, that is to say, the growth of the crack consumes a part of the energy given by the applied forces. Hence, an energy source term describing the total dissipation rate of the body, denoted here by $\Phi(t)$, must be added in the energy balance. Thus, the global balance law for energy can be postulated as

$$\frac{d}{dt}\int_\Omega (W+K)\,dA = \int_{\partial\Omega} \mathbf{TN}\cdot\dot{\mathbf{x}}\,dS - \Phi,\quad \text{for all } t\in I,\ \Omega\in\mathcal{B}_t, \tag{41}$$

where W is the elastic energy density and K is the kinetic energy density, both are defined per unit volume in material configuration.

Localizing eq. (41), we obtain (see eq. (29))

$$\frac{\partial}{\partial t}(W+K) - \mathrm{Div}(\mathbf{T}^T\dot{\mathbf{x}}) = 0,\ \forall\, t\in I,\ \mathbf{X}\in\mathcal{B}_t\setminus C(t), \tag{42}$$

$$[\mathbf{T}^T\dot{\mathbf{x}}]\cdot\mathbf{n} = 0,\ \forall\, t\in I,\ \mathbf{X}\in C(t)\setminus\{\mathbf{Z}(t)\}, \tag{43}$$

$$\Phi = \lim_{\epsilon\to 0}\int_{\partial D_\epsilon}\big((W+K)(\mathbf{V}\cdot\mathbf{N}) + \mathbf{T}^T\dot{\mathbf{x}}\cdot\mathbf{N}\big)dS,\ \forall\, t\in I. \tag{44}$$

It is obvious that eqs. (42) and (43) are the local energy equation and the associated jump condition along the crack, respectively. Also, eq. (44) is the energy flow out of the body and into the crack tip per unit time and if it be divided by the crack propagation velocity V, it will give the well-known, in fracture literature (Freund 1981), dynamic energy release rate G, i.e.,

$$G = \Phi/V. \tag{45}$$

5 BALANCE LAWS IN THE MATERIAL SPACE

The balances which we are dealt within the last section do not exhaust all the relevant quantities involved in our problem. We must further consider balances for the configurational fields, that is, the pseudomomentum and the material angular momentum.

5.1 *The balance of pseudomomentum*

We introduce now the pseudomomentum (or material momentum)

$$\boldsymbol{\mathcal{P}}(\mathbf{X},t) = -\rho\mathbf{F}^T\dot{\mathbf{x}},\ t\in I,\ \mathbf{X}\in\mathcal{B}_t\setminus C(t), \tag{46}$$

a quantity analogous to the physical momentum, concerning changes within the material structure. In a Hamiltonian framework, the pseudomomentum is the dual quantity to the velocity of the inverse motion function, like the physical momentum is the dual of the standard velocity of the body (Maugin and Kalpakides 2002). The contributors to the balance of pseudomomentum will be the material or configurational forces (Maugin 1993). Considering both at a distance and at contact configurational forces, we introduce the configurational body forces $\tilde{\mathbf{f}}$ (source term) and the configurational stress tensor **b** (flux term), respectively . Moreover, we consider a pseudomomentum source term, that is a material force, $\mathcal{F} = \mathcal{F}(t)$, produced by the crack evolution. After all these considerations, we postulate the balance law for the pseudomomentum

$$\frac{d}{dt}\int_{\Omega} \mathcal{P}\, dA = \int_{\partial\Omega} \mathbf{bN}\, dS + \int_{\Omega} \tilde{\mathbf{f}}\, dA + \mathcal{F},\ \forall t \in I,\ \forall \Omega \in \mathcal{B}_t. \tag{47}$$

The local equations, obtained by eq. (47), are given (see eqs. (29)) as follows

$$\frac{\partial \mathcal{P}}{\partial t} - \mathrm{Div}\mathbf{b} - \tilde{\mathbf{f}} = 0,\ \ \forall\, t \in I,\ \mathbf{X} \in \mathcal{B}_t \setminus C(t), \tag{48}$$

$$[\mathbf{b}] \cdot \mathbf{n} = 0,\ \ \forall\, t \in I,\ \mathbf{X} \in C(t) \setminus \{\mathbf{Z}(t)\}, \tag{49}$$

$$\mathcal{F} = -\lim_{\epsilon \to 0} \int_{\partial D_\epsilon} (\mathcal{P}(\mathbf{V}\cdot\mathbf{N}) + \mathbf{bN})\, dS,\ \forall\, t \in I. \tag{50}$$

Eq. (48) is the equation of pseudomomentum, which holds in the smooth part of the body and eq. (49) is the associated jump condition. In addition, eq. (50) represents the material force at the crack tip, which drives the crack evolution. Thus, the quantity $\mathcal{F}$ should be directly related to the energy release rate, *G*. Also, in the static case it holds the following relation,

$$\mathcal{F} = -\lim_{\epsilon \to 0} \mathbf{J}(\epsilon),$$

where **J** is the well-known **J**-integral of Rice (Budiansky and Rice 1973).

In the absence of a crack or any other rearrangement in the material configuration, the last term in the pseudomomentum balance law vanishes and eq. (48) holds all over the body as a simple identity for the solution of the standard elastic problem. In other words, eqs. (48–50) do not make sense in the standard continuum mechanics, where only the motion in physical space is considered. Thus, the balance law (47) must be considered when one studies any kind of evolution of structural defects. From this point of view, it is a configurational balance law.

Remark 1: One can enrich the balance law (47) by considering an additional term of the form $\int_{\gamma\Omega} \mathbf{g}^s\, dl$, accounting for configurational forces acting along the crack curve (Gurtin 2000). In that case, the localization process provides

$$[\mathbf{b}] \cdot \mathbf{n} + \mathbf{g}^l = 0,$$

instead of eq. (49).

Remark 2: The flux term **b**, like the Piola–Kirchhoff stress tensor in standard continuum mechanics, needs a constitutive relation to be further determined. Because the constitutive relations are out of the scope of the present procedure, we adopt without reasoning the relation

$$\mathbf{b} = (W - \frac{1}{2}\rho\dot{\mathbf{x}}^2)\mathbf{I} - \mathbf{F}^T\mathbf{T}, \tag{51}$$

that is, the Eshelby stress tensor for the dynamical case (Eshelby 1995). For the derivation and a discussion about this relationship, viewed as a constitutive relation, we refer to the works of (Gurtin 2000) and (Podio-Guidugli 2002). As concerns the term $\tilde{\mathbf{f}}$, we consider it as a distributed body material force, produced by the material inhomogeneities (Maugin 1993). Moreover, the pseudomomentum source term $\mathcal{F}$, produced by the crack evolution, is referred to by (Maugin 1993; Dascalu and Maugin 1995) as *global material force* and by (Gurtin 2000) as *tip traction.*

5.2 *The balance of material angular momentum*

We proceed to the balance of the material angular momentum, that is, the moment of pseudomomentum, $\mathbf{R} \times \mathcal{P}$, where $\mathbf{R} = \mathbf{X} - \mathbf{0}$ is the position vector of $\mathbf{X}$. The rest contributors to this law should be the moment of material contact and material body forces. Moreover, we consider a material angular momentum source term, $\mathcal{M} = \mathcal{M}(t)$ due to the presence of the crack. We postulate:

$$\frac{d}{dt}\int_{\Omega} \mathbf{R} \times \mathcal{P}\, dA = \int_{\partial\Omega} \mathbf{R} \times \mathbf{bN}\, dS + \int_{\Omega} \mathbf{R} \times \tilde{\mathbf{f}}\, dA + \int_{\Omega} \mathbf{g}\, dA + \mathcal{M}, \quad \forall t \in I,\ \forall \Omega \in \mathcal{B}_t, \tag{52}$$

where $\mathbf{g}(\mathbf{X}, t)$ is a vector field describing the distribution of material couples within the body.

The localization of eq. (52) provides

$$\frac{\partial(\mathbf{R} \times \mathcal{P})}{\partial t} - \mathrm{Div}(\mathbf{R} \times \mathbf{b}) - \mathbf{R} \times \tilde{\mathbf{f}} - \mathbf{g} = 0,\ \forall t \in I,\ \mathbf{X} \in \mathcal{B}_t \setminus C(t), \tag{53}$$

$$\mathcal{M} = -\lim_{\epsilon \to 0} \int_{\partial D_\epsilon} \big(\mathbf{R} \times (\mathcal{P}(\mathbf{V} \cdot \mathbf{N}) + \mathbf{bN})\big)\, dS,\ \forall t \in I \tag{54}$$

and the associated jump condition

$$[\mathbf{R} \times \mathbf{b}] \cdot \mathbf{n} = 0, \quad \forall\, t \in I,\ \forall\, \mathbf{X} \in C(t) \setminus \{\mathbf{Z}(t)\},$$

which holds identically due to the continuity of $\mathbf{X}$ and the jump condition (49).

Eq. (53) is the equation of material angular momentum, which holds in the bulk of the body. If $\tilde{\mathbf{f}} = \mathbf{0}$ and $\mathbf{g} = \mathbf{0}$, then it coincides with the corresponding one of (Golebiewska Herrmann 1982).

If we take into account the equation of pseudomomentum (48), then eq. (53) gives the following relation

$$\mathbf{g} = \mathrm{axl}\mathbf{b} \ \text{ or } \ g_A = -e_{ABC} b_{BC}, \tag{55}$$

where axl**b** denotes the axial vector of **b** (Chadwick 1976). Finally, eq. (53) is written as follows

$$\frac{\partial(\mathbf{R} \times \mathcal{P})}{\partial t} - \mathrm{Div}(\mathbf{R} \times \mathbf{b}) - \mathbf{R} \times \tilde{\mathbf{f}} - 2\mathrm{axl}\mathbf{b} = 0,\ \forall t \in I,\ \mathbf{X} \in \mathcal{B}_t \setminus C(t), \tag{56}$$

which is in accordance with the corresponding one of (Steinmann 2000) for the static case.

Remark 3: If the material is homogeneous and isotropic, then the Eshelby stress tensor **b** is symmetric (Steinmann 2000; Kalpakides and Agiasofitou 2002), which means that $\mathrm{axl}\mathbf{b} = \mathbf{0}$, so eq. (55) gives $\mathbf{g} = \mathbf{0}$.

Remark 6: Adopting the existence of configurational forces distributed along the crack curve as we did in Remark 1, we obtain the jump condition

$$[\mathbf{R} \times \mathbf{b}] \cdot \mathbf{n} + \mathbf{R} \times \mathbf{g}^l = 0.$$

Furthermore, equation (54) gives the form of the *configurational moment* $\mathcal{M}$ *at the crack tip.* Particularly,

$$\mathcal{M}(t) = -\lim_{\epsilon \to 0} \mathcal{M}_\epsilon(t),\ \forall t \in I,$$

where

$$\mathcal{M}_\epsilon(t) = \int_{\partial D_\epsilon} \big(\mathbf{R} \times (\mathcal{P}(\mathbf{V}\cdot\mathbf{N}) + \mathbf{b}\mathbf{N})\big)\, dS. \tag{57}$$

The physical interpretation of $\mathcal{M}_\epsilon$ and its possible connection with the L-integral will be examined in the next subsection.

5.3 *The configurational moment and the L-integral*

It is worth noting that the Eshelby stress tensor used in fracture mechanics literature differs from the one used here. In fracture mechanics, the tensor **b** is defined with the aid of the displacement field $\mathbf{u}(\mathbf{X}, t)$, whereas in our analysis it is defined with the aid of the motion mapping $\mathbf{x}(\mathbf{X}, t)$ (see the relation (51)). If we introduce in eq. (51) the displacement field **u**, we take

$$\mathbf{b} = (W - \frac{1}{2}\rho\dot{\mathbf{x}}^2)\mathbf{I} - \mathbf{F}^T\mathbf{T} = (W - \frac{1}{2}\rho\dot{\mathbf{u}}^2)\mathbf{I} - (\nabla\mathbf{u})^T\mathbf{T} - \tilde{\mathbf{I}}^T\mathbf{T},$$

where $\tilde{\mathbf{I}}$ denotes the two point unit tensor (or the shifter δ_{iA} in a coordinate system). Then, we can write

$$\mathbf{b} = \mathbf{b}^u - \tilde{\mathbf{I}}^T\mathbf{T}, \tag{58}$$

where

$$\mathbf{b}^u = (W - \frac{1}{2}\rho\dot{\mathbf{u}}^2)\mathbf{I} - (\nabla\mathbf{u})^T\mathbf{T}.$$

Notice that using $\mathbf{b}^u$ instead of **b** in pseudomomentum equation and neglecting the configurational body forces $\tilde{\mathbf{f}}$, we obtain, in virtue of eq. (34), an equation of the same form

$$\frac{\partial\mathcal{P}^u}{\partial t} - \mathrm{Div}\mathbf{b}^u = 0,$$

where

$$\mathcal{P}^u = -\rho(\nabla\mathbf{u})^T\dot{\mathbf{u}}. \tag{59}$$

However, under the same manipulation the material angular momentum equation does not retain its form. Indeed, inserting eqs. (58) and (59) into eq. (53), neglecting $\tilde{\mathbf{f}}$, **g** and taking into account the equation of angular momentum, i.e., eq. (35), we obtain

$$\frac{\partial}{\partial t}(\mathbf{R} \times \mathcal{P}^u + \mathbf{u} \times \rho\dot{\mathbf{u}}) - \mathrm{Div}\,(\mathbf{R} \times \mathbf{b}^u + \mathbf{u} \times \mathbf{T}) = 0. \tag{60}$$

Consequently, eq. (60) must be used in any comparison of the present results with the corresponding ones in the linear fracture mechanics. Indeed, if **R** and **T** are replaced by the spatial coordinates **x** and the Caushy stress tensor σ, respectively and ρ and W are defined per unit deformed volume, then eq. (60) coincides with the corresponding one of (Fletcher 1975), for a linear, homogeneous and isotropic elastic body in the absence of body forces.

In addition, doing the same replacements in the integral given by eq. (57), the latter becomes

$$\mathcal{M}_\epsilon(t) = \int_{\partial D_\epsilon} \big((\mathbf{R} \times \mathcal{P}^u + \mathbf{u} \times \rho\dot{\mathbf{u}})(\mathbf{V}\cdot\mathbf{N}) + (\mathbf{R} \times \mathbf{b}^u + \mathbf{u} \times \mathbf{T})\mathbf{N}\big)\, dS \tag{61}$$

In the static case, apart from the contour of integration, the integral $\mathcal{M}_\epsilon$ reduces to the L-integral, as it was given by (Knowles and Sternberg 1972) and (Steinmann 2000) for a nonlinear, homogeneous and isotropic elastic material. Therefore, an integral having the same integrand with $\mathcal{M}_\epsilon$ along an integration path encircling the total crack can be considered as a generalization of L-integral in the dynamical, non-linear case. It is important to remark that, in fracture mechanics literature, the path of the L-integral includes the whole crack, while in our analysis the path ∂D_ϵ is limited around the crack tip. This, on the one hand, justifies the term "configurational moment at the crack tip" and on the other, provides possibly an alternative physical interpretation of $\mathcal{M}$ and $\mathcal{M}_\epsilon$. More specifically, one can conjecture that the quantity $\mathcal{M}$ is related to the tendency of the crack tip (and as a result of the crack) to turn, while the usual interpretation (for instance, see (Golebiewska Herrmann and Herrmann 1981)) of the L-integral concerns the tendency of a stationary straight crack to rotate, as a whole, with respect to its center.

6 THE ENERGY RELEASE RATES AND THE CONFIGURATIONAL FIELDS

In this section, expressions for the energy release rates will be derived. Particularly, the relationship between the dynamical energy release rate with the configurational force at the crack tip as well as the relationship of the rotational energy release rate with the configurational moment at the crack tip, will be established.

6.1 *The energy release rate and the configurational force*

We start with the expression for the rate of energy dissipation, i.e., eq. (44):

$$\begin{aligned}\Phi &= \lim_{\epsilon\to 0}\int_{\partial D_\epsilon}\big[(W+K)(\mathbf{V}\cdot\mathbf{N})+\mathbf{T}^T\dot{\mathbf{x}}\cdot\mathbf{N}\big]dS\\ &= \lim_{\epsilon\to 0}\int_{\partial D_\epsilon}\big[(W-K)(\mathbf{V}\cdot\mathbf{N})\big]dS+\lim_{\epsilon\to 0}\int_{\partial D_\epsilon}\dot{\mathbf{x}}\big[\rho\dot{\mathbf{x}}(\mathbf{V}\cdot\mathbf{N})+\mathbf{TN}\big]dS.\end{aligned}$$

Recalling relation (9), the above equation can be written as

$$\begin{aligned}\Phi &= \lim_{\epsilon\to 0}\int_{\partial D_\epsilon}\left[(W-K)(\mathbf{V}\cdot\mathbf{N})-\mathbf{FV}(\rho\dot{\mathbf{x}}(\mathbf{V}\cdot\mathbf{N})+\mathbf{TN})\right]dS\\ &\qquad +\lim_{\epsilon\to 0}\int_{\partial D_\epsilon}\widetilde{\mathbf{V}}\cdot\left[\rho\dot{\mathbf{x}}(\mathbf{V}\cdot\mathbf{N})+\mathbf{TN}\right]dS\\ &= \lim_{\epsilon\to 0}\int_{\partial D_\epsilon}\mathbf{V}\cdot\left[\left((W-K)\mathbf{I}-\mathbf{F}^T\mathbf{T}\right)\mathbf{N}-\rho\mathbf{F}^T\dot{\mathbf{x}}(\mathbf{V}\cdot\mathbf{N})\right]dS\\ &\qquad +\lim_{\epsilon\to 0}\int_{\partial D_\epsilon}\widetilde{\mathbf{V}}\cdot\left[\rho\dot{\mathbf{x}}(\mathbf{V}\cdot\mathbf{N})+\mathbf{TN}\right]dS.\end{aligned}\qquad (62)$$

Due to eqs. (46) and (51), eq. (62) becomes

$$\begin{aligned}\Phi = &\lim_{\epsilon\to 0}\int_{\partial D_\epsilon}\mathbf{V}\cdot\left[\mathbf{bN}+\mathcal{P}(\mathbf{V}\cdot\mathbf{N})\right]dS\\ &\qquad +\lim_{\epsilon\to 0}\int_{\partial D_\epsilon}\widetilde{\mathbf{V}}\cdot\left[\rho\dot{\mathbf{x}}(\mathbf{V}\cdot\mathbf{N})+\mathbf{TN}\right]dS.\end{aligned}\qquad (63)$$

One can prove that under specific assumptions the second term in eq. (63) vanishes. The essential step to this end is to prove the following

Proposition: *Assume that*

$$\int_{\partial D_\epsilon}|\rho\dot{\mathbf{x}}(\mathbf{V}\cdot\mathbf{N})+\mathbf{TN}|dS,\quad \text{is bounded as}\quad \epsilon\to 0. \qquad (64)$$

Then the following convergence holds

$$\lim_{\epsilon\to 0}\int_{\partial D_\epsilon}\left[\widetilde{\mathbf{V}}(\mathbf{X},t)-\widetilde{\mathbf{U}}(t)\right]\cdot[\rho\dot{\mathbf{x}}(\mathbf{V}\cdot\mathbf{N})+\mathbf{T}\mathbf{N}]\,dS=0. \tag{65}$$

Proof: We have

$$\begin{aligned}&\left|\int_{\partial D_\epsilon}\left[\widetilde{\mathbf{V}}(\mathbf{X},t)-\widetilde{\mathbf{U}}(t)\right]\cdot[\rho\dot{\mathbf{x}}(\mathbf{V}\cdot\mathbf{N})+\mathbf{T}\mathbf{N}]\,dS\right|\\ &\leq\int_{\partial D_\epsilon}\sup_{t\in I}|\widetilde{\mathbf{V}}(\mathbf{X},t)-\widetilde{\mathbf{U}}(t)|\,|\rho\dot{\mathbf{x}}(\mathbf{V}\cdot\mathbf{N})+\mathbf{T}\mathbf{N}|\,dS\\ &\leq\sup_{\mathbf{X}\in\partial D_\epsilon}\left(\sup_{t\in I}|\widetilde{\mathbf{V}}(\mathbf{X},t)-\widetilde{\mathbf{U}}(t)|\right)\int_{\partial D_\epsilon}|\rho\dot{\mathbf{x}}(\mathbf{V}\cdot\mathbf{N})+\mathbf{T}\mathbf{N}|dS.\end{aligned} \tag{66}$$

On the other hand, the condition given by eq. (10) means that

$$\lim_{\mathbf{X}\to\mathbf{Z}(t)}(\sup_{t\in I}|\widetilde{\mathbf{V}}(\mathbf{X},t)-\widetilde{\mathbf{U}}(t)|)=0.$$

¿ From this convergence, taking into account that the boundary ∂D_ϵ shrinks onto $\mathbf{Z}(t)$ as $\epsilon\to 0$, it is implied that the following convergence holds as well

$$\lim_{\epsilon\to 0}\left(\sup_{\mathbf{X}\in\partial D_\epsilon}(\sup_{t\in I}|\widetilde{\mathbf{V}}(\mathbf{X},t)-\widetilde{\mathbf{U}}(t)|)\right)=0.$$

The latter jointly with the assumption (64) gives

$$\lim_{\epsilon\to 0}\left[\sup_{\mathbf{X}\in\partial D_\epsilon}\left(\sup_{t\in I}|\widetilde{\mathbf{V}}(\mathbf{X},t)-\widetilde{\mathbf{U}}(t)|\right)\int_{\partial D_\epsilon}|\rho\dot{\mathbf{x}}(\mathbf{V}\cdot\mathbf{N})+\mathbf{T}\mathbf{N}|dS\right]=0,$$

which, in virtue of inequality (66), completes the proof.

Next, from eq. (65) invoking the balance of physical momentum at the crack tip i.e., eq. (39), we conclude that

$$\begin{aligned}&\lim_{\epsilon\to 0}\int_{\partial D_\epsilon}\widetilde{\mathbf{V}}(\mathbf{X},t)\cdot[\rho\dot{\mathbf{x}}(\mathbf{V}\cdot\mathbf{N})+\mathbf{T}\mathbf{N}]\,dS\\ &=\lim_{\epsilon\to 0}\int_{\partial D_\epsilon}\widetilde{\mathbf{U}}(t)\cdot[\rho\dot{\mathbf{x}}(\mathbf{V}\cdot\mathbf{N})+\mathbf{T}\mathbf{N}]\,dS=0\end{aligned} \tag{67}$$

Therefore, taking into account eq. (67), the energy flux at the crack tip, i.e., eq. (63) becomes

$$\Phi=\mathbf{V}\cdot\lim_{\epsilon\to 0}\int_{\partial D_\epsilon}(\mathbf{b}\mathbf{N}+\mathcal{P}(\mathbf{V}\cdot\mathbf{N}))dS$$

or, due to eq. (50),

$$\Phi=-\mathbf{V}\cdot\boldsymbol{\mathcal{F}}. \tag{68}$$

Using the definition (45) for the energy release rate, eq. (68) gives the following result

$$G=-\boldsymbol{\mathcal{F}}\cdot\mathbf{t}, \tag{69}$$

which confirms that the energy release rate G is the crack driving force.

Remark 5: In the preceding analysis, two relations, which can be viewed as constraints on the singularity order for the velocity and the stress fields at the crack tip, have been arisen. These relations are the condition (64) and the equation (39). Suppose that the independent variables of a function $f(\mathbf{X}, t)$ (say f be the velocity or the stress tensor) can be separated as

$$f(\mathbf{X}, t) = g(r)h(\theta, t),$$

where $r = |\mathbf{X} - \mathbf{Z}(t)|$ and θ are the distance and the angle, respectively, in a polar coordinate system with its origin at the crack tip. Then, assuming that $g(r) = O(r^p)$, $p \geqslant -1$ is sufficient to assure that the condition (64) holds. Furthermore, assuming that $g(r) = O(r^p)$, $p > -1$, we obtain that eq. (39) holds, as well. However, it is well known that for a linear elastic, cracked body, both the near tip velocity and stress fields are of order $O(r^{-1/2})$ (Freund 1981).

6.2 *The rotational energy release rate and the configurational moment at the crack tip*

In this section, we will show that a relationship between the rotational energy release rate and the configurational fields at the crack tip can be established. We start with the relation (57) which can be written as

$$\boldsymbol{\mathcal{M}} = -\lim_{\epsilon \to 0} \int_{\partial D_\epsilon} \mathbf{R} \times \boldsymbol{\mathcal{C}}\mathbf{N}\, dS, \tag{70}$$

where

$$\boldsymbol{\mathcal{C}} = \boldsymbol{\mathcal{P}} \otimes \mathbf{V} + \mathbf{b}^T. \tag{71}$$

Considering that the crack evolves along an arbitrary smooth curve, we introduce the angular velocity of the crack tip

$$\boldsymbol{\omega}(t) = \omega \mathbf{m}, \tag{72}$$

where $\mathbf{m} = \mathbf{t} \times \mathbf{n}$ is the unit normal vector to the plane of the crack. Furthermore, denoting with $\mathbf{a} = a\,\mathbf{n}$ the instantaneous radius of curvature ($\mathbf{n}$ the unit normal to the curve), we can write

$$\omega = \frac{V}{a}. \tag{73}$$

Also, we denote with $\mathbf{R}_Z = \mathbf{Z} - \mathbf{0} = R_Z\mathbf{e}$ the position vector of the crack tip. Then, we can write (see Fig. 4)

$$\mathbf{R}_Z = \mathbf{R} + \boldsymbol{\epsilon}, \quad \boldsymbol{\epsilon} = -\epsilon \mathbf{N}.$$

Therefore, the configurational moment $\boldsymbol{\mathcal{M}}$ becomes

$$\boldsymbol{\mathcal{M}} = -\lim_{\epsilon \to 0} \int_{\partial D_\epsilon} \big(\mathbf{R}_Z \times \boldsymbol{\mathcal{C}}\mathbf{N}\big)\, dS + \lim_{\epsilon \to 0} \int_{\partial D_\epsilon} \big(\boldsymbol{\epsilon} \times \boldsymbol{\mathcal{C}}\mathbf{N}\big)\, dS. \tag{74}$$

One can prove that the last term in eq. (74) vanishes under a particular condition. Indeed,

$$\int_{\partial D_\epsilon} \big(\boldsymbol{\epsilon} \times \boldsymbol{\mathcal{C}}\mathbf{N}\big)\, dS = -\int_{\partial D_\epsilon} \big(\epsilon\mathbf{N} \times \boldsymbol{\mathcal{C}}\mathbf{N}\big)\, dS = -\epsilon \int_{\partial D_\epsilon} \big(\mathbf{N} \times \boldsymbol{\mathcal{C}}\mathbf{N}\big)\, dS.$$

In addition, it holds

$$\Big|\int_{\partial D_\epsilon} \big(\mathbf{N} \times \boldsymbol{\mathcal{C}}\mathbf{N}\big)\, dS\Big| \leq \int_{\partial D_\epsilon} \big(|\mathbf{N}| \times |\boldsymbol{\mathcal{C}}\mathbf{N}|\big)\, dS = \int_{\partial D_\epsilon} |\boldsymbol{\mathcal{C}}\mathbf{N}|\, dS.$$

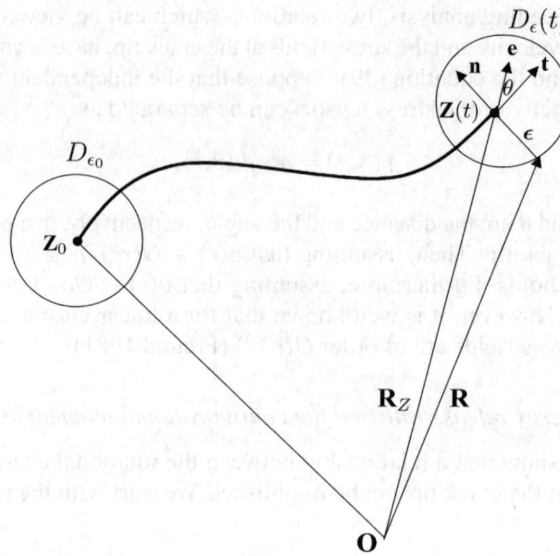

Figure 4. Some geometrical characteristics of the crack.

Thus, assuming that the integral $\int_{\partial D_\epsilon} |\mathcal{C}\mathbf{N}| \, dS$ is bounded as $\epsilon \to 0$, we obtain that the integral $\int_{\partial D_\epsilon} (\epsilon \times \mathcal{C}\mathbf{N}) \, dS$ vanishes as $\epsilon \to 0$.
Consequently, the expression for $\mathcal{M}$ (eq. (74)) becomes

$$\mathcal{M} = -\lim_{\epsilon \to 0} \int_{\partial D_\epsilon} \left(\mathbf{R}_Z \times \mathcal{C}\mathbf{N}\right) dS = -\mathbf{R}_Z \times \left(\lim_{\epsilon \to 0} \int_{\partial D_\epsilon} \mathcal{C}\mathbf{N} \, dS\right) \tag{75}$$

Notice that using the relation (71), the quantity $\mathcal{F}$ is written

$$\mathcal{F} = -\lim_{\epsilon \to 0} \int_{\partial D_\epsilon} \mathcal{C}\mathbf{N} \, dS.$$

So, $\mathcal{M}$ is given by the following simple formula

$$\mathcal{M} = \mathbf{R}_Z \times \mathcal{F}, \tag{76}$$

which confirms the term configurational moment at the crack tip, since, essentially, it is the moment of the configurational force at the crack tip.

In addition, if θ is the angle from the $\mathbf{t}$-axis to the $\mathbf{e}$-axis, then $\mathcal{M}$ is written in terms of $\mathcal{F}$ as follows

$$\mathcal{M} = R_Z\big(\cos\theta \, (\mathcal{F} \cdot \mathbf{n}) - \sin\theta(\mathcal{F} \cdot \mathbf{t})\big)\mathbf{m}. \tag{77}$$

Next, we calculate the product

$$\omega \cdot \mathcal{M} = \omega R_Z\big(\cos\theta \, (\mathcal{F} \cdot \mathbf{n}) - \sin\theta(\mathcal{F} \cdot \mathbf{t})\big) \tag{78}$$

in order to take an expression for the rate of dissipation in terms of the configurational force and the configurational moment at the crack tip

$$\Phi = \frac{a}{R_Z \sin\theta}(\omega \cdot \mathcal{M}) - V \cot\theta(\mathcal{F} \cdot \mathbf{n}). \tag{79}$$

Furthermore, if we denote with

$$G_r = \frac{\Phi}{\omega} \tag{80}$$

the *rotational energy release rate*, that is, the energy flow into the crack tip per unit angle extension of the crack, then from eq. (79) we have

$$G_r = \frac{a}{R_Z \sin\theta}(\boldsymbol{\omega} \cdot \boldsymbol{\mathcal{M}}) - a \cot\theta(\boldsymbol{\mathcal{F}} \cdot \mathbf{n}). \tag{81}$$

Remark 4: Assuming that $\mathbf{Z}(t)$ is a C^2 function, the instantaneous radius of curvature at $\mathbf{Z}(t)$ is related to the instantaneous curvature at $\mathbf{Z}(t)$ by $a = (1/|k|)$ and

$$|k| = \frac{\left|\frac{d\mathbf{Z}}{dt} \times \frac{d^2\mathbf{Z}}{dt^2}\right|}{\left|\frac{d\mathbf{Z}}{dt}\right|^3} = \frac{|\mathbf{V} \times \mathbf{A}|}{|\mathbf{V}|^3}, \tag{82}$$

where $\mathbf{A} = d^2\mathbf{Z}/dt^2$ is the acceleration vector of the crack tip. Therefore, we can write Φ (eq. (79)) in the following alternative form

$$\Phi = \frac{V^2}{R_Z A_n \sin\theta}(\boldsymbol{\omega} \cdot \boldsymbol{\mathcal{M}}) - V \cot\theta(\boldsymbol{\mathcal{F}} \cdot \mathbf{n}), \tag{83}$$

where A_n is the normal component of the vector $\mathbf{A}$.

Remark 7: In the case in which the crack is circular and the origin of the coordinates' system coincides with the center of the circle, the formulas (79) and (81) give

$$\Phi = \boldsymbol{\omega} \cdot \boldsymbol{\mathcal{M}}, \quad G_r = \boldsymbol{\mathcal{M}} \cdot \mathbf{m} = \mathcal{M}. \tag{84}$$

We can see from eq. $(84)_2$ that the rotational energy release rate is simply the magnitude of the vector of the configurational moment at the crack tip. Analogous results to eq. $(84)_1$ have been provided by (Maugin and Trimarco 1995) as well as by (Budiansky and Rice 1973) for disclinations and cavities, respectively. Of course, someone can easily see the analogy between the relations (68), (69) and $(84)_1$, $(84)_2$, respectively.

7 CONCLUSIONS

The objective of this paper was the study of the crack propagation within an elastic medium in the context of configurational mechanics. To this end, we proposed an appropriate kinematics and we formulated the corresponding transport and divergence theorems. In the sequel, we produced a rigorous localization process which has been used to derive the local equations for both the physical and configurational fields.

A significant consequence of the localization process was the expression for the configurational force at the crack tip related to the J-integral as well as the corresponding one for the configurational moment at the crack tip which is related to the L-integral. Based on these expressions, we derived a relationship between the configurational force at the crack tip and the energy release rate, as well as a relation connecting the rotational energy release rate with the configurational moment and force at the crack tip.

In the case of a crack with non constant curvature, the rotational energy release rate depends essentially on the geometrical characteristics of the curve. Therefore, in order to apply the formula (81), the geometrical characteristics of the curve along which the crack will evolve should be a priori known. Such situations appear in delamination cracks, where the crack necessarily follows a particular curve.

REFERENCES

Agiasofitou, E. K. and V. K. Kalpakides (2003). The concept of a balance law for a cracked elastic body and the configurational force and moment at the crack tip. Submitted for publication.

Budiansky, B. and J. R. Rice (1973). Conservation Laws and Energy-Release Rates *J. Appl. Mech.* 40: 201–203.

Chadwick, P. (1976). *Continuum Mechanics, Concise Theory and Problems*, London: George Allen & Unwin.

Dascalu, C. and G. A. Maugin (1995). The Thermoelastic Material-momentum Equation *J. Elasticity* 39: 201–212.

Eischen, J. W. and G. Herrmann (1987). Energy Release Rates and Related Balance Laws in Linear Elastic Defect Mechanics, *ASME J. Appl. Mech.* 54: 388–392.

Eshelby (1995). The elastic energy-momentum tensor. *J. Elasticity* 5: 321–326.

Fletcher, D. C. (1975). Conservation laws in linear elastodynamics. *Arch. Rat. Mech. Anal.* 60: 329–353.

Freund, L. B. (1989). *Dynamic Fracture Mechanics*. Cambridge: Cambridge University Press.

Golebiewska Herrmann, A. and G. Herrmann (1981). On Energy-Release Rates of a plane Crack. *ASME J. Appl. Mech.* 48: 525–528.

Golebiewska Herrmann, A. (1982). Material Momentum Tensor and Path-Independent Integrals of Fracture Mechanics. *Int. J. Solids Stru.* 18: 319–326.

Gurtin, M. E. (1981). *An Introduction to Continuum Mechanics.* Boston: Academic Press.

Gurtin, M. E. (2000). Configurational Forces as Basic Concepts of Continuum Physics. New York: Springer.

Gurtin, M. E. and P. Podio-Guidugli (1996). Configurational Forces and the Basic Laws for Crack Propagation. *J. Mech. Phys. Solids* 44: 905–927.

Kalpakides, V. K. and E. K. Agiasofitou (2002). Configurational balance laws for dynamical fracture. *Theor. Appl. Mech.* 28–29: 205–219.

Kalpakides, V. K. and E. K. Agiasofitou (2002). On material equations in second gradient electroelasticity. *J. Elasticity* 67: 205–227.

Kienzler, R. and G. Herrmann (2000). *Mechanics in Material Space with Applications to Defect and Fracture Mechanics*. Berlin: Springer.

Knowles, J. K. and E. Sternberg (1972). On a class of conservation laws in linearized and finite elastostatics. *Arch. Rat. Mech. Anal.* 44: 187–211.

Maugin, G. A. (1992). *The Thermomechanics of Plasticity and Fracture*. Cambridge: Cambridge University Press.

Maugin, G. A. (1993). *Material Inhomogeneities in Elasticity*. London: Chapman & Hall.

Maugin, G. A. (1995). Material Forces: Concepts and applications. *ASME Appl. Mech. Rev.* 48: 213–245.

Maugin, G. A. and C. Trimarco (1995). Dissipation of configurational forces in defective elastic solids *Arch. Mech.* 47: 81–99.

Maugin, G. A. and V. K. Kalpakides (2002). A Hamiltonian formulation for elasticity and thermoelasticity *J. Phys. A: Math. Gen.* 35: 10775–10788.

Podio-Guidugli, P. (2002). Configurational forces: are they needed? *Mech. Res. Comm.* 29: 513–519.

Steinmann, P. (2000). Application of material forces to hyperelastostatic fracture mechanics. I. Continuum mechanical setting. *Int. J. Solids Stru.* 37: 7371–7391.

Configurational Mechanics, Kalpakides & Maugin (eds)
© 2004, Taylor & Francis Group, London, ISBN 90 5809 667 X

On the use of material forces in the finite element method

K.G. Balassas, V.K. Kalpakides & G.E. Stavroulakis
Department of Mathematics, Division of Applied Mathematics and Mechanics,
University of Ioannina, Greece

ABSTRACT: This work aims at the exploitation of the material forces in the finite element method. The material forces for a homogeneous elastic body in the absence of body forces are identically zero. Nevertheless, the approximate solutions obtained by the finite element method do not fulfil exactly the governing equations of equilibrium. Due to this fact, non-zero material forces appear at the nodes of any discretization. In general, the magnitude and the direction of material forces at any node depend on the error of the approximate solution at that point. Hence, any manipulation which decreases the material forces by rearranging the interior nodes of the mesh improves the finite element solution. A procedure of this type is proposed and is implemented on particular problems of two-dimensional elastostatics.

1 INTRODUCTION

In 1951, J.D. Eshelby, attempting to establish a notion of force acting over a singularity (Eshelby 1951), introduced the energy momentum tensor, now called the Eshelby stress tensor. This paper opened a new topic in continuum mechanics which has attracted the interest of numerous researchers as shown by the great number of papers and monographs published in the last decade (Maugin 1993, Gurtin 2000, Kienzler and Herrmann 2000).

The notion of material (or configurational) force seems to be behind diverse phenomena like phase transitions, crack propagation, dislocations as well as any kind of defects. The material force is the dual quantity to any configurational change (displacements in the material space) like the Newtonian force is the dual of the physical motion (displacements in the Euclidian space). This view justifies the fact that the material forces and the Eshelby stress tensor are the contributors to the pseudomomentum (or material momentum) equation (Maugin 1993, Maugin 1995) in the same manner as the Newtonian body forces and the Cauchy stress tensor are the contributors to the momentum equation. Apart from the pseudomomentum equation, one can enlarge the system of equations concerning the material forces by introducing additional material equations like the material angular momentum equation and material virial equation (Maugin 1993, Steinmann 2000, Maugin and Kalpakides 2002, Kalpakides and Agiasofitou 2002).

It is worth noting that all these new equations, in the absence of inhomogeneities and defects, become simple identities for the solution of the momentum equation, that is, they do not provide any additional information concerning the behavior of the elastic body. In other words, there is no need for using these equations in solving any problem of standard elasticity. Notice that the exact solution to the elastic problem will fulfil the pseudomomentum equation identically. Nevertheless, this is not true when one uses an approximate method as the finite element method (FEM). In this case, the approximate solution will not fulfil the pseudomomentum equation due to the numerical error. Consequently discrete material forces will appear at the nodes of the mesh. These forces depend on the topology of the mesh and can be considered as indicators of the numerical error. Furthermore, one can view these forces as the energy change with respect to the node movement within the mesh.

Thus, any rearrangement of the mesh, which decreases the material forces, will decrease the numerical error as well. This idea was proposed by Braun (Braun 1997) in 1997 and has already been used by other researchers (Steinmann et al 2001, Mueller et al 2002, Mueller and Maugin 2002, Gross et al 2002, Gross et al 2003). In this work a procedure for error minimization through mesh rearrangement, which exploits the concept of material forces is proposed. The mesh rearrangement is carried out automatically with the use of an auxiliary boundary value problem based on a modified momentum equation where the material forces take the position of the Newtonian body forces.

Throughout the paper the Cartesian and the tensorial notation are applied alternatively. A comma followed by the index i means partial derivation with respect to x_i. Some fundamental ideas from material mechanics are discussed in Section 2. The discrete material forces are described in Section 3. The proposed computational scheme and some computational examples are given in Section 4 and 5, respectively. An alternative strategy for the exploitation of the pseudomomentum equation in FEM is discussed in Section 6.

2 THE MATERIAL FORCES IN AN ELASTIC BODY

Consider a smooth elastic body occupying the domain $\Omega \subset IR^2$. In the framework of linear elasticity a strain energy function is assumed in the form

$$w = \hat{w}(\mathbf{x}, \mathbf{e}), \tag{1}$$

where $\mathbf{x} \in \Omega$ and $\mathbf{e}$ is the linear strain tensor written in terms of the displacement vector $\mathbf{u}$ as

$$e_{ij} = \frac{1}{2}(u_{i,j} + u_{j,i}). \tag{2}$$

Also, the standard linear constitutive relation for the Cauchy stress tensor is adopted

$$\sigma_{ij} = \frac{\partial \hat{w}}{\partial e_{ij}}. \tag{3}$$

Assuming that a distribution of body forces f_i acts within the body, the equilibrium equation takes its usual form

$$\sigma_{ij,j} + f_i = 0, \tag{4}$$

Equation (4) is nothing but the standard momentum equation for elastostatics. Furthermore, what is called pseudomomentum (or material momentum) equation holds as an identity (Maugin 1993), too

$$C_{ij,j} + f_i^{mat} = 0, \tag{5}$$

where C_{ij} and f_i^{mat} are the Eshelby stress tensor and the material body forces, respectively given by the relations

$$C_{ij} = w\delta_{ij} - \sigma_{kj}u_{k,i}, \tag{6}$$

$$f_i^{mat} = -f_k u_{k,i} - \frac{\partial \hat{w}}{\partial x_i}. \tag{7}$$

Considering the pseudomomentum equation as a balance of material forces, we remark that there are both boundary material forces (given by the Eshelby stress tensor C_{ij}) and body material forces given by eq. (7). The latter ones are coming from the Newtonian body forces as well as from the inhomogeneity of the strain energy function, thus are present even if the Newtonian body forces do not exist. Consequently, the pseudomomentum equation may contain body (material) forces due to the smooth inhomogeneities of the body. In that case, they are called inhomogeneity material forces (Maugin 1993).

It is worth noting that for a homogeneous body, in the absence of body forces, the pseudomomentum eq. (5) takes the simpler form

$$C_{ij,j} = 0, \tag{8}$$

that is, a conservation law for the equation of motion. That means any analytical solution to the momentum equation fulfil identically the eq. (8). However, this is not true for any approximate solution to the eq. (4) as we discuss in detail in the next section.

3 THE MATERIAL FORCES IN THE DISCRETE CASE

Consider now an approximate solution $\mathbf{u}_e(x, y)$ for the linear elastostatic problem within an element e of a triangular finite element discretization. This should be of the form (Reddy 1984)

$$\mathbf{u}_e(x,y) = \sum_1^3 \mathbf{u}^I N^I(x,y), \tag{9}$$

where $N^I(x, y)$ are the shape functions associated with this particular element and $\mathbf{u}^I$ are the unknown nodal parameters (degrees of freedom) to be determined. Denoting by $\boldsymbol{\eta}$ a test function in the same element, then we can write

$$\eta_i = \sum_I N^I \eta_i^I, \tag{10}$$

$$\eta_{i,j} = \sum_I N_{,j}^I \eta_i^I, \tag{11}$$

where n_i^I are the value of the test function at the node I. Next, we pass to the weak formulation of eq. (5) (Steinmann 2000)

$$\int_\Omega \left(C_{ij,j} + f_i^{mat}\right) \eta_i dV = 0. \tag{12}$$

Applying the divergence theorem this can be written as

$$-\int_\Omega C_{ij}\eta_{i,j}dV + \int_{\partial\Omega} C_{ij}n_j\eta_i dS + \int_\Omega f_i^{mat}\eta_i dV = 0. \tag{13}$$

Taking into account that the boundary of the domain Ω is fixed, we conclude that any variation in the material space should vanish along $\partial\Omega$. In addition, assuming the same discretization with the physical problem and inserting eqs. (10), (11) into eq. (13), we obtain the following equation holding for any element Ω_e

$$\sum_I \left[\int_{\Omega_e} \left(C_{ij}N_{,j}^I - f_i^{mat}N^I\right) d\Omega\right] \eta_i^I = 0. \tag{14}$$

Equation (14) should hold for any admissible test function, thus we obtain

$$\int_{\Omega_e} (C_{ij} N^I_{,j} - f^{mat}_i N^I)\, d\Omega = 0, \tag{15}$$

for any element of the mesh. Notice that the analysis is based on the pseudomomentum eq. (5) which holds for any exact solution of the momentum equation. As we have remarked, eq. (5) is an identity for the solution of the elastic problem. Accounting for the error of the approximate solution, one should expect that the left part of eq. (15) does not vanish. Hence, we can write

$$F^I_i = \int_{\Omega_e} (C_{ij} N^I_{,j} - f^{mat}_i N^I)\, d\Omega, \tag{16}$$

which will be called the material force at the node I, associated with the element Ω_e. In virtue of eq. (7), in the case of a homogeneous body the above formula takes the form

$$F^I_i = \int_{\Omega_e} (C_{ij} N^I_{,j} + f^{phy}_k u_{k,i} N^I) d\Omega. \tag{17}$$

Moreover, in the absence of physical body forces the formula takes the simpler form

$$F^I_i = \int_{\Omega_e} C_{ij} N^I_{,j} d\Omega. \tag{18}$$

Assembling the contribution of the material forces of any neighboring element of the particular node, we obtain the *global discrete material force* (Steinmann et al 2001)

$$\mathbf{F}^G_I = \underset{J=1}{\overset{m}{\mathsf{A}}} \mathbf{F}^J. \tag{19}$$

Here A denotes the assembly operation of finite element analysis and the quantity $\mathbf{F}^J$, represents the material force at the node I, which is calculated with the aid of the J neighboring elements.

4 A SCHEME FOR THE REDUCTION OF MATERIAL FORCES

Figure 1a shows a two-dimensional plane strain problem subjected to a uniform loading q along a part of its upper boundary and fixed boundary conditions along the lower boundary. In Figure 1b, the material forces at the nodes of a sparse triangular mesh (45), are depicted after solving the boundary value problem of elastostatics with the aid of MATLAB package. Constant stress and strain elements are used throughout the paper.

The magnitude of the material forces is related with the error of the approximate solution at this node. Moreover, the direction of the material forces shows the opposite direction to which we must move the nodes in order to reduce the approximation error. Consequently, the next step is to rearrange the mesh according to what the material forces indicate. Mueller et al (Mueller et al 2002) and Mueller and Maugin (Mueller and Maugin 2002) solve this problem by using an appropriate, but otherwise arbitrary, constant c in the following iteration scheme

$$\mathbf{P}^{new}_I = \mathbf{P}^{old}_I - c\mathbf{F}^G_I, \tag{20}$$

where $\mathbf{P}_I^{new}$, $\mathbf{P}_I^{old}$ denote the position of the node I after and before its movement, respectively. Considering that the material constants should participate in the "motion" of the mesh, we formulate an auxiliary elastostatic problem in which the distribution of the material forces, in the opposite

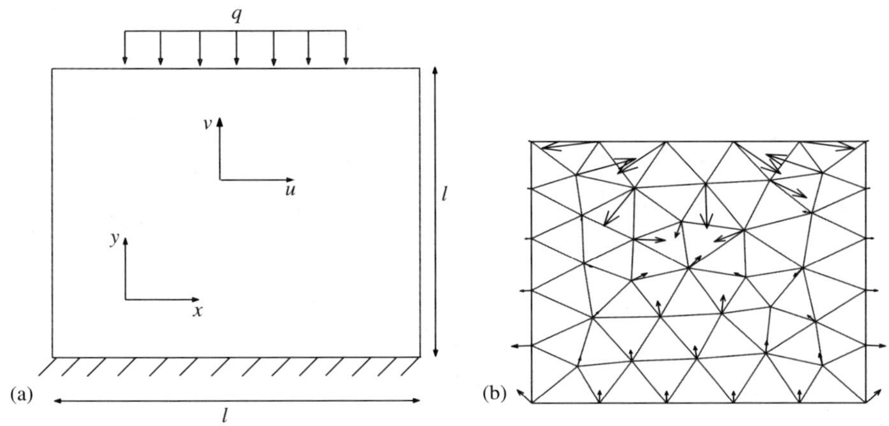

Figure 1. (a) Homogeneous plate under pressure (b) Distribution of material forces.

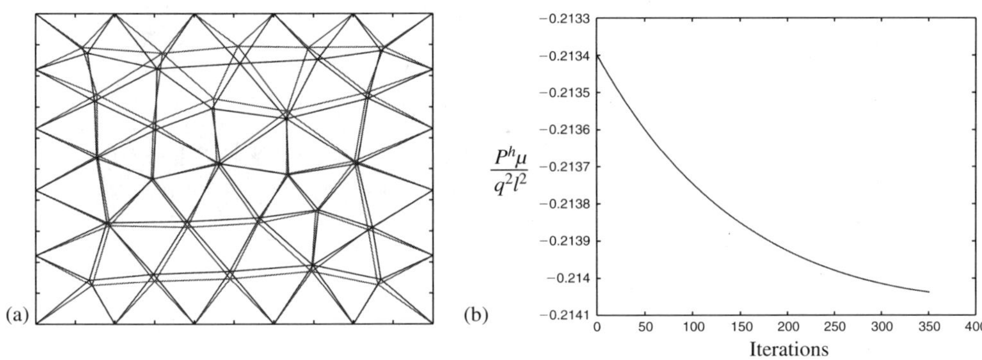

Figure 2. (a) Initial and updated mesh (b) Change of total potential energy during update iterations.

direction, is used as external body loading. On the boundary we impose Dirichlet boundary conditions to prevent the body to change its initial configuration along it. This is equivalent to the assumption that an auxiliary elastic body is deformed, under the influence of the material forces. Denoting with $\mathbf{u}_I^\psi$ the solution of the auxiliary problem at the node I, we propose the following iteration scheme

$$\mathbf{P}_I^{new} = \mathbf{P}_I^{old} + \mathbf{u}_I^\psi. \tag{21}$$

The finite element solution with the same mesh topology but with the new position of the nodes approximates better the exact solution of the physical problem, thus the new distribution of the material forces will be of smaller magnitude and so on. Thus, it seems that the difference between our scheme and the scheme eq. (20) lies in the fact that we let the material to choose the constant c which changes automatically at each step. In Figure 2a, the new mesh after 350 iterations is shown and in Figure 2b the relaxation of the total potential energy with respect to the number of iterations.

Recall that a discrete displacement-based finite element solution overestimates the potential energy of the continuous problem. A systematic mesh refinement or, as in the present case a suitable rearrangement approximate asymptotically the value of the continuous problem from above. Figure 3 depicts the vertical component of the approximate solution versus the iteration number. It is remarked that to obtain an analogous accuracy with the standard FEM, the same package (PDETOOL of MATLAB) needs a fine mesh of 8865 nodes instead of 45 that are used in our example.

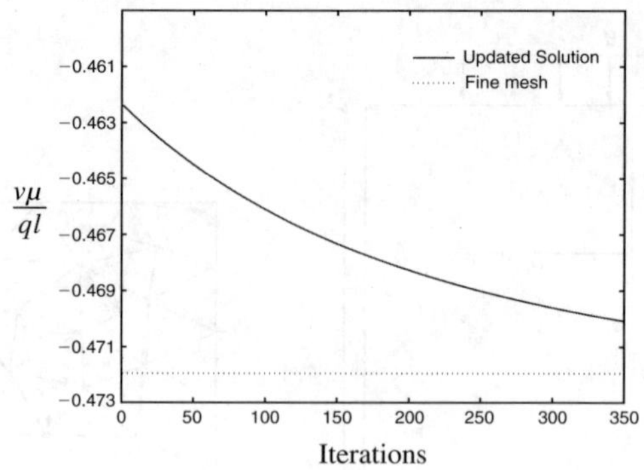

Figure 3. Change of displacement v during update iterations.

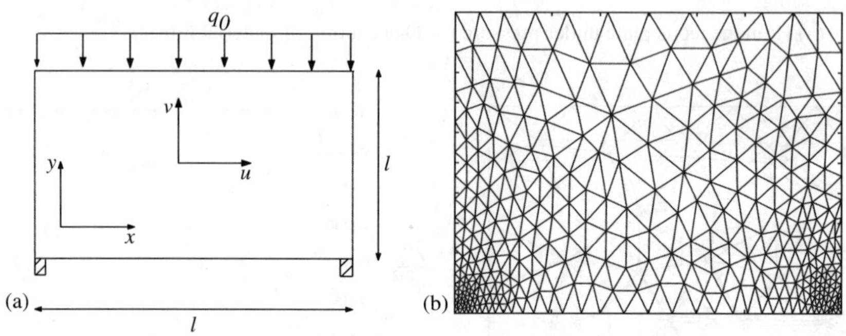

Figure 4. (a) Homogeneous plate supported at the bottom under pressure (b) Initial mesh.

5 EXAMPLES

5.1 *A plane strain problem*

Consider the plane strain problem depicted in Figure 4a, where a continuous and uniform load q_0 is applied on the upper part, while it is supported at the bottom. The body is homogeneous, linear elastic, isotropic and the equilibrium equation is considered. We start with an initial discretization of 353 nodes and 624 elements as shown in Figure 4b. With the aid of this mesh we take the finite element solution for the displacement and stress fields, which are used for the calculation of the material forces at all nodes. Looking at the distribution of material forces, we remark that the higher values appear at the part of the boundary where Dirichlet conditions have been imposed. This can be explained by the fact that the higher values for the stresses and strains appear at these nodes as well. Over the neighbouring region the material forces are small, but in any case greater than of the corresponding ones at the central nodes.

Next, solving the auxiliary problem we let the interior nodes move according to the scheme eq. (21) in the direction which reduces the material forces. Notice that due to the Dirichlet conditions of the auxiliary problem the boundary nodes cannot move. Figure 5a shows the decrease of the total potential energy during the iterations of the proposed algorithm. It is obvious that the scheme leads to the relaxation of the energy in a smooth manner. The solution of the physical problem is also improved by the iterations as one can see in the Figure 5b, where the displacement v of the middle node on the upper edge is showed. It is noticed that the updated solution is compared with the corresponding one obtained by a fine mesh of 21313 nodes and 41984 elements. Finally, in Figures 6a and 6b we show the distribution of the stresses σ_{yy} in successive restructuring of the mesh. The enhancement of the stress approximation with the rearrangement of the mesh is remarkable.

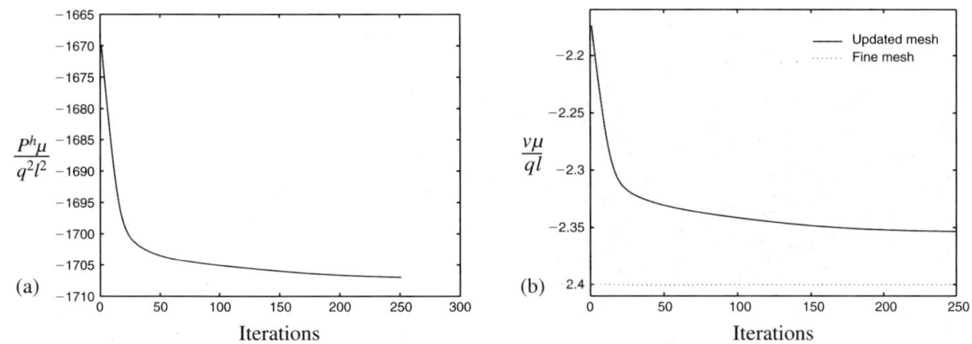

Figure 5. (a) Change of total potential energy during update iterations (b) The displacement v in comparison with the solution of a fine mesh.

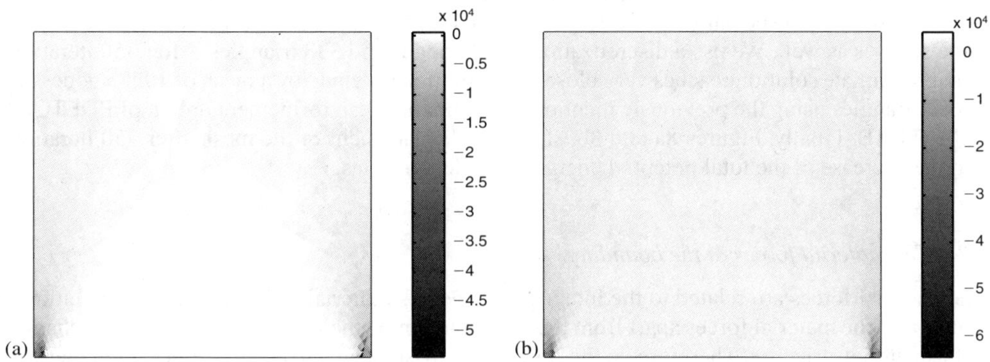

Figure 6. (a) σ_{yy} distribution to the initial mesh (b) σ_{yy} distribution to the updated mesh.

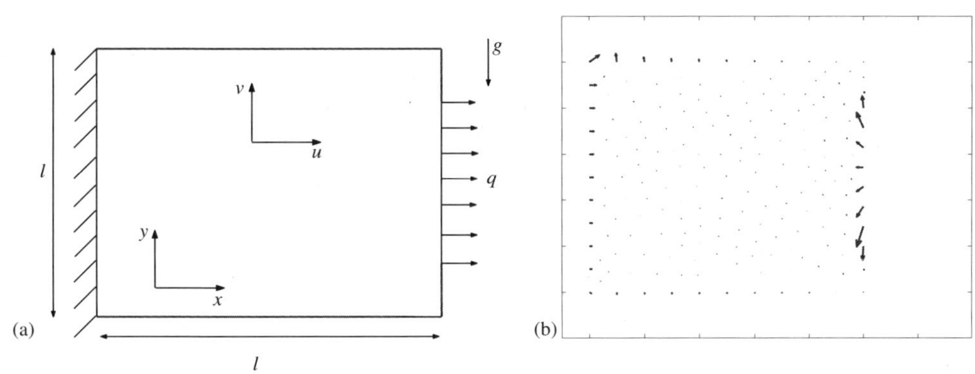

Figure 7. (a) Homogeneous plate in tension (b) Distribution of Material Forces.

5.2 *Plane strain problem in the presence of the body forces*

In this example we consider a beam clamped at the left edge and loaded over a part of the right edge by a uniform load q as it is shown in Figure 7a. The body is under the influence of the gravity g, which provides a distribution of body forces. In this case, use of eq. (17) is made for the calculation of the material forces. An interpretation of the material forces is given in Figure 7b.

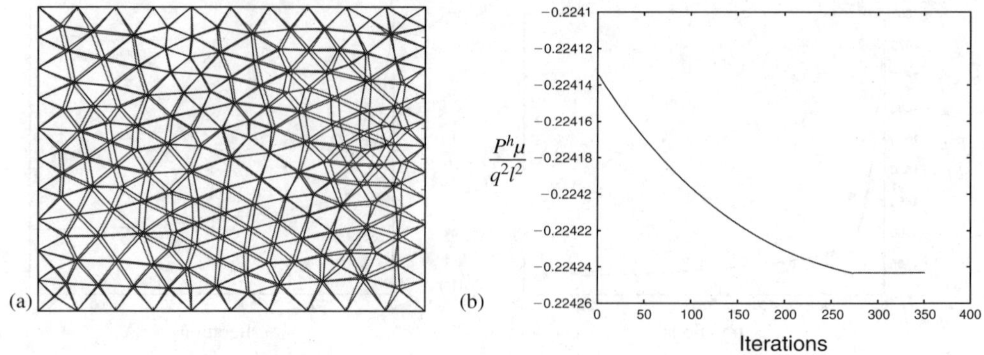

Figure 8. (a) Initial and updated mesh (b) Behaviour of total potential energy during update iterations.

In the interior of the body, the material forces almost vanish, while large material forces appear over the boundary, especially on the loaded part and at the upper left corner where one expects large stresses as well. We use a discretization of 191 nodes and 338 triangles. After 350 iterations our approximate solution reaches very closely the solution obtained by a mesh of 10985 nodes and 21632 triangles using the previously mentioned, automatic mesh refinement option of PDETOOL in MATLAB. Finally, Figures 8a and 8b, show the displacement of the mesh after 350 iterations and the decrease of the total potential energy with the iterations.

5.3 *The material forces at the boundary*

The material forces are related to the integrity of a body's material structure. Thus, as mentioned above, the true material forces apart from the case of inhomogeneities, are connected with changes in the material structure. Therefore, in the case of a body subjected to elastic deformation there are no true material forces in the interior of the body. On the contrary, we believe that along the boundary true material forces do exist in order to maintain the shape of the body's structure. Recalling eq. (6), we confirm that the Eshelby stress tensor depends on the stress and strain distribution and the strain energy function. Consequently it vanishes everywhere in the body and along its boundary for the case of an unloaded elastic body. In order to catch the material forces at the boundary, we must substitute the strain energy function in eq. (6) with the internal energy function accounting for the additional energy needed to keep the material points at their position within the material structure. Then the Eshelby stress tensor will take the form

$$C_{ij} = \epsilon\delta_{ij} - \sigma_{kj}u_{k,i}. \tag{22}$$

Assuming that the internal energy has the form

$$\epsilon = w + w_0, \tag{23}$$

where w_0 is the extra energy. In the case of an unloaded body, we conclude that the Eshelby stress tensor takes the simpler form

$$C_{ij} = w_0\delta_{ij}. \tag{24}$$

We calculate now the material forces for a homogeneous elastic block without any loading using the Eshelby stress tensor given by eq. (23) instead of eq. (6). An arbitrary constant is used for the value of w_0. The result of this calculation is presented in Figure 9, where the expected uniform distribution of material forces at the boundary is shown.

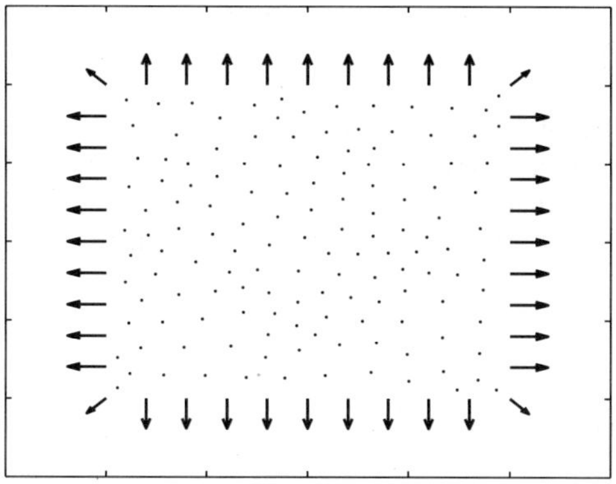

Figure 9. Distribution of material forces along the boundary.

6 A REMARK ON AN ALTERNATIVE MANIPULATION

After having computed the material forces for an approximate solution of a BVP, the main question is how one must move the nodes in order to reduce them. From the previous analysis, it is apparent (see eq.(6)) that the material forces are contributors to the pseudomomentum equation. Therefore, the simplest answer to this question could be: By solving a new BVP concerning the pseudomomentum equation. In this perspective, it is convenient to set up the problem in terms of the deformation $\mathbf{x} = \mathbf{x}(\mathbf{X})$ which is an invertible function instead of the displacement $\mathbf{u} = \mathbf{u}(\mathbf{X})$ used in the previous analysis. It seems that the non-linear setting is more appropriate. The idea is to formulate the momentum equation in terms of the deformation mapping $\mathbf{x} = \mathbf{x}(\mathbf{X})$ and the pseudomomentum equation in terms of the inverse deformation mapping $\mathbf{X} = \mathbf{X}(\mathbf{x})$. In the case of Dirichlet boundary conditions, a BVP for the inverse deformation mapping can be established with the aid of pseudomomentum equation. Theoretically, in the continuous case the standard BVP for the deformation function (momentum equation) and the proposed here BVP for the inverse deformation function (pseudomomentum equation) are equivalent in the sense that the solution to the former is the inverse of the solution to the latter (Maugin 1993, Maugin and Kalpakides 2002). In the approximate case, the two solutions diverge each other due to the inserted error. However, solving the momentum equation, the material forces indicating the error magnitude will contribute in the pseudomomentum equation as source terms. Inversely, solving the pseudomomentum equation approximately, the numerical error will cause “physical body forces” which will contribute in the momentum equation. Thus, a FEM solution for the inverse deformation mapping can be used to optimize an initially uniform discretization for the momentum equation (direct formalism). Generally speaking, the solution for the inverse deformation can improve the discretization for the direct problem and vice versa. Consequently, a new computational strategy could be used whereby the BVP for the momentum equation – formulated, classically, in terms of direct deformation – and the pseudomomentum equation – formulated in terms of the inverse deformation mapping – are solved numerically in succession improving in each cycle the finite element solution. An implementation of this procedure will be the objective of a future work.

7 CONCLUSIONS

A systematic procedure for automatic mesh rearrangement based on the material force concept was presented. The main idea for the reduction of the material forces was based on the formulation of an auxiliary BVP which successfully move the nodes in the direction of reducing the material forces.

The scheme was implemented in plane strain problems of homogeneous linear elastostatics. The presented examples demonstrated that a rearrangement of the mesh, without disturbing its topology, in the direction which minimizes the material forces according to the proposed automatic algorithm leads to a reduction of the potential energy and gives a better FE solution for the displacements and the stresses.

Finally, although the idea of minimizing the material forces gives successful numerical as was shown here and in other works (Braun 1997, Mueller et al 2002, Mueller and Maugin 2002, Steinmann 2000), we believe that the problem of mesh optimization could be ultimately solved in an alternative setting by the use of the inverse deformation function.

REFERENCES

Braun, M. (1997). Configurational forces induced by finite-element discretization. *Acad. Sci. Phys. Math.* 46: 24–31.

Eshelby, J.D. (1951). The force on an elastic singularity. *Philos. Trans. Roy. Soc. London* A 244: 87–112.

Gross, D., Mueller, R. and Kolling, S. (2002). Configurational forces–morphology evolution and finite elements *Mech. Res. Comm.* 29: 529–536.

Gross, D., Kolling, S. and Mueller, R. (2003). Configurational forces and their application in solid mechanics *Eur. J. Mech. A/Sol.* Article in press.

Gurtin, M.E. (2000). Configurational Forces as Basic Concepts of Continuum Physics. New York: Springer.

Kalpakides, V.K. and Agiasofitou, E.K. (2002). On material equations in second gradient electroelasticity. *J. Elasticity*. 67: 205–227.

Kienzler, R. and Herrman, G. (2000). *Mechanics in Material Space with Applications to Defect and Fracture Mechanics*. Berlin: Springer-Verlag.

Maugin, G.A. (1993). *Material Inhomogeneities in Elasticity*. London: Chapman & Hall.

Maugin, G.A. (1995). Material Forces: Concepts and applications. *ASME Appl. Mech. Rev*. 48: 213–245.

Maugin, G.A. and Kalpakides, V.K. (2002). A Hamiltonian formulation for elasticity and thermoelasticity. *J. Phys. A: Math. Gen*. 35: 10775–10788.

Mueller, R. and Maugin, G.A. (2002). On material forces and finite element discretization. 67: 205–227.

Mueller, R., Kolling, S. and Gross, D. (2002). On configurational forces in the context of the finite element method. *Int. J. for Numerical Methods in Engineering*. 53(7):1557–1574.

Reddy, J.N. (1984). *Energy and Variational Methods in Applied Mechanics with an Introduction to the Finite Element Method*. New York: John Willey & Sons.

Steinmann, P. (2000). Application of material forces to hyperelastostatic fracture mechanics. I. Continuum mechanical setting. *Int. J. Solids Stru*. 37: 7371–7391.

Steinmann, P., Ackermann, D. and Barth, F.J. (2001). Application of material forces to hyperelastostatic fracture mechanics. II. Computational setting. *Int. J. Solids Stru*. 38: 5509–5526.

Configurational Mechanics, Kalpakides & Maugin (eds)
© 2004 Taylor & Francis Group, London, ISBN 90 5809 667 X

Author index